Dionysius Lardner

Handbook of Natural Philosophy

Mechanics

Dionysius Lardner

Handbook of Natural Philosophy
Mechanics

ISBN/EAN: 9783337330743

Printed in Europe, USA, Canada, Australia, Japan

Cover: Foto ©berggeist007 / pixelio.de

More available books at **www.hansebooks.com**

PREFACE.

THIS work is intended for all who desire to attain an accurate knowledge of Physical Science, without the profound methods of Mathematical investigation. Hence the explanations are studiously popular, and everywhere accompanied by diversified elucidations and examples, derived from common objects, wherein the principles are applied to the purposes of practical life.

It has been the Author's especial aim to supply a manual of such physical knowledge as is required by the Medical and Law Students, the Engineer, the Artisan, the superior classes in Schools, and those who, before commencing a course of Mathematical Studies, may wish to take the widest and most commanding survey of the field of inquiry upon which they are about to enter.

Great pains have been taken to render the work complete in all respects, and co-extensive with the actual state of the Sciences, according to the latest discoveries.

Although the principles are here, in the main, developed and demonstrated in ordinary and popular language, mathematical symbols are occasionally used to express results more clearly and concisely. These, however, are never employed without previous explanation.

The present edition has been augmented by the introduction of a vast number of illustrations of the application of the various branches of Physics to the Industrial Arts, and to the practical business of life. Many hundred engravings have also been added to those, already numerous, of the former edition.

For the convenience of the reader, the series has been divided into Four Treatises, which may be obtained separately.

MECHANICS . . .	. One Volume.
HYDROSTATICS, PNEUMATICS, and HEAT	. One Volume
OPTICS and ACOUSTICS . .	. One Volume.
ELECTRICITY, MAGNETISM, and METEOROLOGY	. One Volume.

The Four Volumes taken together will form a complete course of Natural Philosophy, sufficient not only for the highest degree of School education, but for that numerous class of University Students who, without aspiring to the attainment of Academic honours, desire to acquire that general knowledge of these Sciences which is necessary to entitle them to graduate, and, in the present state of society, is expected in all well educated persons.

CONTENTS.

BOOK I.

Properties of Matter.

CHAPTER I.

MAGNITUDE, FIGURE, AND STATE OF AGGREGATION.

Sect.		Page
1. Most important properties of matter	- - -	1
2. Magnitude classified	- - -	ib.
3. Linear magnitude	- - -	2
4. Superficial magnitude	- -	ib.
5. Solid magnitude	- - -	ib.
6. Solid magnitude defined	- -	ib.
7. On what the form of a solid depends	- - -	ib
8. Boundaries of solid bodies	- -	ib.
9. Boundaries of surfaces	- -	ib.
10. 11. Magnitude unlimited	- -	2-3
12. Bodies consist of parts	- -	3
13. Solid, liquid, and gaseous states	-	ib.
14. Solid state	- - - -	ib.
15. Liquid state	- - - -	ib.
16. Pulverised state	- - -	4
17. Gaseous state	- - -	ib.
18. Atmospheric air an example	-	ib
19. Other examples	- - -	ib.
20. Same body may exist in either state	- - - -	ib.
21. Water an example	- - -	ib.

CHAP. II.

IMPENETRABILITY.—DIVISIBILITY.

22. All matter impenetrable	- -	5
23. Gaseous bodies impenetrable	-	ib.
24. Air an example	- - -	ib.
25. Examples of apparent penetrability explained	- - - -	ib.
26. Walking through the air	- -	ib.
27. Solid plunged in liquid	- -	ib.
28. Divisibility unlimited	- -	ib.
29. Water, its ultimate atoms compound	- - - -	6
30. Water divisible without practical limit	- - - -	ib.

Sect.		Page
31. Other bodies likewise so divisible	6
32. Examples	- - - -	ib.
33. Pulverised marble	- -	ib.
34. Polished surfaces covered with asperities—Diamond	-	7
35. Gold visible on touchstone	-	ib.
36. Minuteness of tubular filaments of glass	- - - - -	ib.
37. Such a filament might penetrate the flesh without pain	-	ib.
38. Wollaston's micrometric wire	-	8
39. Illustration of its extreme minuteness	- - - - -	ib.
40. Minuteness of organised filaments	9
41. Thinness of a soap-bubble	-	ib.
42. Thinness of insects' wings	-	ib.
43. Thinness of leaf gold	- -	ib.
44. 45. Wire used in embroidery	-	10
46. Composition of blood	-	ib.
47. Magnitude and form of its corpuscles	- - - -	ib.
48. Number of corpuscles in a drop of blood	- - - -	11
49. Minuteness of animalcules; their organisation and functions	-	ib
50. Unlimited divisibility by solution of solids in liquids	- -	12
51. Minute divisibility proved by colour	- - - -	ib.
52. Divisibility of musk	- -	ib.
53. Fineness of spiders' web	-	ib.
54. Divisibility shown by the taste	-	ib.
55. Effect of strychnine dissolved in water	- - - -	13
56. Effect of salt of silver dissolved in water	- - - -	ib.
57. Effect of sugar dissolved	-	ib.
58. Is matter, therefore, infinitely divisible?	- - - -	ib.
59. Existence of ultimate molecules may be inferred	- -	ib.
60. Crystallisation indicates their existence	- - -	ib.
61. Process of crystallisation	-	14
62. Existence of ultimate molecules inferred	- - -	ib.
63. Planes of cleavage	- -	15

CONTENTS.

Sect.		Page
64.	Planes of cleavage indicate the forms of the ultimate molecules	15
65.	Inference that all bodies consist of ultimate atoms of determinate figure	ib.
66.	Those molecules too minute to be the subjects of direct observation	ib.
67.	Ultimate molecules of matter indestructible	ib.
68.	No matter destroyed in combustion	16
69.	No matter destroyed in evaporation	ib.
70.	Destructive distillation	17
71.	General conclusion	ib.

CHAP. III.

POROSITY, DENSITY, COMPRESSIBILITY, AND DILATABILITY.

Sect.		Page
72.	Component molecules of a body not in contact	17
73.	Spaces which separate them, called pores	18
74.	Proportion of mass to pores determines density	ib.
75.	Porosity and density correlative terms	ib.
76.	Uniform diffusion of pores constitutes uniform density	ib.
77.	Pores different from cells	ib.
78.	Pores sometimes occupied by more subtle matter	ib.
79.	Examples of porosity and density	19
80.	Porosity of wood	ib.
81.	Buoyancy of wood	ib.
82.	Densest substances have pores	ib.
83.	Filtration	20
84.	Petrifaction	ib.
85.	Porosity of mineral substances	ib.
86.	Porosity of mineral strata	ib.
87.	Compression diminishes bulk and augments density	ib.
88.	All bodies compressible	21
89.	Compressibility increases with porosity	ib.
90.	Compression of wood	ib.
91.	„ of stone	ib.
92.	„ of metal	ib.
93.	„ of liquds	ib.
94.	„ of gases	22
95.	Contractibility of liquids	ib.
96.	Elastic and inelastic bodies	ib.
97.	Elasticity of gases	ib.
98.	Elasticity of liquids	ib.
99.	Expansibility of gases	23
100.	Elasticity of solids	ib.
101.	Examples of elasticity of solids	ib.
102.	Examples—Ivory balls	ib.
103.	„ Caoutchouc balls	ib.
104.	„ Elasticity of steel springs	24

Sect.		Page
105.	Limits of elastic force	24
106.	Elasticity of torsion	ib.
107.	Dilatation by elevation of temperature	25
108.	Dilatation of liquids in thermometers	ib.
109.	Useful application of dilatation and contraction	ib.
110.	General effects of dilatation and contraction	26

CHAP. IV.

INERTIA.

Sect.		Page
111.	All matter inert	26
112.	Inertia, inability to change state of rest or motion	ib.
113.	Vis inertiæ, term leading to erroneous conclusions	27
114.	Erroneous supposition that matter more inclined to rest than motion	ib.
115.	Why motion of bodies in general retarded	28
116.	Astronomy supplies proof of law of inertia	29
117.	Examples of inertia	ib.
118.	„ Effect of sudden change of speed on horseback	ib.
119.	„ Leaping from carriages in motion	30
120.	„ Coursing	ib.

CHAP. V.

SPECIFIC PROPERTIES.

Sect.		Page
121.	Properties general and specific	31
122.	Elasticity and hardness	ib.
123.	Relative hardness of metals	ib.
124.	Hardness of a metal may be modified	ib.
125.	Effects of elasticity	32
126.	No body perfectly elastic or perfectly inelastic	ib.
127.	Elasticity not proportional to hardness	ib.
128.	Vibratory metals	33
129.	Hardness and elasticity of metals affected by their combination	ib.
130.	Flexibility and brittleness	ib.
131.	Malleability	ib.
132.	Malleability varies with temperature	34
133.	Process of annealing	ib.
134.	Welding	ib.
135.	Ductility	35
136.	Tenacity	ib.
137.	Tenacity	ib.
138.	Table of relative tenacities of metals	ib.
139.	Table of tenacity of fibrous textures	36
140.	Chemical properties	ib.

BOOK II.

Of Force and Motion.

CHAPTER I.

COMPOSITION AND RESOLUTION OF FORCES.

Sect. Page
141. Force produces, destroys, or changes motion - - - 38
142. Force expressed by weight - ib.
143. Direction of force - - - ib.
144. Effect of forces acting in same direction - - - - ib.
145. Resultant of forces in same direction - - - - - 39
146. Resultant of opposite forces - ib.
147. Resultant of forces in different directions - - - - ib.
148. Composition of forces - - 40
149. Resultant and components mechanically interchangeable - 41
150. Resultant of any number of forces ib.
151. Composition of forces applied to different points - - - ib.
152. Resultant of parallel forces - 42
153. Composition of parallel forces acting in same direction - - 43
154. Composition of parallel forces acting in opposite directions - ib.
155. Couple - - - - - 44
156. Mechanical effect of couple - ib.
157. Equilibrium of couples - - 45
158. Condition under which two forces admit a single resultant - - ib.
159. Mechanical effect of two forces in different planes - - - 46
160. Suspension bridges - - 48

CHAP. II.

COMPOSITION AND RESOLUTION OF MOTION.

161. Direction and velocity of motion 50
162. Direction of motion in a curve - ib.
163. Velocity defined - - ib
164. Table of velocities of certain moving bodies - - - 51
165. Uniform velocity - - - 52
166. Principles of composition and resolution of force equally applicable to composition and resolution of motion - - - ib.
167. Resultant of two motions in one direction - - - - 53
168. Resultant of two motions in opposite directions - - - ib.
169. Resultant of two motions in different directions - - - ib.
170. Examples of composition and resolution of motion - 54
171. „ swimming across a stream - ib.
172. „ effect of wind and tide on a ship - 56

Sect. Page
173. Examples of rowing a boat - 57
174. „ motion of fishes, birds, &c. - - ib.
175. „ effect of wind on sailing vessels - - ib.
176. „ tacking - - - 58
177. „ ball dropped from ship's topmast - 59
178. „ object let fall from railway carriage - 60
179. „ billiard-playing - 61
180. „ example proving the diurnal rotation of the earth - - 62
181. Motion absolute and relative - 63
182. Example — Person walking on deck of ship - - ib.
183. „ Gymnastic and equestrian feats - - 64
184. „ Flying kite - 65
185. Force of moving mass - - 66
186. Momentum of solid masses - ib.
187. „ liquids - - ib.
188. „ air - - - ib.
189. Conditions which determine force of moving mass - - - 67
190. Moving force augments with velocity - - - - - ib.
191. When velocity same, moving force augments with mass moved - ib.
192. Example — Force necessary to project stone by hand - - 68
193. „ Force necessary to row a skiff - - ib.
194. Arithmetical expression for momentum - - - - ib.
195. General condition of equality of moving forces - - - 69

CHAP. III.

COMMUNICATION OF MOMENTUM BETWEEN BODY AND BODY.

196. Effects of collision - - 69
197. Case of inelastic masses - ib.
198. Effect of moving body striking body at rest - - - ib.
199. Example—Boat towing ship - 71
200. Collision of two bodies moving in same direction - - ib.
201. Equality of action and reaction - 72
202. Example—Mutual action of two boats - - - - ib.
203. Effect of collision of two bodies moving in contrary directions - ib.
204. Effect verifies law of action and reaction - - - 73
205. Collision of equal masses with equal and opposite velocities - 74

Sect.		Page
206.	Examples of railway trains and steam-boats	74
207.	Pugilism	ib
208.	Collision of masses moving in different directions	75
209.	How action and reaction are modified by elasticity	ib.
210.	Perfect and imperfect elasticity	76
211.	Rebound of an ivory ball	ib.
212.	Oblique impact of elastic body	ib.
213.	When striking body perfectly elastic, angles of incidence and reflection equal	77
214.	When striking body imperfectly elastic, angle of incidence less than angle of reflection	ib.
215.	Apparatus to illustrate collision experimentally	78
216.	Laws of motion	80
217.	Meaning of these laws	81
218.	Maxim that there is always the same quantity of motion in the world explained	ib.
219.	Cannot a living agent produce new motions?	82
220.	Spontaneous motion of an animal explained	ib.
221.	Motion of railway train	ib.
222.	Earth a great reservoir of momentum	83
223.	Would spontaneous progressive motion be possible in absence of a mass like the earth to react upon?	ib.

Sect.		Page
243.	Atwood's machine for illustrating phenomena of falling bodies	92
244.	Morin's apparatus	97
245.	Heights from which body falls proportional to squares of times of fall	ib.
246.	Formulæ, expressing rate, heights, velocities, and times	98
247.	Calculation of height from which body falls in one second	100
248.	Method of calculating all circumstances of descent of falling body	ib.
249.	Force with which body falls as square root of height	101
250.	Why fall from great height not so destructive as might be expected	ib.
251.	Case of person falling down coalpit	102
252.	Retarded motion of bodies projected upwards	ib.
253.	Motion down inclined plane	ib.
254.	Motion of projectiles	105
255.	Case of projectile shot horizontally	ib.
256.	Case of oblique projection	106
257.	Projectiles move in parabolic curves	ib
258.	These conclusions modified by resistance of the air	108
259.	Application of projectiles to gunnery	ib.
260.	Effect of a hammer upon a nail	109
	Influence of time on the effect of force	111

CHAP. IV.

TERRESTRIAL GRAVITY.

CHAP. V.

CENTRE OF GRAVITY.

224.	Plumb line	84
225.	Vertical direction	ib.
226.	Level surface	ib.
227.	Terrestrial gravity indicated by these facts	ib.
228.	Earth attracts all bodies towards its centre	85
229.	Bodies fall in vertical lines	ib.
230.	What is reaction, corresponding to action of falling body	ib.
231.	All bodies fall with the same velocity	ib.
232.	Effects incompatible with this explained	86
233.	Guinea and feather experiment	ib.
234.	Weight of bodies proportional to their quantity of matter	87
235.	Motion of a falling body accelerated	ib.
236.	Force of fall not proportional to height	88
237.	Analysis of the motion of a falling body	ib.
238.	Velocity augments with time of fall	89
239.	Uniformly accelerated motion and force	ib.
240.	Body falling acquires speed which would in same time carry it over twice the space	90
241.	Analysis of heights fallen through in successive seconds	ib.
242.	Tabular analysis of motion of falling body	91

261.	Weight of body is the aggregate of weights of its molecules	112
262.	Effect of cohesion on gravity of molecules	ib.
263.	Resultant of gravitating forces of molecules	113
264.	Experimental method of determining resultant of gravitating forces of molecules	114
265.	Different resultant for every different point of suspension	115
266.	All these resultants have common point of intersection	ib.
267.	Another experimental proof of this	116
268.	Common point of intersection called centre of gravity	117
269.	When centre of gravity supported, body will remain at rest	ib.
270.	Body having regular figure, centre of magnitude centre of gravity	ib.
271.	Centre of gravity of sphere	ib.
272.	Body having symmetrical axis, centre of gravity will be on it	ib.
273.	Centre of gravity not always within body	118
274.	Has same properties	ib.
275.	Centre of gravity takes lowest position	119
276.	Centre of gravity at rest must be always above or below a point of support	ib
277.	Centre of gravity not supported, oscillates	ib.

Sect.	Page
278. Pendulum - - - - -	119
279. Conditions determining stability of body - - - -	120
280. Stability of pyramid - -	121
281. Case in which line of direction falls outside base - - -	ib.
282. Leaning towers of Pisa and Bologna - - - - -	ib.
283. Case in which line of direction falls upon edge of base -	122
284. Stability of loaded vehicle -	ib.
285. Drays - - - - -	123
286. Stability of table - - -	ib.
287. Gestures and motions of animals governed by direction of centre of gravity - - - -	124
288. Motion of centre of gravity of person walking - - -	ib.
289. Use of knee-joint - - -	ib.
290. Position of centre of gravity changes with change of posture	125
291. Porter carrying load - -	ib.
292. Walking up or down hill -	ib.
293. Rising from chair - -	ib.
294. Case of quadruped - -	ib.
295. Cylinder rolled on level plane	126
296. Elliptic body on level plane	127
297. Stable, unstable, and neutral equilibrium - - -	128
298. Criterion of stable equilibrium	ib.
299. Criterion of unstable equilibrium	ib.
300. Criterion of neutral equilibrium	ib.
301. Example 1. Children's toys	ib.
302. „ II. Feats of public exhibitors - - -	129
303. „ III. Spinning top - -	ib.
304. „ IV. Object spinning on point of sword -	130
305. „ V. Cases in which centre of gravity seems to ascend - -	ib.
306. „ VI. Case of globe rolled up inclined plane -	131
307. Magic clock - - -	132
308. Centre of gravity of fluids -	ib.
309. Centre of gravity of two separate bodies - - - -	133

CHAP. VI.

CENTRIFUGAL FORCE.

310. Force consequent on rotatory motion - - -	133
311. Centrifugal force - - -	ib.
312. Centrifugal force a consequence of inertia - - -	134
313. Method of calculating centrifugal force - - - -	ib.
314. Rule to calculate centrifugal force, weight, velocity, and radius of rotation being given -	135
315. Rule to calculate centrifugal force, weight, radius of rotation, and number of revolutions per second being given - -	136
316. Application of whirling-table to illustrate theory - -	ib.
317. Centrifugal force of bodies revolving in same time round common centre of gravity equal -	138
318. Examples of centrifugal force	139

Sect.	Page
319. Example I. Turning rapidly round corner -	139
320. „ II. Horse moving round circus - -	ib.
321. „ Method of calculating inclination of horse towards centre -	140
322. „ III. Carriage turning corner - - -	141
323. „ IV. Stone in a sling	142
324. „ V. Glass of water in a sling - - -	ib.
325. „ VI. Water whirling	143
326. Centrifugal drying machine	144
327. Case of body moving down convex surface - - -	146
328. Centrifugal forces of solid body revolving on fixed axis -	ib.
329. First case, in which centrifugal forces in equilibrium -	147
330. Second case, in which they have a single resultant - -	ib.
331. Third case, in which they are equivalent to a couple -	ib.
332. Fourth case, in which they have effect of single force and couple combined - - - -	ib.
333. Examples of application of these principles - - -	148
334. Example I. Ring revolving round its centre in its own plane - -	ib.
335. „ II. Flat circular plate -	ib.
336. „ III. Cylinder revolving round its geometrical axis - -	ib.
337. „ IV. Solid of revolution revolving round its geometrical axis -	ib.
338. Case of solids which have a symmetrical axis - - -	150
339. Case of rectangular prism -	151
340. Case of pyramid - - -	ib.
341. Case of axis of revolution passing through centre of gravity - -	ib.
342. Centrifugal forces round such an axis are either in equilibrium or equivalent to a couple -	152
343. Axes of centrifugal equilibrium called principal axes -	ib.
344. Case in which all lines through centre of gravity are principal axes - - - -	ib.
345. Case of axis of revolution parallel to principal axis through centre of gravity - -	153
346. Such axis called also principal axis	ib.
347. Three principal axes at right angles to each other pass through each point - - -	ib.
348. Effect of revolution round a line which is not a principal axis -	ib.
349. Experimental illustration of these properties - - -	ib.

CHAP. VII.

MOLECULAR FORCES.

350. Pores of bodies the region of molecular forces - -	155
351. Attraction of cohesion -	ib.
352. Attraction of adhesion -	156

Sect.		Page
353.	Capillary attraction	156
354.	Chemical affinity	ib.
355.	Atoms of bodies manifest attraction and repulsion	ib.
356.	Cohesion manifested in solids and liquids	157
357.	Example of cohesion	ib.
358.	Shot manufacture	ib.
359.	Why liquids form spherical drops	ib.
360.	Earth and planets once fluid	158
361.	Mutual repulsion of atoms of gases	ib.
362.	Gases may be reduced to liquid and solid states	ib.
363.	Existence of attraction of cohe-	

Sect.		Page
	sion between their atoms inferred	158
364.	Mutual repulsion ascribed to influence of heat	ib.
365.	Same body may exist in solid, liquid, or gaseous state	159
366.	Adhesion of solids	ib.
367.	Adhesion of wheels of locomotive to rails	ib.
368.	Effect of lubricants	160
369.	The bite in metal working explained	ib.
370.	Effect of glues, solders, and like adherents	ib.
371.	Silvering mirrors	ib.

BOOK III.

Theory of Machinery.

CHAPTER I.

GENERAL PRINCIPLES.

Sect.		Page
372.	What constitutes a machine	161
373.	Moving power	ib.
374.	Working point	ib.
375.	Weight	ib.
376.	Various resistances and physical conditions omitted provisionally in exposition of theory of machines	162
377.	Physical science consists of a series of approximations to truth	163
378.	This gradual approximation to truth not peculiar to mechanical science	ib.
379.	How machine provisionally regarded	ib.
380.	Mechanical truths improperly invested with the appearance of paradox	164
381.	Effect of fixed points or props	ib.
382.	An indefinitely small power raising an indefinitely great weight involves no paradox	ib.
383.	Effects of a machine under different relations of power and weight	165
384.	When power and weight in equilibrium, rest not necessarily implied	ib.
385.	Equilibrium infers either absolute rest or uniform motion	166
386.	When the power more than equilibrates, accelerated motion ensues	ib.
387.	All machines are in this state when started	ib.
388.	Analysis of the effect in this case	ib.

Sect.		Page
389.	When the power is too small to equilibrate, motion is retarded	167
390.	Proper functions of a machine	168
391.	No machine can really add to the mechanical energy of the power	169
392.	Method of expressing mechanical energy of power and weight	170
393.	Moments of power and weight	ib.
394.	Relation between these moments determines state of machine	ib.
395.	Equality of these moments determines equilibrium	171
396.	When moment of power exceeds that of weight, motion accelerated in direction of power	ib
397.	When moment of power less than moment of weight, motion retarded	ib.
398.	In equilibrium, velocity of power is to that of weight as weight is to power	ib.
399.	Power always gained at expense of time	ib.
400.	All paradox thus removed from mechanical theorems	ib.
401.	Utility of machines not limited to make small power overcome great resistance	172
402.	Case in which point of application of power and working point do not move in direction of power and weight	ib.

CHAP. II.

SIMPLE MACHINES.

| 403. | Machines, simple and complex | 174 |
| 404. | Classification of simple machines | ib. |

CONTENTS. xiii

Sect. Page
405. Condition of equilibrium of machine having fixed axis - 174
406. Condition of equilibrium of flexible cord - - - 175
407. Condition of equilibrium of weight upon hard inclined surface - ib.
408. Mechanic powers - - - ib.

THE LEVER.

409. Levers, 1st, 2nd, and 3rd kinds - 176
410. Condition of equilibrium - - ib.
411. Effect of power or weight varies as leverage - - - 177
412. Power or weight oblique to lever ib.
413. Relation of power and weight in levers of 1st, 2nd, and 3rd kinds 178
414. Examples of levers of 1st kind — Balance - - - - ib.
415. Sensibility of balance - - - 179
416. Steelyard - - - - 180
417. Letter balance - - - - 181
418. Spring balance - - - - ib.
419. Crowbar, scissors, &c. - - ib.
420. Examples of 2nd kind, — Oar, rudder, &c. - - - 182
421. Examples of 3rd kind,—Limbs of animals, &c. - - - 183
422. How to determine pressure on fulcrum of lever - - ib.
423. Rectangular lever - - - 184
424. Effect of beam resting on two props - - - - - ib.
425. Condition of equilibrium in compound lever - - - 185
426. Weighing machines - - 186
427. Knee lever - - - - 187
428. Beautiful example of complex leverage in the mechanism which connects the key and hammer in Erard's pianoforte - - ib.
429. Power of machine how expressed 189
430. Equivalent lever - - - ib.
431. Complex machine may be represented by equivalent compound lever - - - - ib.

WHEEL-WORK.

432. Wheel and axle - - - ib.
433. Condition of equilibrium - 190
434. Various ways of applying power ib.
435. Windlass - - - - ib.
436. Capstan - - - - ib.
437. Treadmill, &c. - - - 191
438. French quarry wheels - - ib.
439. Case in which power or resistance variable - - - 192
440. Method of augmenting ratio of power without complicating machine - - - - 193
441. Compound wheels and axles analogous to compound levers - ib.
442. Various methods of communicating force between wheels and axles - - - - 194
443. By endless bands or cords - ib.
444. By rough surfaces in contact - 195
445. By teeth - - - - 196
446. Formation of teeth - - ib.
447. Methods of computing condition of equilibrium in wheel work - ib.
448. Spur wheels - - - - ib.
449. Crown wheels - - - 197
450. Bevelled wheels - - - ib.

Sect. Page
451. Rack and pinion - - 197
452. Ratchet wheel - - - 198

PULLEYS.

453. Ropes not perfectly flexible nor perfectly smooth - - - ib.
454. Hence necessity of sheaves - 199
455. Fixed pulley, useful to change direction of power - - ib.
456. Case in which power and resistance parallel - - - 200
457. Power and weight equal when a single rope used with fixed pulley - - - - ib.
458. Mechanical convenience of changing direction of power - ib.
459. Fire-escape - - - 201
460. Single movable pulley - - ib.
461. Case in which cords not parallel ib.
462. Movable block with several sheaves - - - - 202
463. Condition of equilibrium of such a block - - - - ib.
464. Effect of attaching end of rope to movable block - - - ib.
465. Smeaton's and White's pulleys - ib.
466. Systems of pulleys consisting of several ropes and movable blocks - - - - 203
467. Practical effect of pulleys varies considerably from their theoretical effect - - - 204

INCLINED PLANE,—WEDGE AND SCREW.

468. Effect of inclined surface on weight - - - 205
469. Inclined plane — Condition of equilibrium - - 206
470. Apparatus to illustrate experimentally inclined plane - ib.
471. Inclined roads - - - ib.
472. Inclined planes on railways - 207
473. Case in which power acts in direction inclined to plane - - ib.
474. Examples - - - 208
475. Case of double inclined plane - ib.
476. Case of self-acting planes on railways - - - - 209
477. Wedge - - - - ib.
478. Wedges consist of two inclined planes - - - - 210
479. Theory of wedge practically inapplicable - - - ib.
480. Power applied to wedge principally percussion - - - ib.
481. Practical use of wedge - - ib.
482. Practical examples—cutting and piercing instruments - - 211
483. Utility of friction in application of wedge - - - ib.
484. Screw - - - - ib.
485. Power applied to screw by means of lever - - - - 212
486. Methods of transmitting power to resistance - - - ib.
487. Condition of equilibrium - 213
488. Great mechanical force of screw explained - - - ib.
489. Various methods of connecting screw and nut - - - ib.
490. Various examples of application of screw - - - 214

CONTENTS.

Sect.		Page
491.	Manner of cutting screw	215
492.	Method of augmenting force of screw	ib.
493.	Hunter's screw	ib
494.	Micrometer screw	216
495.	Endless screw	ib.

CHAP. III.

PENDULUM, BALANCE, AND FLY.

496.	The pendulum	217
497.	Oscillation of pendulous mass	ib.
498.	Isochronism of the pendulum	219
499.	Time of oscillation independent of weight of pendulum	ib.
500	How time of oscillation affected by length	ib.
501.	Experimental illustration	220
502.	Time of vibration being given, to find length of pendulum	221
503.	Time of oscillation varies with attraction of gravity	ib.
504.	Law of this variation	222
505.	Pendulum indicates variation of gravity in different latitudes	ib.
506	Analysis of motion of a pendulous mass of definite magnitude	ib.
507.	Centre of oscillation	224
508.	Centres of oscillation and suspension interchangeable	ib.
509.	Variations of pendulum consequent on change of temperature	225
510.	Compensation pendulums	ib.
511.	Pendulum a measure of time	ib.
512.	Pendulum measure of force of gravity	ib.
513.	Pendulum indicates form and diurnal rotation of the earth	226
514.	Pendulum measures velocity of falling bodies	ib.
515.	Curious properties of a swing	227
516.	Actual length of pendulum which vibrates seconds	229
517.	Balance wheel	ib.
518.	The fly	ib.

CHAP. IV.

REGULATION AND MODIFICATION OF FORCE AND MOTION.

519.	Regularity of motion necessary in machinery	232
520.	Causes of irregular motion	ib.
521.	When varying power is opposed to uniform resistance	233
522.	When uniform power opposed to varying resistance	ib.
523.	When variation of power not proportional to variation of resistance	ib.
524.	When effect of power unequally transmitted to resistance	234
525.	Regulators — General principle of their action	ib.
526.	The governor	ib.
527.	Fusee in watchwork	235

Sect.		Page
528.	When efficacy of machine to transmit power to working point varies	236
529.	Effect of crank	ib.
530.	Regulating effect of fly wheel	238
531.	Methods of augmenting or mitigating energy of moving power	239
	Case of watchwork moved by mainspring	ib.
532.	Case of clockwork moved by weight	240
533.	Cases in which energy of power augmented	ib.
534.	Analysis of action of hammers, sledges, &c.	ib.
535.	Inertia supplies means of accumulating force	241
536	Effect of weapon called life preserver, flails, &c.	ib.
537.	Loaded lever of screw press	242
538.	This accumulation of force involves no paradox	ib.
539.	Rolling and punching mills — Effect of fly wheel	ib.
540.	Position of fly wheel	243
541.	Method of cutting open work in metal	ib.
542.	Modification of motion	ib.
543.	Continued rectilinear motion	245
544.	Reciprocating rectilinear motion	246
545.	Continued circular motion	248
546	Universal joints	250
547.	Alternate circular motion	251
548.	Joints	257
549.	Gimbals	ib.
550	Ball and socket	258
551.	Cradle joint	259
552.	Hinge	ib.
553.	Trunnions	260
554.	Axles	ib.
555.	Telescope joint	ib.
556	Bayonet joint	ib.
557.	Clamps and adjusting screws	261
558.	Couplings	ib.

CHAP. V.

RESISTING FORCES.

559.	Forces which destroy but cannot produce motion	264
560.	Effects of friction	265
561.	Friction aids power in supporting weight, but opposes power in moving it	266
562.	How this modifies conditions of equilibrium	ib
563.	Cases in which friction is the whole power	ib.
564.	Great use of friction in economy of nature and art	267
565.	Friction of sliding and rolling	ib.
566.	Laws of friction empirical but still useful	ib.
567.	Methods of measuring sliding friction	268
568.	Angle of repose	269
569.	Friction proportional to pressure	ib.
570.	Effect of crossing the grains	ib.
571.	Pivots of wheels	ib.
572.	Selection of lubricants	ib.
573.	Rolling friction	270

CONTENTS.

Sect.		Page
574	Sledges	270
575.	Use of rollers and wheels	271
576.	Friction rollers	272
577.	Friction wheels	ib.
578.	Effect of magnitude of wheels	ib
579.	Best line of draught	273
580.	Friction of wheel carriages on roads	ib.
581.	Railway	ib.
582.	Brakes	274
583.	Friction a point of resistance	275
584.	Effects of imperfect flexibility of ropes	276
585.	Resistance of fluids	277
586.	Resistance depends on frontage of moving body	278
	Also affected by form of front	ib.
587.	Increases in high ratio with speed	ib
588.	Advantages of ponderous missiles	279
589.	Resistance of air to motion of falling bodies	ib.

CHAP. VI.

STRENGTH OF MATERIALS.

590.	Strength of solid bodies	280
591.	Difficulty of ascertaining its laws	ib.
592.	Several ways in which strength may be manifested	281
593.	Strength to resist direct pull	ib.
594.	Method of experimentally measuring it	282
595.	Table showing most recent results of such experiments	ib
596.	Iron the most tenacious — Effect of alloys	ib.
597.	Effect of heat	283
598.	Strength of timber	ib.
599.	Strength of cordage	ib.
600.	Animal substances	ib
601.	Strength to resist pressure or direct thrust	ib.
602.	Tables showing the strength of columns	284
603.	Results of Hodgskinson's researches	ib.
604.	Strength to resist torsion	285
605.	Strength to resist transverse strain	ib.
	Case of a beam supported at one end	ib.
606.	Case of a beam supported at both ends	287
607.	Case of rectangular beams	288
608.	How strength of beam affected by form of its transverse section	289
609.	Transverse strength of solid and hollow cylinders	290

Sect.		Page
610.	Examples in the structure of animals and plants	290
611.	Tabular statements of the average transverse strength of beams	291
612.	Tredgold's table of the transverse strength of metals and woods	ib.
613.	Barlow's table of the transverse strength of woods	292
614.	Effect of distribution of the weight on beam	ib.
615.	Strength of a beam increased by partially sawing it transversely and inserting a wedge	293
616.	Why strength of structure diminished as magnitude increased	ib.
617.	Strength on the large scale not to be judged by that of the model	ib.

CHAP. VII.

MOVING POWERS.

618.	Mechanical agents	294
619.	Dynamical unit	ib.
620.	Machines and tools	ib.
621.	Animal power	295
622.	Relation between the load and speed	ib
623.	Force of a man varies with the way it is applied	296
624.	Man working at capstan	ib.
625.	Man working by his weight	ib.
626.	Apparatus for this method	297
627.	Spade labour	298
628.	Earthwork on railways	299
629.	Horse power	ib.
630.	Horse power compared with that of other animals	301
631.	Various estimates of animal power	ib.
632.	Peron's estimate	ib.
633.	Tredgold's estimate of horse work	ib.
634.	Comparative force of different animals	302
635.	Steam horse	ib.
636.	Water power	303
637.	Wind power	ib.
638.	Steam power	ib.
639.	Consumption of fuel	304
640.	Mechanical effects of fuel	ib.
641.	Springs and weights	ib.
642.	Forces produced by heat	305
643.	By congelation	ib.
644.	Electro-magnetic force	ib.
645.	Chemical agency	306
646.	Gun cotton	307
647.	Capillary attraction	ib.
648.	Perpetual motion	ib.

BOOK IV.

Illustrations of the Application of Mechanical Principles in the Industrial Arts.

CHAPTER I.

CRANES.

Sect.		Page
649.	Cranes in general - -	310
650.	Movable cranes - - -	ib.
651.	Cranes for building - -	311
652.	Manufactory and railway cranes	312
653.	Fixed cranes - - -	313
654.	Drops for loading and unloading vessels - - - -	315

CHAP. II.

ENGINES WORKING BY IMPACT.

655.	Pile engines - - -	317
656.	Stamping engines - -	320
657.	Sledge hammer - - -	321
658.	Coining press - - -	323

CHAP. III.

BORING, PLANING, SAWING, AND SCREW AND TOOTH-CUTTING ENGINES.

659.	Boring tools - - -	327
660.	Planing machine - - -	329
661.	Saw mills - - -	330
662.	Screw-cutting engine - -	335
663.	Engine for cutting the teeth of wheels - - - -	336

CHAP. IV.

FLOUR MILLS.

664.	Process of grinding - -	337
	Description of flour mills -	ib.
	Form and structure of mill stones - - - -	ib.
	Separation of the flour and bran	ib.

CHAP. V.

CLOCK AND WATCH WORK.

665.	Time measurer a necessity of civilised life - - - -	339
666.	Sun-dials - - -	340
667.	Position of gnomon - -	341
668.	Clepsydra - - -	342
669.	Sand-glass - - -	ib.
670.	Mercurial time measurer -	343
671.	Clocks and watches - -	ib
672.	Pendulum - - -	344

Sect.		Page
673.	Compensation pendulums -	345
674.	Harrison's gridiron pendulum	347
675.	Mercurial pendulum - -	348
676.	Another compensation pendulum	ib.
677.	Connection between pendulum and escapement - -	349
678.	Motion of hand intermitting	350
679.	How the motion of the pendulum is sustained - - -	351
680.	Its action upon the escapement -	ib.
681.	Motion of hands, how proportioned	352
682.	Method of cutting wheels and pinions - - - -	ib.
683.	Moving power, weight -	353
684.	Mainspring - - -	354
685.	Fusee - - - -	356
686.	Balance wheel - - -	359
687.	Isochronism - - -	360
688.	Compensation balances -	ib.
689.	Movement imparted to the hands of a watch - - -	ib.
690.	Movement of the hands of a clock	363
691.	To regulate the rate - -	366
692.	Recoil escapement - -	367
	Cylindrical escapement -	ib.
693.	Duplex escapement - -	370
	Lever escapement - -	371
694.	Detached escapement - -	ib.
695.	Maintaining power in clocks	373
696.	Maintaining power in watches	374
697.	Cases in which weights and springs are used - -	376
698.	Watches and chronometers, marine chronometers - -	377
699.	Observatory clocks - -	ib.
700.	Striking apparatus - -	ib.

CHAP. VI.

THE PRINTING PRESS.

701.	Setting the types - -	381
702.	Old method of printing -	ib.
703.	Inking rollers - - -	382
704.	Modern printing presses -	ib.
705.	Arrangement of the form -	383
706.	Single printing machines -	ib.
707.	Double printing machines -	384
708.	Applegath and Cowper's double printing machine - -	385
709.	Printing machines on this principle erected for the *Times* in 1814	388
710.	Further improvement in this machine, increasing its power -	ib.
711.	*Times* printing machine of 1850 -	390
712.	Marinoni's newspaper printing machine - - -	394
713.	Marinoni's book printing machine - - - -	400

Frank Southern (handwritten signature)

ELEMENTARY COURSE

OF

MECHANICS.

BOOK THE FIRST.
PROPERTIES OF MATTER.

CHAPTER I.
MAGNITUDE, FIGURE, AND STATE OF AGGREGATION.

1. THE material world, the bodies which compose it, and their qualities and properties, form the subjects of contemplation and inquiry to the natural philosopher.

Among these properties, the most important are the following: —

1. MAGNITUDE and FORM.
2. STATES OF AGGREGATION.
3. IMPENETRABILITY.
4. DIVISIBILITY.
5. POROSITY and DENSITY.
6. COMPRESSIBILITY and CONTRACTIBILITY.
7. ELASTICITY and DILATABILITY.
8. INACTIVITY.
9. SPECIFIC PROPERTIES.

We shall, therefore, in the present book, explain and illustrate these qualities.

2. The idea of magnitude is so simple that any attempt to explain it by definition would only tend to obscurity.

B

Magnitude is either *linear*, *superficial*, or *solid.*

3. *Linear magnitude* is length or distance. Length or distance is expressed numerically by means of some unit; the unit adopted being, as a matter of convenience, small or great, according to the magnitude of the length or distance to be expressed. Thus, we say, the side of this leaf is so many *inches*, the length of the room is so many *feet*, the distance from London to York is so many *miles.*

4. *Superficial magnitude*, sometimes called *area*, has length and breadth. Its quantity is expressed by so many *square inches*, so many *square feet*, or so many *square miles*. Thus, we say, the area of this page of paper is so many *square inches*, that of the floor of this room so many *square feet*, and the entire surface of the earth is so many *square miles.*

5. *Solid magnitude* has length, breadth, and depth, height or thickness.

Its quantity, sometimes called its *volume*, is expressed by so many *cubic inches*, so many *cubic feet*, or so many *cubic miles*. Thus, we say, the volume of this book is so many *cubic inches*, the volume of this room is so many *cubic feet*, and the volume of the terrestrial globe is so many *cubic miles.*

6. The *solid magnitude* or *volume* of a body, is the space which that body occupies or fills.

7. The *form*, *figure*, or *shape* of a line, surface, or solid, depends upon the relative position of its parts or limits.

Two lines may be of equal length, while they differ in form, shape, or figure. Thus, the arc of a circle and a straight line may both measure a foot in length, while their figure or shape is different. Again, two lines may have the same form or figure, but very different lengths: thus, the two straight lines which *form* the edges of this page have the same form, both being *straight*, but they have different *lengths*. In like manner, two different arcs of the same circle will have like forms but different lengths, and two arcs of different circles may have equal lengths, and will both be curved and both circular, but they will, nevertheless, have different forms, inasmuch as the curvature of one will be greater than that of the other.

8. *Solid* bodies are bounded by *surfaces*. Thus, a globe is included within its spherical surface, a cube is included within six plane square surfaces.

9. *Surfaces* are bounded by *lines*. Thus, the surface of this page is bounded by four straight lines, forming its edges; the surface of a circle is bounded by the line called its circumference.

10. *Magnitude is unlimited as well in its increase as in its diminution.* — There is no magnitude so great, that we cannot conceive

a greater; and none so small, that we cannot conceive a smaller. The diameter of the earth measures about 8000 miles; but it is very small compared to the diameter of the sun, which measures nearly 900000 miles; and this, again, is itself very small compared with the distance between the earth and the sun, which measures little less than 100,000000 of miles; and even this last space, great as it is, vanishes to nothing, compared with the distance between the sun and the fixed stars.

11. In like manner, there is no limit to the diminution of which magnitude is susceptible. There is no line so small, that may not be bisected, or halved, or reduced, in any desired proportion. Let P I (*fig.* 1.) be a line of any proposed length, as, for example, the tenth of an inch.

Draw P O at right angles to it, and a line I, 10 parallel to P O.

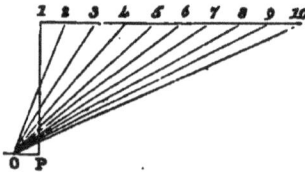

Take upon this line distances 1, 2, 3, 4, &c. successively equal to P O: the line O 2 will cut off half P I; O 3 will cut off one-third of P I; O 4 will cut off one-fourth of it. O 5 one-fifth, and so on. Now, as there is no limit to the number of equal parts that may be taken successively on the parallel 1, 10, so

Fig. 1.

there is no fraction so minute, that a more minute one may not be cut from the line P I.

12. All bodies consist of component parts, similar in their qualities to the whole, and into which, as will presently be shown, they are separable or divisible. Thus, a mass of metal may be reduced to powder, by filing, grinding, and a variety of other expedients; each particle of this powder will have the same qualities as the entire mass of metal of which it constituted a part.

13. These component parts of bodies are observed to exist in three distinct states of aggregation, which are distinguished in mechanical science by the terms *solid, liquid,* and *gaseous.*

14. **Solid state.** — A solid body is one of which the component parts cohere with such force, that it maintains its figure, unless submitted to some action more or less violent, by which it will be fractured, bruised, or otherwise changed in form. Thus, a solid body laid upon a plane surface will rest upon it in any position, without dropping asunder by the tendency of the weight of its component parts to separate them.

Stones, metals, wood, are examples of solids.

15. **Liquid state.** — A liquid body is one of which the component parts do not cohere with sufficient force to prevent their separation by the mere influence of their weight. Thus, a mass of

liquid placed upon a plane will separate by the separate weights of its particles, and will spread itself in a film, more or less thin, over the surface of the plane.

If placed in a vessel within which it is confined by a bottom and sides, it will flow into all the inferior parts of the vessel, and its surface will become level; for if any part of such surface were more elevated than another, the particles forming such elevation would fall down by their own weight to the level of the others.

16. Solids in a state of pulverization must not be confounded with liquids, in the properties of which they do not participate. Sand or dust consists of a great number of *small solids*, each of which has the qualities of a solid as definitely as the largest mass of rock. If such particles be examined with a microscope, they will be found to have different forms, to be bounded by surfaces and lines, and to maintain their figure in virtue of cohesion.

17. **Gaseous state.** — A body in the gaseous or aeriform state is one whose component particles not only are not held together by mutual cohesion, but have towards each other a *repulsion*, in virtue of which they will separate, so that the whole mass has a power of expansion to which there is no known limit.

18. *Atmospheric air* is the most familiar example of this state of body. If we suppose a quantity of air to be confined in a cylinder, under a piston which moves air-tight, such piston being drawn upwards, the air will not rest in the bottom of the cylinder, as the same volume of liquid would do, leaving vacant the space above it, but it will expand in virtue of the repulsive force already mentioned, which prevails among its particles, and it will fill the entire space under the piston. To this expansion there is no practical limit; such a piston may be indefinitely raised, producing an indefinitely-increased space in the cylinder under it, and the air will still continue to expand, so as to fill this space to whatever extent it be increased.

19. Atmospheric air, as will be seen hereafter, is by no means the only example of bodies in the gaseous form. Innumerable varieties of such bodies are found existing in the material world, and still greater varieties result from the experimental operations of the natural philosopher and the chemist.

20. There are numerous bodies which may exist in any of these three states of solid, liquid, or gas, according to certain physical conditions, which will be explained hereafter; and analogy renders it probable that all bodies whatever may assume, under different conditions, these several states.

21. **Water** affords a familiar example of this: as ice it exists in the solid state, as water in the liquid state, and as steam in the gaseous state.

CHAP. II.

IMPENETRABILITY. — DIVISIBILITY.

22. IMPENETRABILITY is the quality in virtue of which a body occupies a certain space, to the exclusion of all other bodies. This idea is so inseparable from matter, that some writers affirm that it is nothing but matter itself; that is, that when we say that a body is impenetrable, we merely say that it is a body.

However this be, the existence of this quality of impenetrability is so evident as to admit of no other proof than an appeal to the senses and the understanding. No one can conceive two globes of lead, each a foot in diameter, to occupy precisely the same place at the same time. Such a statement would imply an absurdity, manifest to every understanding.

23. Even bodies in the gaseous form, the most attenuated state in which matter can exist, possess this quality of impenetrability as positively as the most hard and dense substances.

24. If we invert a common drinking-glass, and plunge its mouth in water, the water will be excluded from the glass, in spite of the pressure produced by the weight of the external water; because the air which filled the glass at the moment of immersion is still in it, and its presence is incompatible with that of any other body. It is true, however, that in this case the water will rise a little above the mouth of the glass; but this effect arises not from the penetrability of the air, but from another quality, viz. its compressibility, which we shall presently explain.

25. The numerous examples of apparent penetration which will present themselves to all observers are only cases of displacement.

26. If we walk through the atmosphere, our bodies may be said in one sense to penetrate it; but they do so only in the same manner as they would penetrate a liquid in passing through it. They displace, as they move, as much air as is equal to their own bulk.

27. If a cambric needle be plunged in a glass of water, it might appear that penetration took place; but it is evident that the needle displaces a quantity of water precisely equal to its own bulk; and if we had the means of measuring with the necessary precision the position of the surface of the water in the glass, we should find that on plunging the needle in the liquid the surface rises through a space corresponding precisely to a quantity of water equal in volume to the bulk of the needle.

28. **Divisibility unlimited.** — As all bodies consist of parts which are similar in their qualities to the whole, a question arises whether there is any limit to their subdivision or comminution. Are there, in short, any ultimate particles at which the process of division must cease? or, to speak more correctly, are there any ultimate particles so constituted that any further division would resolve them into parts differing in quality from the entire mass?

29. To make our meaning understood, let us take the example of the most common of all substances, water. The discoveries of modern chemistry have disclosed the fact that water is a compound body, formed by the combination of two gases, called *oxygen* and *hydrogen*. These gases, in their sensible properties and appearance, are similar to atmospheric air, and have never separately assumed the liquid form; but, by certain means which will be explained in a future part of this work, they may be made to combine and coalesce, and when so combined they form water.

Hence it is certain that the liquid, water, consists of ultimate particles, or molecules, as they are sometimes called; which molecules are composed of particles of the two gases above mentioned combined together.

30. Now, the question is, can water be practically so minutely divided that its particles admit of no further subdivision, save and except that which would resolve them into their constituent gases, oxygen and hydrogen?

To this it is answered, that, although the process of subdivision may be carried to an extent which has no practical limit, yet that by no process of art or science have we ever even approached to those ultimate constituent atoms which admit of no other division save decomposition.

31. Nor is this unlimited divisibility and comminution peculiar to water. It is common to all substances, whether solid, liquid, or gaseous. They may all be reduced to particles of the most unlimited minuteness, and yet each of these minute particles will possess the same qualities as the largest mass of the same substance.

32. **Examples.** — As this quality of unlimited divisibility involves conditions of the most profound interest, as well in the sciences as in the arts, we shall offer here several examples in illustration of it.

33. The most solid bodies are capable of unlimited comminution, by a variety of mechanical processes, such as cutting, filing, pounding, grinding, &c. If a mass of marble be reduced to a fine powder by the process of grinding, and this powder be then

purified by careful washing, its particles, if examined by a powerful microscope, will be found to consist of blocks having rhomboidal forms, and angles as perfect and as accurate as the finest specimens of calcareous spars. These rhomboids, minute as they are, may be again broken and pulverised, and the particles into which they are divided will still be rhomboids of the same form and possessing the same character. The particles of such powder being submitted to the most powerful microscopic instruments, and the process of pulverization being pushed to the utmost practical extreme, it is still found that the same forms are reproduced.

34. The polish of which the surfaces of certain bodies, such as steel, the diamond, and other precious stones, are susceptible, is an evidence at once of the limited sensibility of our organs, and the unlimited divisibility of matter. This polish is produced, as is well known, by the friction of emery powder or diamond dust, and consequently each individual grain of such powder or dust must leave a little trench or trace upon the surface submitted to such friction. It is evident, therefore, that after this process has been completed, the surface which presents to the senses such brilliant polish, and apparently infinite smoothness, is in reality covered with protuberances and indentations, the height and depth of which cannot be less than the diameter of the particles of powder by which the polish has been produced.

35. In the detection of matter in a state of extreme comminution, the sense of sight is infinitely more delicate than that of touch. If we rub a piece of gold upon a touchstone, we plainly see the particles of matter which are left upon the surface of the stone. The touch, however, cannot detect them.

36. In the preceding examples the comminution, however great, cannot be easily submitted to actual measurement. Certain processes, however, in the arts enable us to obtain exact numerical estimates of a minute divisibility, which without them might appear incredible. If a thin tube of glass, being held before the flame of a blow-pipe until the glass be softened and acquire a white heat, be drawn end from end, a thread of glass may be obtained so fine that its diameter will not exceed the two thousandth part of an inch. This filament of glass will have all the fineness and almost all the flexibility of silk, and yet a bore proportional to that which passed through the original tube will still pass through its centre. The presence of this bore may be rendered manifest by passing a fluid through it.

37. It has been ingeniously conjectured, that if a filament of this degree of fineness could be obtained of a material which would retain sufficient inflexibility, it might be made to penetrate the

flesh without producing either pain or injury, inasmuch as its magnitude would be much less than that of the pores of the integuments.

38. In the application of the telescope to astronomical purposes, the distance between objects which are present at one and the same time within the field of view of the instrument, is measured by fine threads which are extended parallel to each other across the field of view, and which may be moved towards and from each other until they are made to pass through the objects between which we desire to measure the distance. An experiment, then, which determines the distance between these threads measures the distance between the objects.

But these threads, being placed before the eye-glass of the telescope, and therefore necessarily magnified in the same manner as the objects themselves, would, unless such filaments were of an extreme degree of tenuity, appear in the field of view like great broad bands, and would conceal many of the objects which it might be necessary to observe. It was therefore necessary to resort to the use of filaments of extraordinary minuteness for this purpose. The threads of spiders' webs are used with more or less success; but the late Dr. Wollaston invented a beautiful expedient by which metallic threads of any degree of fineness might be obtained.

Let us suppose a piece of platinum wire, one-hundredth of an inch in diameter, a fineness easily obtainable by the process of wire-drawing, to be extended along the axis of a cylindrical mould, one-fifth of an inch in diameter, the wire being thus the twentieth part of the diameter of the mould. Let the mould be then filled with silver in a state of fusion. When this is cold we shall have a cylinder of silver, having in its axis a thread of platinum the twentieth part of its diameter. This compound cylinder is then submitted to the common process of wire-drawing, during which the platinum in its centre is drawn with the silver, the proportion of their diameters being still maintained. When the wire is drawn to the greatest degree of fineness practicable, a piece of it is plunged in nitric acid, by which the surrounding silver is dissolved, and the platinum wire remains uncovered.

39. By this process Dr. Wollaston obtained platinum wire so fine, that thirty thousand pieces, placed side by side in contact, would not cover more than an inch. It would take one hundred and fifty pieces of this wire bound together to form a thread as thick as a filament of raw silk. Although platinum is the heaviest of the known bodies, a mile of this wire would not weigh more than a grain. Seven ounces of this wire would extend from London to New York.

40. The natural filaments of wool, silk, and fur, afford striking examples of the minute divisibility of organized matter. The following numbers show how many filaments of each of the annexed substances placed in contact, side by side, would be necessary to cover an inch : —

Coarse wool · - · · · · · 500
Fine Merino wool · · · · · 1250
Silk - · · - · · - 2500

The hairs of the finest furs, such as beaver and ermine, hold a place between the filaments of Merino and silk, and the wools in general have a fineness between that of Merino and coarse wool. All these objects are sensible to the touch.

It will be remembered that they are compound textures, having a particular structure, each containing very different elements, which are prepared by the processes of nutrition and secretion.

41. The optical investigations of Newton disclosed some astonishing examples of the minute divisibility of matter. A soap-bubble as it floats in the light of the sun reflects to the eye an endless variety of the most gorgeous tints of colour. Newton showed, that to each of these tints corresponds a certain thickness of the substance forming the bubble ; in fact, he showed in general, that all transparent substances, when reduced to a certain degree of tenuity, would reflect these colours.

Near the highest point of the bubble, just before it bursts, is always observed a spot which reflects no colour and appears black. Newton showed that the thickness of the bubble at this black point was the 2,500000th part of an inch! Now, as the bubble at this point possesses the properties of water as essentially as does the Atlantic Ocean, it follows, that the ultimate molecules forming water must have less dimensions than this thickness.

42. The same optical experiments were extended to the organic world, and it was shown, that the wings of insects which reflect beautiful tints resembling mother-of-pearl owe that quality to their extreme tenuity. Some of these are so thin that 50000 placed one upon the other would not form a heap of more than a quarter of an inch in height!

43. In the process of gold-beating the metal is reduced to laminæ or leaves of a degree of tenuity which would appear fabulous, if we had not the stubborn evidence of common experience in the arts as its verification. A pile of leaf-gold the height of an inch would contain 282000 distinct leaves of metal! The thickness, therefore, of each leaf is in this case the 282000th part of an inch. Nevertheless, such a leaf completely conceals the object which it is used to gild ; it moreover protects such object from the action of external agents as effectually as though it were plated with gold an inch thick.

44. In the manufacture of embroidery, fine threads of silver gilt are used. To produce these, a bar of silver, weighing 180 oz., is gilt with an ounce of gold ; this bar is then wire-drawn until it is reduced to a thread so fine that 3400 feet of it weigh less than an ounce. It is then flattened by being submitted to a severe pressure between rollers, in which process its length is increased to 4000 feet. Each foot of the flattened wire weighs, therefore, the 4000th part of an ounce. But as in the processes of wire-drawing and rolling the proportion of the two metals is maintained, the gold which covers the surface of the fine thread thus produced consists only of the 180th part of its whole weight. Therefore the gold which covers one foot is only the 720000th part of an ounce, and consequently the gold which covers an inch will be the .8,640000th part of an ounce. If this inch be again divided into one hundred equal parts, each part will be distinctly visible without the aid of a microscope, and yet the gold which covers such visible part will be only the 864,0000000th part of an ounce.

But we need not stop even here. This portion of the wire may be viewed through a microscope which magnifies 500 times ; and by these means, therefore, its 500th part will become visible.

45. In this manner, therefore, an ounce of gold may be divided into 432000,000000 parts, and each part will still possess all the characters and qualities found in the largest mass of the metal. It will have the same solidity, texture, and colour, will resist the same chemical agents, and will enter into combination with the same substances. If this gilt wire be exposed to the action of nitric acid, the silver within the coating will be dissolved, but the hollow tube of gold which surrounds it would still cohere and remain suspended.

46. The organic world affords most interesting and striking examples of the minute divisibility of matter. None can be selected more remarkable than that presented by the blood of animals, which is not, as it appears to the naked eye, a uniform red liquid, but consists of a transparent colourless fluid called *lymph*, in which innumerable small red corpuscles of solid matter float.

47. In different species these red corpuscles differ both in form and size. In the human blood, and in that generally of animals who suckle their young, they are circular discs, their surfaces being slightly concave, like the spectacle glasses used by short-sighted persons. In birds, reptiles, and fishes, they are generally oval, the oval being more or less elongated in different species. The surface of the discs in these species, instead of being concave, are convex, like the spectacle glasses used by weak-sighted persons. The thickness of these discs varies from one third to one quarter of their diameter. Their diameter in human blood is the

3500th part of an inch ; they are smallest in the blood of the *Napu* musk-deer, where they measure only the 12000th of an inch. It would require 50000 of these discs as they exist in the human blood to cover the head of a small pin, and 800000 of the disks of the blood of the musk-deer to cover the same surface.

48. It follows from these dimensions that in a drop of human blood which would remain suspended from the point of a fine needle there must be about 3,000000 of disks, and in a like drop of the blood of the musk-deer there would be about 120,000000, yet these corpuscles are rendered not only distinctly visible to the senses by the aid of the microscope, but their forms and dimensions are rendered apparent. Small as they are, they are divisible, and can be resolved into their elements by chemical agents.

49. But these diminutive globules are exceeded in minuteness by innumerable creatures whose existence the microscope has disclosed, and whose entire bodies are inferior in magnitude to the globules of blood.

Microscopic research has disclosed the existence of animals, a million of which do not exceed the bulk of a grain of sand, and yet each of these is composed of members as admirably suited to their mode of life as those of the largest species. Their motions display all the phenomena of vitality, sense, and instinct. In the liquids which they inhabit they are observed to move with the most surprising speed and agility; nor are their motions and actions blind and fortuitous, but evidently governed by choice and directed to an end. They use food and drink, by which they are nourished, and must, therefore, be supplied with a digestive apparatus. They exhibit a muscular power far exceeding in strength and flexibility, relatively speaking, the larger species. They are susceptible of the same appetites, and obnoxious to the same passions, as the superior animals, and, though differing in degree, the satisfaction of these desires is attended with the same results as in our own species.

Spallanzani observes that certain animalcules devour others so voraciously that they fatten and become indolent and sluggish by over-feeding. After a meal of this kind, if they be confined in distilled water so as to be deprived of all food, their condition becomes reduced, they regain their spirit and activity, and once more amuse themselves in pursuit of the more minute animals which are supplied to them. These they swallow without depriving them of life, as, by the aid of the microscope, the smaller, thus devoured, have been observed moving within the body of the greater.

The microscopic researches of Ehrenberg have disclosed most surprising examples of the minuteness of which organized matter is susceptible. He has shown that many species of infusoria exist which are so small that millions of them collected into one mass would not exceed the bulk of a grain of sand, and a thousand might swim side by side through the eye of a needle. The shells of these creatures are found to exist fossilized in the strata of the earth in quantities so great as almost to exceed the limits of credibility.

By microscopic measurement it has been ascertained that in the slate found at Bilin, in Bohemia, which consists almost entirely of these shells, a cubic inch contains forty-one thousand millions ; and as a cubic inch weighs two hundred and twenty grains, it follows that one hundred and eighty seven millions of these shells must go to a grain, each of which would consequently weigh the 187,oooooooth part of a grain.

All these phenomena lead to the conclusion that these creatures must be supplied with an organization corresponding in beauty with those of the larger species.

50. If a grain of salt be dissolved in 1000 grains of distilled water, each grain of the water will contain the 1000th part of the grain of the salt ; and if a grain of this water be mixed with 1000 grains of distilled water, the 1000th part of a grain of salt which it holds in solution will be uniformly diffused through the latter, so that each grain of the latter solution will contain the millionth part of a grain of salt. The presence of the salt in this second solution can be detected by certain chemical tests. It is evident that this process may be continued to a still greater extent.

51. A grain of sulphate of copper, dissolved in a gallon of water, will impart to the whole mass of the liquid a plainly perceptible tinge of blue ; and a grain of carmine will give its peculiar red to the same quantity of water. It follows, therefore, that a minute drop of such water will contain such a proportion of either of these substances as the drop bears to the gallon.

52. The sense of smelling, although it does not inform us of the mechanical qualities of minute masses of matter, determines, nevertheless, their presence · thus, it is known that a grain of musk will impregnate the atmosphere of a room with its odour for a quarter of a century, or more, without suffering any considerable loss in its weight. Every particle of the atmosphere which produces the sense of the odour must contain a certain quantity of the musk.

53. A thread of a spider's web, measuring four miles, will weigh very little more than a single grain ! Every one is familiar with the fact, that the spider spins a thread, or cord, by which his own weight hangs suspended. It has been ascertained that this thread is composed of about six thousand filaments.

54. The sense of taste, like that of smelling, may determine the

presence of matter, without manifesting, by direct evidence, anything concerning its mechanical qualities.

55. A portion of strychnine, so minute as to be scarcely perceptible to the sight, if dissolved in a pint of water, will render every drop of the water bitter. Now, it is evident that in this case, the strychnine being uniformly diffused through the water, the minute portion of it above mentioned is subdivided into as many parts as there are drops of water in a pint.

56. In like manner, a single grain of the salt of silver, called ammoniacal hyposulphite, will impart a flavour of sweetness to a gallon of water. Now, a gallon of water will weigh about 70000 grains; and as the flavour of the salt is perceptible in each grain of the water, it follows that one grain of this salt is thus divided into 70000 equal parts.

57. A small lump of sugar, dissolved in a cup of tea measuring half a pint, will sweeten the whole perceptibly. In this half-pint of tea there are 31000 drops. Each drop, therefore, must contain the 31000th part of the sugar dissolved, and each such drop is perceptibly sweet. But if the point of a needle be inserted in one of these drops, and withdrawn from it, a film of moisture will remain upon it, and the drop will not be visibly diminished. Yet this film of moisture will still be sweet, and will, therefore, contain a fraction of the 31000th part of the lump of sugar, too minute to admit of numerical estimation.

58. It may be asked, whether we are then to conclude, from these various facts, that matter is infinitely divisible, and that there are no original constituent atoms of determinate magnitude and figure, at which all subdivision must cease. Such an inference, however, would be unwarranted, even if we had no other means of deciding the question except those of direct observation, as we should thus impose those limits on the operations of nature which she has imposed upon our powers of observing them.

59. Although we are unable, by direct observation, to perceive the existence of molecules, or material atoms of determinate figure, yet there are many observable phenomena which render their existence in the highest degree probable, if not positively certain.

60. **Crystallization.**—The most remarkable of such phenomena are observed in the crystallization of salts. When salt is dissolved in distilled water, as in the preceding example, the mixture presents the appearance of a transparent liquid like water itself, the salt altogether disappearing from sight and touch. The presence of the salt in the water, however, can be established by weighing the solution, which will be found to exceed the original weight of the water by the exact amount of the weight of the salt dissolved.

Now, if this solution be heated to a sufficient temperature, the water will gradually evaporate; but this process of evaporation not affecting the salt, the remaining water will still contain the same quantity of salt in solution, and it will consequently become, by degrees, a stronger and stronger saline solution, the water bearing, consequently, a less and less proportion to the salt. The water will at length be diminished, by evaporation, to that point, that a sufficient quantity does not remain to hold in solution the entire quantity of salt contained in it. When this has taken place, each particle of water which is evaporated leaving behind it the salt which it held in solution, and this salt not being capable of being dissolved by the water which remains, it will float in such water in its solid and natural state, undissolved, just as particles of dust, or other matter not soluble in the water, would do. But the saline particles which thus remain floating in the liquid undissolved, will not collect in irregular solid pieces, but will exhibit themselves in regular figures, terminated by plane surfaces, always forming regular angles, these figures being invariably the same for the same species of salt, but different for different species. There are several circumstances attending the formation of these crystals which merit attention.

61. If one of these be detached from the others, and the gradual progress of its formation be submitted to observation, it will be found to grow larger, always preserving its original figure. Now, since its increase must be produced by the continual accession of saline molecules, disengaged by the water evaporated, it follows that these molecules, or atoms, must have such a shape, that, by attaching themselves successively to the crystal, they will maintain the regularity of its bounding planes, and preserve the angles which these planes form with each other unvaried. In fact, they must be so shaped, that the structure of the crystal they form may be built up by their regular aggregation into the form which it assumes. If one of these crystals be taken from the liquid during the process of its formation, and be broken, so as to destroy the regularity of its form, and then restored to the liquid, it will be observed soon to recover its regular form, the atoms of salt, successively dismissed by the evaporating water, filling up the irregular cavities produced by the fracture.

62. Two consequences obviously follow from this phenomenon : First, that the atoms of the salt dismissed by the water evaporated have such a form, as enables them, by combination, to give to the crystals the shape which they exhibit; and, Secondly, that the atoms which are successively attached to the crystals in the process of formation, attach themselves in a particular position, to explain which it is necessary to suppose that corresponding sides

of the crystals have attractions for each other, so that the atoms of salt not only attach themselves to the sides of the crystals, but place themselves there in a particular position. In a word, we must suppose that the walls of the crystal are built with these atoms in the same manner, and with the same regularity, as the walls of a building are formed with bricks.

All these, and many similar details of the process of crystallization, are, therefore, very evident indications of a determined figure in the ultimate atoms of the substances which are crystallized.

63. There are certain planes called planes of *cleavage*, in the direction of which natural crystals are easily divided. In substances of the same kind, these planes have always the same relative position; but they differ in different substances.

64. We must conclude, therefore, that these planes of cleavage are parallel to the sides of the constituent atoms of the crystals, and their directions therefore form so many conditions for the determination of the shape of these atoms.

65. It follows, therefore, from these effects, and the reasoning established upon them, that the substances which are susceptible of crystallization consist of ultimate atoms of different figure. Now, all solid bodies whatever are included in this class, for they have severally been found in, or are reducible to a crystallized form. Liquids crystallize in freezing: several of the gases have been already reduced to the liquid and solid forms, and analysis justifies the conclusion that all are capable of being reduced to this form.

Hence it appears reasonable to presume that all bodies whatever are composed of ultimate atoms, having determinate shape and magnitude; that the different qualities with which we find different bodies endued, depend upon the shape and magnitude of these atoms; that these atoms cannot be disturbed or changed so long as the body to which they belong is not decomposed into other elements, as we find the qualities which depend on them unchangeably the same under all the influences to which they have been submitted.

66. We must conclude also that these atoms are so minute in their magnitudes that they cannot be observed by any means which human art has yet contrived, but nevertheless that such magnitudes still have limits.

67. **Matter indestructible.**—The extreme division to which bodies are subjected in many natural and artificial processes, and especially when exposed to the application of heat or fire, has naturally suggested to minds not habituated to the rigid process of scientific reasoning, the idea that bodies are destructible. The

ancients, instead of the modern practice of inhumation, disposed of the bodies of their dead by burning them, upon the supposition that their component parts were by such operation destroyed.

The more exact reasoning of modern philosophy, however, teaches us that a power to destroy matter would be as inconceivable in a finite agent as a power to create it. It is certain that the quantity of matter which exists upon and in the earth has never been diminished by the annihilation of a single atom. Matter is in fact indestructible by any agency short of divine power. It may be asked, then, what becomes of the matter composing a body which, being subjected to the action of fire, gradually and completely disappears. The answer is, that in this, as well as in all other cases of the apparent destruction of matter, nothing takes place except its subdivision and the change of its form and position.

68. When a body is subjected to the action of heat, its elements are decomposed, and its constituent particles separated, many of them combine with other particles of matter, and form new substances possessing other qualities. Thus, when coal or other fuel is burned, the carbon enters into combination with one of the constituents of the atmosphere called oxygen, and forms a gaseous substance called carbonic acid, which rises into and mixes with the atmosphere. Another element, hydrogen, combines with the same constituent of the atmosphere and forms vapor, which also disperses in the atmosphere.

Sulphur, which is also occasionally present in fuel, combines with the same constituent of the air, forming a gas called sulphureous acid, which also escapes into the atmosphere. Thus the entire matter of the fuel, with the exception of a small portion of incombustible matter which falls into the ash-pit, is dispersed in the air, and no destruction or annihilation takes place.

That no portion of the matter of the fuel is destroyed or annihilated can be established by the incontrovertible experimental proofs of the chemist, for by the expedients of his science all the products of the combustion which have been just mentioned can be preserved, weighed, and decomposed. The oxygen which has entered into combination with each element of the fuel can be reproduced, as well as the constituents of the fuel itself, the latter of which being weighed, as well as the incombustible ash, the weight of the whole is found to be precisely equal to the weight of the fuel which was burned and apparently destroyed.

69. Liquids when subjected to heat are converted into vapor, and this vapor disperses in the atmosphere, so that the liquid seems to be boiled away; but if the vapor be preserved, as it may

be in a separate vessel, and exposed to cold, it will return to the liquid form, and its weight and measure will be found to be precisely the same as that of the liquid evaporated.

70. **Destructive distillation.** — There is a process in chemistry which is called destructive distillation. The term is objectionable because it implies a destruction where no destruction takes place. If a piece of wood, being previously weighed, be placed in a close retort and submitted to what is called destructive distillation, it will be found that water, a certain acid, and several gases will issue from it, all of which may be preserved, and mere charcoal will remain in the retort at the end of the process. If the water, acid, and gases which thus escape be weighed with the charcoal, the weight of the whole will be found to be precisely equal to that of the wood which was subjected to destructive distillation.

71. **General conclusion.** — Thus various forms of matter may be fused, evaporated, or submitted to combustion; animals and vegetables may die, organized bodies may be dissolved and decomposed, but in all cases their elementary and constituent parts maintain their existence. The remains of our own bodies after death are deposited in the grave, and enter into innumerable combinations with the materials of the soil, with the vegetation which covers it, and the air which circulates above it.

Consequently, these parts enter into an infinite series of other combinations, forming parts of other organized bodies, animal and vegetable, and which, after having discharged their functions, are thrown off again, mixing with the soil, the air, or organized matter, and once more running through the round of physical combinations.

The constituent atoms of matter are thus constantly performing a circle of duties in the economy of nature with infinitely more certainty and regularity than is observed in the most disciplined army or in the best regulated manufactory.

CHAP. III.

POROSITY, DENSITY, COMPRESSIBILITY, AND DILATABILITY.

72. THE volume or magnitude of a body is, as has been already explained, the space which is included within its external surfaces. The mass of a body, or the quantity of matter of which it consists, is the collection of atoms or molecules which compose it.

c

If these atoms were in actual contact, the volume would be completely occupied by the mass; but numerous results of observation prove that this is never the case. There is no body so solid or compact that it cannot, by various processes to be explained hereafter, be forced into less dimensions. Now, if its atoms were in absolute contact, and had no unoccupied spaces between them, this compression could not take place. It follows, therefore, that between the atoms or molecules which form the mass of a body there are vacant spaces in the magnitude or volume, which therefore consists partly of the spaces occupied by the atoms, and partly of the spaces which intervene between these atoms.

73. **Pores.** — These intervening spaces which separate the constituent atoms of a mass of matter are called pores; and the quality of bodies in virtue of which their constituent atoms are thus separated by vacant spaces is called *Porosity*.

74. **Density.** — In bodies of different species, the pores bear a greater or less proportion to the whole volume; or, in other words, the component atoms of the mass are placed more or less closely together. This circumstance determines what is called the *density* of bodies.

75. *Porosity* and *density* are therefore correlative terms. The greater the porosity, the less the density; and the greater the density, the less the porosity. Gold and platinum are bodies of great density; cork is one of much less density; and air or the gases still less. When one body is heavier than another under the same bulk, it is concluded that its density is proportionally greater than that of the other, and consequently its porosity proportionally less.

76. When the pores are uniformly diffused throughout the dimensions of a body, the body is said to be uniformly dense.

77. The porosity of bodies is sometimes illustrated and explained by a sponge, which allows the cavities which pervade it to be filled with water or other fluid; but such an illustration is not strictly apposite. The cavities of a sponge are not its pores, any more than are the cells of a honeycomb the pores of wax. Cellular structures in general present peculiar modifications of matter totally different from what is understood by porosity.   •

78. The pores of a body are sometimes filled with another material substance of a more subtle nature. Thus, if the pores of a body be greater than the atoms of air, such body being surrounded by the atmosphere, the air will pervade its pores. This is found to be the case, for example, with certain sorts of wood, with chalk, sugar, and many other substances. If a piece of such wood, chalk, or sugar be pressed to the bottom of a vessel filled with water, the air which fills the pores will be observed to escape in bubbles,

and to rise to the surface, the water pervading the pores, and taking their place.

79. **Examples.** — The following examples will illustrate the qualities of porosity and density : —

80. If a cylinder of wood be inserted in the bottom of a cup which is attached to the mouth of a glass receiver placed upon the plate of an air-pump, the cup thus placed being filled with mercury, on withdrawing the air from the interior of the reservoir the pressure of the external atmosphere will force the mercury through the pores of the wood, and it will be observed to fall in a shower of silver within the receiver.

81. Wood in general is lighter, bulk for bulk, than water, and will therefore float in it; but this comparative lightness is in some cases not a property of the wood, but of the air which fills its pores. To prove this, let a piece of such wood be held beneath the surface of water contained in a vessel placed under the receiver of an air-pump. On exhausting the receiver, the air contained in the piece of wood will force its way out by reason of its elasticity, and will rise in bubbles to the surface of the water. If, when the air has been thus expelled from the pores, the pressure of the atmosphere be made to act again upon the surface of the water by opening the cock which admits air to the receiver, the water will be forced into the pores of the wood and will fill the spaces deserted by the air, and the wood will then sink to the bottom.

82. There is no substance so dense as to be divested of pores. The celebrated Florentine experiment, performed at the Academia Del Cimento in 1661, and often repeated since that time with the same result, showed that gold itself has pores sufficiently large to admit the particles of water to pass through them. A globe of gold, being completely filled with water, was closed by a screw, and submitted to a severe pressure. As a globe is the figure which within the same surface contains the greatest possible volume, any change produced in its figure by external pressure must necessarily diminish its volume. When the globe, therefore, thus filled with water, is submitted to a pressure which changes it to a form slightly elliptical, or turnip-shaped, it would necessarily contain less liquid, and either of two effects must ensue, viz., the globe must burst, or a portion of the liquid must force its passage through the pores of the gold. The latter effect ensued; and as the globe changed its form, the water was seen collecting in a dew on the external surface of the metal. This proved that the particles of water found their way through the pores of the gold without tearing, rupturing, or otherwise doing violence to its general structure.

83. The process of filtration, so extensively used in the arts and sciences, depends on the quality of porosity. The substance through which a liquid is filtrated has pores large enough to allow the particles of the liquid to pass, but too small to permit the passage of the foreign matter suspended in the liquid, and of which it is intended to purify the liquid by the process of filtration. The most ordinary filters are soft stone, paper, and charcoal.

84. Animal and vegetable petrifactions furnish striking examples of porosity, since the stony substance which petrifies them must have been infiltrated through their mass, so as to penetrate all their fibres.

85. Mineral substances are all more or less porous. Opaque stones are in general more porous than transparent ones. Chalk and marble are formed of the same constituents, with different degrees of porosity. If water be poured on chalk, it is instantly observed passing into its innermost pores; if it be poured on marble, it rests on the surface without penetrating. Stones, however, which resist the admission of water to their pores under ordinary circumstances, will be penetrated by the liquid provided an intense pressure be used for a sufficient length of time. Thus, stones taken from the bottom of the sea, especially if the depth be considerable, are found penetrated by water to their very centre. Among siliceous stones, such as agate and flint, there is one called Hydrophane, whose porosity is attended by a singular phenomenon. A piece of this substance, in its common state, is nearly opaque; but if it be plunged in water, it is found, on withdrawing it from the liquid, to be nearly as transparent as glass. In this case the water penetrates the stone exactly as oil penetrates paper; bubbles of air are disengaged from its pores, which are filled with the water absorbed, the presence of which gives the transparency.

86. Large mineral masses existing naturally in the strata of the earth present examples of porosity still more striking. Water percolates through the sides and surfaces of caverns and grottos, and, being impregnated with calcareous and other earths, forms stalactites, or pendulous protuberances, presenting curious appearances, with which every one is familiar.

87. **Compression.**—The quality in virtue of which a body allows its volume to be diminished without diminishing its quantity of matter is called *compressibility*, when the effect is produced by the application of external mechanical force; and *contractibility* when produced by change of temperature, or any other agency not mechanical. When the volume of a body is diminished, whether by compression or contraction, its constituent atoms are brought into closer contiguity, its pores are consequently diminished, and its density proportionally increased.

88. All known bodies, whatever be their nature, are capable of having their dimensions reduced without diminishing their mass; and this is one of the most conclusive proofs that all bodies are porous.

89. It is evident in general that the more porous a body is the more easy is its compression. This truth is manifested by innumerable examples derived from organized bodies, especially those of a fibrous texture. All those whose porosity is such as to allow them to be easily penetrated by fluids can be diminished by the application of pressure; and in this case, if they have been previously filled with fluids, these fluids are expelled by the pressure exactly as water is squeezed from a piece of sponge. Innumerable processes in the arts supply examples of this.

90. Wood of even the hardest kind, in its natural state, is so porous as to absorb both air and water in considerable quantities. When such wood is used in the arts, in cases where extreme hardness is required, it is previously submitted to severe pressure, by which the fluids absorbed are expelled from the pores, the volume diminished, and the density increased. The wooden wedges used in fastening the rails of railways in their chairs are prepared in this manner.

91. Even the most solid stone, when loaded with a considerable weight, is found to be compressed. The foundations of buildings, and the columns which sustain incumbent weights in architecture, supply numerous proofs of this.

92. Malleable metals are compressed by percussion or hammering: they become thus more compact and dense. In the process of coining, medals and pieces of money are struck by a severe pressure, by which they are made to receive the impression and characters upon them more accurately than softened wax would from the pressure of the hand. Under the blow of the press they not only change their form, accommodating themselves to the characters and figures sunk upon the die, but they are at the same time compressed and rendered more dense, so that the coin or medal has a volume sensibly less than the blank piece had before it was struck.

93. Liquids in general are less easily compressed than solids; so much so, that in practical science they are regarded as incompressible.

They are, however, strictly speaking, capable of a slight compression under the operation of considerable mechanical force. In the year 1761, the compressibility of water and other liquids was established. It was found that water, submitted to a mechanical pressure, amounting to fifteen pounds on a square inch, would be diminished in its volume by forty-five parts in a million,

c 3

that is to say, a million of cubic inches would be reduced to about forty-five cubic inches less.

In more recent experiments, a quantity of water was enclosed in a piece of cannon and submitted to a mechanical pressure amounting to fifteen thousand pounds per square inch. Under this pressure it was diminished by one-twentieth of its volume, and the cannon enclosing it was burst.

94. But if liquids are so little compressible, they are, in a very high degree, susceptible of contraction.

If a quantity of water coloured with ink or other colouring matter be included in a glass bulb connected with a tube of small bore, it will be found, that when the bulb is exposed to cold, the level of the coloured water in the tube will descend. This is an effect of the contraction which the liquid undergoes in consequence of its diminution of temperature. This contraction by cold is an universal quality of matter, which will be explained more fully in a subsequent part of this work.

95. **Compression of gases.** — Of all forms of matter the gases are the most susceptible of compression. This quality has already been briefly noticed. There appears to be no practical limit to the compression of which this form of matter is susceptible, its volume being diminished in the exact proportion of the compressing force applied to it.

96. **Elasticity.** — This is the quality in virtue of which a body, after having been compressed, recovers its former dimensions, on being relieved from the force which compresses it. Bodies which retain their compressed state after the force ceases to act, and do not resume their original dimensions, are said to be inelastic.

97. The class of bodies which affords the most striking examples of elasticity are the gases and aeriform bodies. If a quantity of air be included in a syringe under a piston, and be compressed by a force applied to the piston, on the removal of that force the air, by virtue of its elasticity, will force the piston upwards until it resumes the position from which it had been driven by the compressing force.

98. All liquids, when compressed, immediately recover their original dimensions when relieved from the compressing force, and therefore may be said to be perfectly elastic. The play of the compressive and elastic principle, however, in the case of liquids, is so extremely limited, that for all practical purposes this form of body is treated as both incompressible and inelastic. All the theorems of those parts of physical science called *Hydrostatics*, *Hydrodynamics*, *Hydraulics*, &c., are based upon the principle that liquids are incompressible and inelastic; for although it be true, as has been stated, that within certain very minute limits they are

both compressible and elastic, yet these limits are so small as to produce no appreciable effects under ordinary circumstances.

99. Gaseous bodies are not only compressible and elastic without any practical limit, but also endued with unlimited dilatability. Thus, if a quantity of gas be included in any given volume, and that this volume be augmented in any required proportion, the gas will spontaneously, and without the application of any external agency, dilate itself so as to fill the augmented volume, and this expansion will go on, no matter to what extent the volume be augmented.

100. The quality of elasticity is manifested in solid bodies, but in a less decided manner than in gases. Caoutchouc, or elastic gum, is perhaps of all bodies that which has most elasticity. This quality, combined with the methods recently discovered of varying the form of this substance, has extended considerably the application of it to the useful purposes of life.

101. **Examples.** — The following examples will illustrate the quality of elasticity as found in solid bodies.

102. If a flat and hard surface be smeared with a thin coating of oil, and an ivory ball be allowed to drop upon it, the ball will rebound by reason of its elasticity. On examining that part of the surface of the ball which struck the flat surface from which it rebounded, it will be found that a somewhat extensive circular space will have been stained with the oil. If the ball be brought gently into contact with the flat surface, a minute space only would be stained with the oil. Why, then, it may be asked, did a larger space receive a stain when the ball was allowed to drop with a certain force upon the surface? The answer to this is, that the force of the impact *flattens* the surface of the ball to a certain extent; that, in virtue of its elasticity, the ball recovers its spherical figure; and that the force with which it recovers this figure causes the rebound. The extent of the surface stained by the oil is a little, but not much greater than the extent of the circle flattened by the impact.

If the ball be let fall from several different heights, it will be found that the circular space stained by the oil will be greater, the more elevated the point from which the ball is allowed to depart. This effect is only what might have been anticipated; the greater the height from which the ball falls, the greater will be the force of the impact, and consequently the greater will be the extent over which its surface will be flattened, and the greater, consequently, will be the elastic force which produces the rebound.

103. If such an experiment be made with a ball composed of a substance softer than ivory, and equally elastic, the flattening may be rendered directly perceptible to the senses. This may be made

C 4

evident by the large caoutchouc balls inflated with air, used in the plays of children. When they strike the ground, they are flattened at the surface over a circle of very considerable magnitude, and which flattening may be exhibited by pressing them on the ground by the force of the hand. This is only an exaggeration of what would actually take place in the case of a ball of ivory or glass.

104. Elasticity in bodies is sometimes manifested by their disposition to recover their form when disturbed by a force which does not affect their volume. For example, a plate of steel when bent would have the same dimensions which it had before the pressure, yet its elasticity will be rendered apparent by its immediately recovering its original form after the force which bends it had ceased to act. The play of springs of every form affords examples of this. When a straight bar of steel is bent into a curve, both compression and expansion of its molecules take place. The molecules which compose that side which becomes convex are forcibly drawn asunder, and those which form the surface which becomes concave are forcibly compressed. This is evident, inasmuch as the convex side becomes longer, and the concave shorter, by the change of form. The tendency of the molecules, by virtue of their elasticity, to recover their original position, causes those on the convex surface to contract, and those on the concave surface to expand, the combined effects of such contraction and expansion being the restoration of the bar to its original form.

105. **Limits of the elastic force.** — As elasticity results from a derangement of the component molecules of bodies, it will be easily understood that there must be limits, beyond which such derangements cannot be produced without a permanent change in the form of the body; and there are consequently limits to the play of the elastic principle. These limits will be obviously different in different bodies.

In the case of the most elastic class of bodies, such for example as caoutchouc, these limits are very extensive. In ivory they are more extensive than glass, for ivory will recover its figure after a compression which would cause the fracture of glass. These limits are narrow in the case of such metals as lead; for although considerable compression will not cause the fracture of lead as it would that of glass, yet the derangement which such compression produces amongst the molecules of that metal is greater than their feeble elasticity can resist, and the metal accordingly takes permanently any form given to it.

106. **Torsion.** — Elasticity is sometimes manifested by torsion or twisting. Thus, let us suppose a filament of raw silk stretched by a weight attached to it. If this weight be made to revolve several times in the same direction, so as to twist the silk and then

be disengaged, the fibre of silk in virtue of its elasticity will un-
twine itself, causing the weight to revolve in a contrary direction;
and this process of untwining will continue until the filament recovers its original position; but the twisting may have been continued to such an extreme as to exceed the limits of the elasticity
of the silk; and in that case a permanent derangement of the
molecules of the silk will take place, and it will not recover its
original form.

The same effects would ensue if the weight had been suspended
to a fine wire of copper, silver, or any other metal, but the limit
at which the twisting would produce a permanent derangement of
form, or, in other words, the limit of play of the elastic principle,
would be different.

107. **Dilatability.** — When the extension or augmentation
of the volume of a body is produced by any physical agency,
such, for example, as heat, not coming under the denomination
of mechanical force, it is called *dilatation*. All bodies whatever,
when submitted to the action of heat, are susceptible of having
their dimensions enlarged; and to this augmentation of magnitude
or dilatation by increase of temperature there is no practical
limit.

Innumerable examples of the operation of this principle in the
arts and sciences may be produced.

108. In the thermometer the dilatation of a liquid is used as the
measure of the degree of heat which produces it. This instrument consists of a glass bulb attached to a tube of small bore.
The bulb and part of the tube are filled with a liquid. As the
temperature to which the instrument is exposed is increased or
diminished, the liquid affected by it expands or contracts in a
much greater degree than does the glass in which the liquid is
contained. The consequence of this is, that in order to find room
for its increased volume, a portion of the liquid in the bulb is
forced into the tube. The column in the tube consequently becomes longer, and its increase of length, measured by a scale
attached to the tube, becomes a measure of the increased temperature.

109. The dilatation and contraction of metal consequent upon
change of temperature has been applied some time ago in Paris to
restore the walls of a tottering building to their proper position.
In the *Conservatoire des Arts et Métiers*, the walls of a part of the
building were forced out of the perpendicular by the weight of the
roof, so that each wall was leaning outwards. M. Molard conceived the notion of applying the irresistible force with which
metals contract in cooling to draw the walls together. Bars of
iron were placed in parallel directions across the building, and at

right angles to the direction of the walls. Being passed through the walls, nuts were screwed on their ends outside the building. Every alternate bâr was then heated by lamps, and the nuts screwed close to the walls. The bars were then cooled; and the lengths being diminished by contraction, the nuts on their extremities were drawn together, and with them the walls were drawn through an equal space. The same process was repeated with the intermediate bars, and so on alternately, until the walls were brought into a perpendicular position.

110. **General effects of dilatation and contraction.** — Since there is a continual change of temperature in all bodies on the surface of the globe, it follows that there is also a continual change of magnitude. The substances which surround us are constantly swelling and contracting under the vicissitudes of heat and cold. They grow smaller in winter, and dilate in summer. They swell their bulk on a warm day, and contract it on a cold one. These curious phenomena are not noticed only because our ordinary means of observation are not sufficiently accurate to appreciate them. Nevertheless, in some familiar instances, the effect is very obvious. In warm weather the flesh swells, the vessels appear filled, the hand is plump, and the skin distended. In cold weather, when the body has been exposed to the open air, the flesh appears to contract, the vessels shrink, and the skin shrivels.

CHAP. IV.

INERTIA.

111. **All matter inert.** — The quality of matter which stands foremost in importance in all mechanical inquiries, forming the basis of the whole theory of force and motion, is *inactivity* or *inertia;* and important as this quality is, there is perhaps nothing which has given rise to so many erroneous conceptions.

These errors have chiefly arisen from the adoption of a vicious phraseology on the part of many writers on Natural Philosophy

112. **Inertia.** — At any given moment of time a body, mechanically considered, must be in one or other of two states, rest or motion. Inertia or inactivity is the total absence of all power in the body to change its state. If the body be at rest, it cannot put itself in motion; if the body be in motion, it can neither change that motion nor reduce itself to rest. Any such change must be produced from some external cause independent of the body.

INERTIA 27

113. The phrase *vis inertiæ*, or force of resistance, used in many treatises on Natural Philosophy, has been a fertile source of error. Such a phrase implies a disposition in matter to resist being put in motion when at rest. Now no such disposition is found to exist; and if it did exist, it would be as utterly incompatible with the quality of inactivity as is the power to produce spontaneous motion.

114. Innumerable effects which fall daily under our observation prove to us the inability of mere matter when at rest to put itself in motion, or when in motion to augment its speed; but, on the other hand, we have not the same direct and manifest evidence of its inability to destroy or diminish any motion which it may have received; and it happens, therefore, that while few will deny to matter the former effect of inertia, many will at first doubt or fail to comprehend the latter.

Philosophers themselves, so late as the epoch signalised by the writings of Bacon, held it as a maxim that matter is more inclined to rest than to motion; and this being so, we cannot be surprised to find those who have not been familiar with physical science still slow to believe that a body once put in motion would continue for ever to move in the same direction and with the same speed, unless stopped by some external cause.

But a careful examination of the circumstances which affect the movement of the bodies around us with which we are most familiar will soon convince us, that in every case in which we observe the motion of those bodies gradually diminished, or entirely destroyed, such effects arise, not, as has been erroneously supposed, from any natural disposition of the bodies themselves to be retarded or brought to rest, but from the operation of causes of which there is no difficulty in rendering an account.

In some of the modern popular works on Natural Philosophy, a trifling experiment is mentioned as an example of the effect of inertia, and explained on principles somewhat erroneous. A card being placed on the top of the finger, and a coin placed on the card, a sudden blow being given with the back of the nail to the edge of the card, it will be projected from its place between the coin and the finger, the coin remaining unmoved on the finger.

This has been explained by stating that the inertia of the coin is comparatively so great, that the friction produced between it and the card is insufficient to move it from its place.

If by these words it be understood that the coin resists the force exerted upon it by means of the friction, it is erroneous, and would be incompatible with that quality of inertia to which the effect is ascribed.

The correct explanation of the experiment is as follows. A part

of the momentum given to the card by the blow is communicated
to the coin in consequence of the resistance to the motion of the
card produced by the friction which takes place between it and
the coin. But the coin contains comparatively so much matter,
that this moving force, when distributed among its component
particles, which it will necessarily be, will give to the whole coin a
velocity in the direction of the motion of the card incomparably
smaller than that of the card, and so small that the resultant of the
forces produced by the weight of the coin upon the finger and this
force scarcely deviates from the direction of the weight of the
coin: consequently, although the coin remains on the finger, it
does not remain precisely in the same position over the finger
which it had when it rested on the card; its position will be
changed in a slight degree in the direction of the motion of the
card.   •

115. When a stone is rolled along the surface of the ground,
the inequalities of its form, as well as those of the ground on which
it moves, present impediments which gradually retard its movement, and soon bring it to rest. Render the stone round and
smooth, and the ground level, and the motion will be considerably
prolonged; a much longer interval will elapse, and a much greater
space will be traversed, before it will come to rest. But asperities more or less considerable will still remain on the surface of
the stone, and on the surface of the ground. Substitute for it a
ball of highly polished metal, moving on a highly polished steel
plane truly level, and then the motion will continue for a very
long time.

But, even in this case, asperities will remain on the surface of
the moving body, as well as on the surface on which it moves,
which will gradually destroy the motion, and ultimately bring it
to rest.

But, independently of the obstructions to the motion of bodies
arising from the friction of the surfaces which move in contact
with each other, all motions which take place on or near the surface of the earth are necessarily made in the fluid medium of the
atmosphere. This fluid, however attenuated, still offers considerable resistance to the motion of bodies through it. An extensive
flat surface spread at right angles to the direction of the motion
will thus meet a powerful resistance. This resistance arises from
the body moved being compelled to push out of its way a volume
of air proportional to the extent of the surface which the body
presents in the direction of the motion. If on a calm day an open
umbrella be carried with its concave surface presented in the
direction in which we are moving, a powerful resistance will be
encountered, which will increase with every increase of speed.

116. **Proofs of inertia.**—As these causes of resistance to the motion of bodies are everywhere present on and near the surface of the earth, we are unable, by direct experiment, to establish the proposition that a body when once put in motion would continue for ever to move in the same direction, and with the same speed, if undisturbed; but astronomical observations supply an immense mass of evidence to established this principle. In the heavens we find a vast apparatus, every movement of every part of which establishes incontrovertibly the inertia of matter, inasmuch as the reasoning by which all these motions are explained, and by which all these phenomena are predicted, is based upon the fundamental principle of the complete inertia of matter. The celestial bodies, removed from all the casual obstructions and resistances on the surface of the globe which disturb our reasoning, roll on in their appointed paths with unerring regularity, preserving undiminished all that motion which they received at their creation from the hand which launched them into space. These phenomena alone, unsupported by other reasoning, would be sufficient to establish the quality of inertia; but, viewed in connection with the other circumstances already mentioned, and with the whole superstructure of mechanical science, leading to innumerable truths verified by daily and hourly experience, no doubt can remain that this important principle is a universal law of nature.

117. **Examples.**—The following examples will illustrate the quality of inertia :—

118. If a horse or a vehicle of any kind moving with considerable speed be suddenly stopped by any cause which does not at the same time affect the rider or those who are transported by the vehicle, then the body of the rider, or those who are transported, still retaining the progressive motion of which the horse or vehicle is suddenly deprived, will be projected forwards; and unless some means of resistance be adopted, the rider will be thrown over the head of the horse, and the passengers thrown forwards from the vehicle.

In the same manner, if a horse or vehicle being at rest be suddenly started forwards with considerable speed, the rider, or the persons placed upon the vehicle, not being as suddenly affected by the same forward motion will be thrown backwards.

In both these cases the effects are the consequence of the quality of inertia. In the one case they manifest the tendency of the bodies to continue the motion they have already received, and in the other they manifest the disposition of the same bodies to continue at rest.

119. If a passenger in a carriage which moves with considerable speed leap to the ground, he will fall in the direction in which the

carriage is moving; for in descending to the ground his entire body will still retain all the progressive motion which it had in common with the carriage. When his feet touch the ground, they and they alone will be suddenly deprived of this progressive motion, which being retained by the remainder of his body, he will fall as if he were tripped by some object impeding his motion, in the direction of the carriage.

120. The sport of coursing presents many amusing and instructive examples of the force of inertia. From the movements of the hare, one might suppose that it is, indeed, an expert mechanical philosopher. The hound which pursues it being a comparatively heavy body, and moving at the same or a greater speed, cannot suddenly arrest its course, because, in virtue of its inertia, it has a tendency to proceed forward in the same straight line.

The hare, a comparatively light body, and moreover being prepared for the evolution, first gradually retards its motion so as to diminish the effects of inertia, and at the moment when the hound, urged to its extreme speed, is in the act of seizing the game, the hare dexterously turns at an acute angle to its former course, leaving the hound propelled forwards in the direction in which it was previously moving.

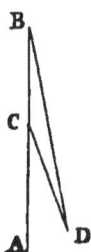

Thus, if the line A B (*fig.* 2.) represent the direction in which the hound was pursuing the hare, the hare, having arrived at the point c, suddenly turns in the direction c D; while the hound, unprepared for the trick, and hurried forward by the inertia of its motion, is carried on in the direction A B to the point B, while the hare has passed along the line c D to the point D. The distance now between the hound and the hare is the line B D, the base of the obtuse angle formed by two lines, c B and c D, simultaneously moved over by the hound and the hare.

Fig. 2.

CHAP. V.

SPECIFIC PROPERTIES.

121. THE qualities of matter which have been illustrated and explained in the preceding chapters, are those which are common, in a greater or less degree, to all bodies, in whatever form or under whatever circumstances they may exist. There remains to be noticed another group of properties which may be denominated for distinction, specific properties, being found in some species of matter and not in others, or at least varying in degree so extremely in different sorts of bodies as to give them specific characters.

122. **Elasticity and hardness.**—Although the property of elasticity in its general sense may be considered as one which, in various degrees, is common to all bodies, yet it is manifested in so peculiar a manner in bodies of different forms, that it may be not incorrectly considered as giving them a specific character. It is intimately connected with another mechanical quality which may be called *hardness*. This quality consists in a certain degree of coherence, by which the constituent molecules of a body keep their relative position so as to resist any force which tends to change the figure of the body.

123. *Hardness* is distinct from *density*, as we frequently find the most dense bodies possess this quality in a much less degree than lighter substances. Glass, for example, is harder than gold, or even than platinum, which is still harder and denser than gold. A piece of glass will scratch the surface of gold or platinum, an effect which shows that the particles of gold or platinum yield and are displaced more easily than those of glass.

Again, in comparing different species of metals one with another, their hardness is evidently independent of their densities. Gold and platinum, the most dense of metals, are softer than iron or zinc, which are much lighter. Among the hardest of the metals are iron, zinc, copper, manganese, nickel, titanium, and pelagium. The softest of the common metals is lead, but the new metals developed by chemical enquiries, such as potassium and sodium, are so soft as to yield under the finger like putty.

124. Some metals are capable of having their structure modified without the combination of any other substance with them, so as to render them harder or softer within certain limits at pleasure. Thus steel, when heated, and then suddenly cooled by being plunged in cold water, becomes harder than glass; but if it be cooled more gradually, then it becomes soft and flexible.

125. **Elasticity.**—Elasticity manifests itself in various ways according to the form and character of the body to which it belongs. The elasticity of a flat and thin bar of steel is manifested by the force with which it will recover its figure when bent by lateral pressure; the elasticity of an ivory ball is manifested by the force with which it will recover its figure when flattened by impact against some hard surface. In the case of a steel spring, the body yields to a slight pressure, readily changing its form; in the case of an ivory ball, the body does not yield to mere pressure, and requires the force of impact to produce change of form.

When the force with which a body recovers its form is equal to the force by which its form has been changed, the elasticity is said to be perfect. Thus, if a bent spring recover its position when relieved from the force which bent it with an energy equal to such bending force, then the spring is said to be perfectly elastic; but when the restoring force is less than the bending force, the elasticity is imperfect. In the same manner, if an ivory ball flattened by a blow recover its form with a force equal to that of the blow which flattens it, the elasticity is perfect, but otherwise imperfect.

126. It is more exact to say that no body whatever is, in an absolute sense, either perfectly elastic or perfectly inelastic. All bodies possess some degree of elasticity, however small, but some bodies, such as the gases, for example, possess elasticity in so high a degree, and others, such as the liquids, in a degree comparatively so small, that not only, in popular language, is the one considered elastic and the other inelastic, but it has been found convenient to assume hypothetically these two qualities of perfect elasticity and perfect inelasticity as the bases of those divisions of physical science, in which the laws which regulate the phenomena of liquids and gases are developed.

It has been already shown, however, that liquids themselves admit of some compression, and it may be added that they recover their volume with a force sensibly equal to the compressing force; and to this extent, and in this sense, they are therefore almost perfectly elastic.

127. The quality of elasticity is intimately connected with that of hardness; so much so, that it has sometimes been said that one quality is proportional to the other. This is, however, erroneous. Many of the gums, and eminently that called caoutchouc, are highly elastic, and yet these substances are among the softest of the solids. The elasticity of caoutchouc is nearly perfect, and yet this substance, especially when it is warm, has great softness. On the other hand, glass, flint, marble, ivory, afford examples of solids in which hardness is combined with great elasticity.

Putty, wet paste, moist clay, and similar bodies afford examples of substances nearly deprived of elasticity. The figure of any of these may be changed by pressure or by impact, and no tendency to recover the figure so changed is perceptible.

128. Sound, as will be explained hereafter, is produced by vibration imparted to the air by some solid body which is itself in a state of sympathetic vibration. It is obvious, therefore, that the metals best suited for bells, and other forms of matter intended to produce sound, must be those which are most elastic.

129. The hardness and elasticity of metals are affected in a striking manner by their combination. It often happens that two metals, neither of which is eminently hard or elastic, produce by their combination in certain proportions, one which possesses these qualities in a high degree. Thus, bells formed of pure copper, or of pure tin, will have little sonorous quality; but if these two metals be united in a certain proportion, their combination will give a beautiful musical sound. The compounds of different metals which have this quality are accordingly known as *bell metal.*

130. **Flexibility and brittleness.**—When a body easily yields, and changes its form in obedience to a force exerted at right angles to its length, as, for example, when a bar being supported at the middle is pressed upon the ends, it is said to be *flexible*; but if, upon the action of such a force, instead of yielding and changing its form, it breaks, it is said to be *brittle.*

Flexibility and brittleness are specific qualities which bodies possess in an infinite variety of degrees.

In general, brittleness is connected with hardness, nor is it, as might at first appear, at all inconsistent with certain forms of elasticity. Glass, for example, which is highly elastic, is also the most brittle of known substances. Brittleness, like hardness and elasticity, is a quality which the same body may acquire, or be deprived of, according to certain conditions to which it may be subjected. Thus, the metals, iron, steel, brass, and copper, if they be heated and suddenly cooled, by being plunged in cold water, will become brittle; but if, when heated, they are buried in a hot sand-bath, and allowed to cool very gradually, then they will lose their brittleness, and acquire the contrary quality of flexibility.

131. **Malleability.** — This is a quality by which the metals in general are eminently distinguished, but which they possess in extremely different degrees. It is a property in virtue of which a substance admits of being reduced to thin plates or leaves under the blow of a hammer, or the intense pressure of rollers. No process is of more extensive use in the arts. In

D

large iron works, great lumps of metal at a white heat, but still solid, are taken from the furnace, stuck upon the end of a long bar of iron, and placed under a sledge hammer of enormous weight, which rapidly strikes them, and reduces them to an elongated form approaching to that of an iron bar. The metal, being still red hot, is then passed between rollers, which are formed to the shape of the transverse section of the rails used on our railways. When pressed between and drawn through these rollers, the rail has acquired its proper form, but is still red and soft; and when received from the rollers is so flexible, that it bends by its own weight like a rod of wax. It is then laid on a flat surface, where it cools and hardens, and assumes the condition of the rails on which we travel.

The malleability of bodies depends on the combination in them of the qualities of tenacity and softness. Without softness, they could not yield either to the impact of the hammer or the pressure of the roller; without tenacity, they would be fractured by the severe process of their fabrication.

The most malleable of the metals are gold, silver, iron, and copper.

132. The malleability of a metal varies in degree according to its temperature. There are certain temperatures in which this quality exists in the highest degree. Iron is most malleable when it first attains the white heat which follows the red; zinc becomes malleable at a much lower temperature, possessing this quality in the greatest degree between 300° and 400°. Some metals possess the quality of malleability in so slight a degree as to be in this respect specifically different from metals in general: among them may be mentioned antimony, arsenic, bismuth, and cobalt, all of which are brittle. The metals may be rendered brittle as they are rendered hard, by being heated and then suddenly cooled, in which case they lose their malleable quality. This quality, however, may always be restored by again heating them and cooling them gradually, as before described.

133. **Annealing.**—This process of gradual cooling, which is of great importance in the arts, is called *annealing*.

Metals are also rendered brittle, and deprived of their malleability, by constant hammering. Thus, a bar of iron may be hammered until it entirely loses its flexibility. In this case, as before, the malleability may be restored by heating and annealing.

134. **Welding.**—Metals which are highly malleable admit of being united, piece to piece, by the process called *welding*. In this process, the two pieces of metal are raised to that heat at which they are most malleable, and the ends being laid one upon

the other, are rapidly beaten by a welding-hammer. The particles are thus driven into such intimate contact, that they cohere, and form one uniform mass. Different metals may in some cases be thus welded together.

135. **Ductility.**—The property in virtue of which a metal admits of being drawn into wire, is called *ductility*. This quality is also eminently specific, being possessed by some sorts of metal in a very high degree, while others are entirely destitute of it.

136. Ductility is a quality which must not be confounded with malleability; for the same metals are not always ductile and malleable, or, at least, do not possess these qualities in the same degree. Iron possesses ductility in a much greater degree than it possesses malleability, for it admits of being drawn into extremely fine wire, though it cannot be beaten into extremely thin plates. Tin and lead, on the other hand, are highly malleable, being capable of being reduced to extremely attenuated leaves; but they are not ductile, since they cannot be drawn into small wire. Gold and platinum possess both ductility and malleability in a high degree. Gold has been drawn into wire so fine, that 180 yards' length of it did not weigh more than one grain, and an ounce weight would consequently extend over fifty miles.

137. **Tenacity.**—The property in virtue of which a body resists the separation of its parts, by extension in the direction of its length, is called *tenacity*. This manifestation of strength must be carefully distinguished from that of which the absence or feebleness is expressed by brittleness. The one form of strength may exist in the highest degree in a body in which the other is in the lowest degree. A thin rod of glass, if laid at its middle point on any support, will be broken by the slightest force pressing on its ends; but the same rod, if suspended by one end in a vertical position, will sustain an immense weight attached to the lower end without being broken. It has at once great brittleness and great tenacity; while its longitudinal strength is considerable, its lateral strength is almost nothing.

Different bodies vary extremely in their tenacity. Experiments have been made on an extensive scale for determining the tenacity of those bodies most used in the arts. The tenacity of metals has been tested by suspending a weight from the end of a wire.

138. In the following table, the greatest weights are given which were found to be supported by wires of the different metals, having a diameter of $\frac{1}{10000}$ths of an inch.

	Weights supported.		Weights supported.
Iron	- 549·250 lbs.	Gold	- 150·173 lbs.
Copper	- 302·278 „	Zinc	- 109·540 „
Platinum	- 274·320 „	Tin	- 34·630 „
Silver	- 187·137 „	Lead	- 27·621 „

The process of annealing, which improves the malleability and ductility of metals, is found in some cases, as, for example, in iron, copper, and the combinations of zinc and copper, to diminish their tenacity. In organized substances, those which possess a fibrous texture have greater tenacity than those of cellular tissue. Hence, we find that cotton has much less tenacity than thread, rope, or silk.

139. L'Abbé Labillardière found that threads of the following substances, having the same diameter, were capable of supporting weights in the proportion of the annexed numbers : —

Silk	•	•	• - 3400	Flax (common) - - • 1175
New Zealand Flax	:	: 2380	Ditto (Plta) (*Agave Americana*) 700
Hemp	-	•	- 1633	

140. **Chemical properties.**—There is an endless variety of specific properties of bodies, the exposition and investigation of which belong properly to chemistry. It will be sufficient here to notice briefly the distinctions between these qualities and those which form the proper subjects of Natural Philosophy commonly so denominated.

If two substances, being mixed together, retain respectively their separate qualities, the combination is said to be a mechanical mixture. Thus, if blue and yellow powders be mingled together, the mixture will appear green ; but on examining it with a microscope, it will appear to consist of large blocks of matter, of the colours blue and yellow, these being the particles of the separate powders retaining their distinctive qualities, which are mechanically mingled. The combination produces upon the eye a green colour, the effect of the separate particles being too minute to be separated by unassisted vision.

There are two gases, called oxygen and hydrogen, which have the common mechanical properties of atmospheric air. If one ounce weight of hydrogen be mingled with eight ounces of oxygen, the gases will be interfused and mingled; but the entire mass will retain the same mechanical qualities as before, and the separate particles will remain side by side in the mixture, exactly as did the particles of blue and yellow powder in the preceding example. But if an electric spark be imparted to this mixture, a striking change will take place. The mixture will in an instant be reduced to water, or rather to vapour, which. being cooled, will be soon converted into water. In fact, the oxygen has in this case united with the hydrogen, and the mass has lost the mechanical qualities which it possessed, and has acquired those of the liquid water. This is a *chemical phenomenon.*

There is a metal called *sodium*, and a gas called *chlorine*, each of which, separately, is poisonous, and destructive of life, if taken into the stomach or lungs. If these two substances be brought together, they immediately explode, and burst into flame. If the substance resulting from this phenomenon be preserved and cooled, it will be found to be common kitchen-salt, one of the most wholesome condiments, and highly antiputrescent. Thus, two ingredients possessing the most noxious properties, when combined chemically, lose those properties, and produce a substance wholly different in form and *quality*.

The investigation of this and all similar phenomena belongs to the province of chemistry.

BOOK THE SECOND.

OF FORCE AND MOTION.

~~~~~~~~~~~~

CHAPTER I.

THE COMPOSITION AND RESOLUTION OF FORCES.

141. ANY agency which, applied to a body, imparts motion to it, or produces pressure upon it, or causes both of these effects together, is called in mechanics *a force*. To determine a force, therefore, with precision, three things are necessary:

First, the point of the body to which it is applied, technically called its *point of application; Secondly*, its *intensity* or *quantity*; and *Thirdly*, its *direction*.

142. **Force expressed by weight.**—It is in general convenient and customary to express the intensity or quantity of forces by equivalent weights. Weight is the sort of force with which we are most familiar. Every one is acquainted with the effect produced by the pressure of a given weight; and whatever be the force whose intensity or quantity it is required to express, a weight may be named which would produce the same effect. Thus, if a piece of iron, attracted by a magnet, be resisted by any surface, it will press against this surface with a certain force. A weight may in this case be named, which, being placed in the dish of a balance, would press upon the surface of the dish with the same force. The intensity of the attraction of the magnet on the iron would then be expressed by the amount of such an equivalent weight.

143. **Direction.** — When a force applied to any point of any body causes that point to move, the direction of its motion is the direction of the force. If the force do not produce motion, but mere pressure, then the direction of the force is that in which the pressure is directed, and in which the point would move in obedience to the force, if it were free.

144. **Forces in same direction.** — If two or more forces act upon the same point, and in the same direction, their effect will be equivalent to a single force which is equal to their sum. This is so self-evident that it scarcely needs demonstration.

If a vehicle be drawn by three horses, one placed before the other, one horse pulling with a force of 80, another with a force of 60, and the third with a force of 40 lbs., then the combined action of the three horses upon the vehicle will be equal to the action of a single horse which should pull with a force of 180 lbs., which is equal to 80 lbs. + 60 lbs. + 40 lbs.

145. A single force acting on a body which would thus produce the same motion or pressure as several forces acting together, is called technically the *resultant* of these forces. Thus, in this preceding example, the force of 180 lbs. acting on the vehicle in the same direction as the three independent forces of 80 lbs., 60 lbs., and 40 lbs., is the resultant of those three.

146. **Opposite forces.** — If two forces act upon a body in opposite directions, then the lesser of these forces will neutralize so much of the greater as is equal to its own quantity, and an effective force will remain in the direction of the greater, equal to their difference. This is also self-evident. If, for example, a vehicle be pulled backwards by a weight of 100 lbs. acting over a pulley, and that it be drawn forwards by a horse acting with a force of 150 lbs., then 100 lbs. of the horse's force will be neutralized by the weight which draws the vehicle backwards, and an effective force of 50 lbs. will remain in the direction of the horse's traction.

This principle is stated generally by saying that the resultant of two forces applied to the same point in opposite directions is equal to their difference and in the direction of the greater.

If any number of forces act upon the same point, some in one direction, and the others in the direction immediately opposed to it, then the resultant of such a combination of forces will be found by taking the difference between the sum of all the forces which act in the one direction, and the sum of all the forces which act in the other direction, the direction of such resultant being that of the forces whose sum is the greater.

147. **Forces in different directions.** — When two forces applied to the same point act in the direction of different and diverging

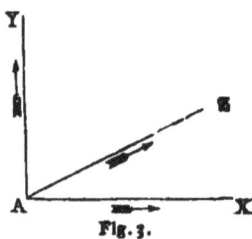

Fig. 3.

straight lines, such as A x and A y (*fig.* 3.), then the direction and quantity of their resultant is not so evident as in the case just mentioned. It is indeed apparent that the combined effect of two such forces on the point A must be in some direction, such as A z, intermediate between A x and A y; but how this direction A z divides the angle formed by the two components is not apparent.

The following example, in which, as usual, weights are used to represent the forces in question, will, however, elucidate this.

Let two weights A and B (*fig.* 4.) be attached to the extremities of a flexible cord which passes over two pulleys, M and N. Let another cord be knotted to this at any intermediate point, such as P; and let a third weight c be suspended from it. The weight c will then draw the cord which unites A and B into an angle M P N. The system after some oscillations will come to rest, and when it is at rest it will be evident that the point P is solicited by three forces; 1st, by the weight A acting in the direction of the line P M; 2ndly, by the weight B acting in the direction P N; and 3rdly, by the weight c acting in the direction of the line P c.

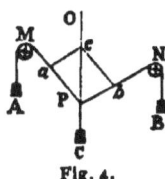

Fig. 4.

Now, it is evident that the weight c acting in the direction P c would equilibrate with an equal force acting in the opposite direction P c. Since, then, the weight c would precisely counterpoise an equal weight in the direction of P c, and that it is also in equilibrium with the weights A and B, which act in the directions P M and P N respectively, it follows that the resultant of the forces A and B acting in the directions P M and P N will be a single force equal to c acting in the direction P c.

It now remains to show in what manner this direction of the resultant of the two diverging forces M and N is connected with their quantities.

Let us suppose, for example, that the weight A is 6 oz., the weight B 4 oz., and the weight c 6¼ oz. If, then, we take upon the line P o a distance P c of 6¼ inches, and if we draw two lines, one c a parallel to P N, and the other c b parallel to P M, so as to form a parallelogram P a c b, we shall find, on measuring the side P a, that it is 6 inches, and on measuring the side P b, that it is 4 inches.

148. **Composition of forces.**—Hence it appears, that while the diagonal P c consists of as many inches as there are ounces in the resultant of the two forces, the sides of the parallelogram which are in the direction of these two forces respectively consist of as many inches as there are ounces in these two forces. This result may be enunciated in general terms as follows:

If two forces acting upon the same point be represented in quantity and direction by two lines drawn through that point, then the resultant of such forces will be represented in quantity and direction by the diagonal of the parallelogram of which these lines are the sides.

But it may be objected that the result we have obtained by the use of three particular weights may be accidental, and that it may not always happen that the third weight c, which balances the other two, will throw the cords into such directions as would give the remarkable result here obtained.

The validity of such an objection may be easily tested by varying the weights at pleasure, and by submitting the position of the string to the same process of measurement as we have given above. It will then be found that in whatever manner the three weights may be varied, the knot which unites the three strings P M, P N, and P C will invariably establish itself in such a position, that while the diagonal P c will measure as many inches as there are ounces in the resultant c, the sides P a and P b will measure as many inches as there are ounces in the components A and B.

The proposition which we have here established is of the utmost importance in all mechanical investigations, and is known as the principle of the *composition of forces.*

149. In virtue of this principle, whenever two forces in different directions act upon the same point of a body, a single force determined as above by the diagonal can be substituted for them without changing the mechanical state of the body; or, on the other hand, if a single force act upon any point of a body, two forces acting on the same point may be substituted for them, provided such forces can be represented, in quantity and direction, by the sides of a parallelogram whose diagonal represents in quantity and direction the single force for which they are substituted.

150. **Resultant of any number of forces.**— If any number of forces whatever act upon the same point of a body, and in any directions whatever, a single force can always be assigned which will be mechanically equal to them, and will therefore be their resultant. After what has been established, nothing is more easy than the solution of this question. Let the several single forces supposed to act upon the point in question be expressed by A, B, C, D, E, &c. 1st. Let the resultant of A and B be found by the principle of the parallelogram of forces explained above, and let this resultant be A'. 2nd. Let the resultant of A' and c be found by the same principle, and let this resultant be B'. 3rd. Let the resultant of B' and D be found, and let this resultant be c'; and so on. In this way we shall finally arrive at the determination of a single force, which will be equivalent to, and will therefore be the resultant of the entire system.

151. In what precedes, we have supposed the forces whose combined effects are to be determined as applied to the same point of the body on which they act. It often happens, however, that the

forces are applied to different points. We shall therefore now proceed to consider this case; and, first, we shall take the more simple condition under which the force acts in parallel directions.

152. **Parallel forces.**—As before, we shall consider the forces represented by weights.

Let P and P' (*fig.* 5.) be the points to which the two forces in question are applied, and let these two forces be represented in direction by parallel cords P M and P' M' passing over pulleys, and let them be represented in quantity by two weights A and A' suspended from these cords. Now the resultant of these two weights, A and A', or the single force which would be equal to them, may be determined by means precisely similar to those which we have adopted in the case of diverging forces, by ascertaining where a single force may be applied and what will be its amount, so as to balance the two forces A and A'.

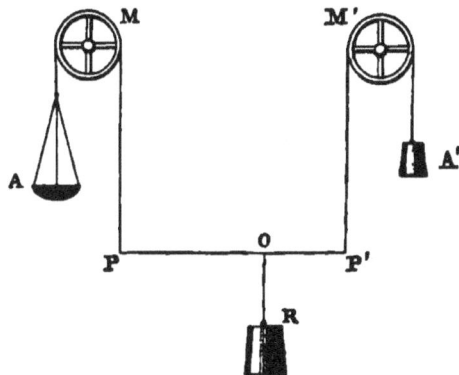

Fig. 5.

For this purpose, let us suppose a weight, R, to be suspended from a point, o, between P and P'.

Instead of suspending a determinate weight from the string carried over the pulley M, let us suppose the dish of a balance to be suspended there, capable of receiving any heavy matter which may be placed in it. Things being thus arranged, let sand be poured into the dish A, until it is found that the three weights A, A', and R are in equilibrium. Let us suppose, for example, that the weight A' is 6 oz., and the weight R 10 oz. If the weight of the sand and of the scale which bears it at A be ascertained, it will be found to be 4 oz.; and it therefore follows that the sum of the two weights at A and A' being 10 oz., is equal to the weight R. Hence it follows that the resultant of the two parallel forces A and A' is in this case a force equal to their sum.

Now if the experiment be varied in any manner, the same result will still be obtained. Thus, if the weight A' be 8 oz. and the weight R be 20 oz., then the weight of the sand in the dish A will be found to be 12 oz.; the sum of A and A', 12+8, being still equal to R, which is 20. In a word, in whatever manner the weights A and R may be varied, so long as R is greater than A, the weight A' will invariably be their difference ; and we therefore conclude in general that *the resultant of two parallel forces acting in the same direction upon two different points of the same body, is a force parallel to their direction, and equal to their sum acting at some intermediate point.*

Now, it remains to determine what is the intermediate point between P and P' at which this resultant will act.

After having established an equilibrium by pouring the quantity of sand into the dish A, if we measure the distances P o and P' o, we shall find that they are invariably in the *inverse proportion* of the two weights A and A' ; that is to say, if the weight A' be 8 oz· and the weight A 12 oz., then the proportion of P o to P' o will be 8 to 12, and this will be found to be invariably the case. If the position of the string supporting the weight R be varied, as it may be, it will always be found that the ratio of the two weights A and A', which establish an equilibrium in the system, will be inversely as the distance of the points P and P' from the point o, where the resultant is applied ; while the distance P o represents in quantity the component A', the distance P' o will represent in quantity the component A.

153. This general principle, which is of great importance in mechanics, may be enunciated as follows : —

The resultant of two forces which act on different points of the same body in parallel lines, and in the same direction, is a single force equal to their sum acting parallel to them, and in the same direction, at an intermediate point which divides the line joining the two points of application of the components in the inverse proportion of the quantities of those components.

If the forces A' and R be considered as components, the force A may be considered as the opposite of their resultant. It consequently follows that the resultant of A' and R is a force equal in quantity to their difference, and applied at P, acting in a line parallel to them, and in the direction of the greater force R.

154. **Parallel forces in opposite directions.** — This general principle may be enunciated as follows : —

The resultant of two forces which act on different points of the same body in parallel lines in opposite directions, will be a single force equal to their difference, and acting at a point beyond the greater of the two forces, and so situated that the point of application of the greater of the two forces will divide the distance between the lesser and the resultant in the inverse proportion of the quantities of the lesser and of the resultant.

Having the means of determining the resultant of two parallel forces, we can determine the resultant of any number of such forces by taking them respectively in pairs, as we have done in the case of diverging forces. Thus, let any two forces of such a system be taken, and the resultant found. Then, considering such resultant as a component, let it be combined with a third component, and their resultant found; and so on.

155. There is a case of parallel forces which does not admit of a single resultant, and which is of considerable importance in mechanical inquiries. This case is that in which two equal forces act upon two points of a body in parallel and opposite directions. The effect of such forces cannot be represented by any single force. In fact, such a combination of forces has no tendency to produce in a body any progressive motion, but has a tendency to cause it to revolve round a point intermediate between the direction of the two forces. Such a system of forces is called a *couple*.

156. The mechanical effect of such a system depends, consequently, on the intensity of the forces, the perpendicular distance between their lines of direction, and on the direction of the plane which passes through their lines of direction.

If p and p′ (*fig.* 6.) be the points of application, and one of the forces act in the direction of p m, while the other acts in the

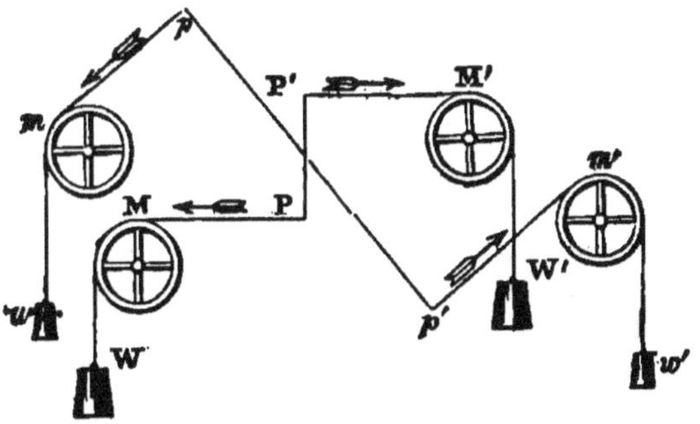

Fig. 6.

direction of p′ m′, then their effect will depend on their intensity, on the length of the perpendicular distance p p′ between their directions, and on the direction of the plane in which the lines p m and p′ m′ lie.

Representing such forces by weights, as before, let us suppose strings attached to the points p and p′ carried over the pulleys m and m′, and supporting the two equal weights, w w′.

The obvious tendency of these weights is to turn the line p p′

round in the direction in which the hands of a clock would move. Now this tendency cannot be counteracted by any single force, but it may be resisted by another *couple* applied to other two points of the body. Let us suppose, for example, that $p\ p'$ be two other points of the same body, either situate in the same plane as the lines P M and P' M', or in any parallel plane, and that strings be applied to them extended by weights, in the same manner as in the former case, the strings lying in such parallel plane.

Let $p\ m$ and $p'm'$, carried over pulleys, support weights w and w', but let them be so applied to the line $p\ p'$ that they shall have a tendency to turn the body round *contrary* to the motion of the hands of a clock, and therefore contrary to the effect of the former couple. Now let the weights w and w' be so adjusted by trial, that this second couple· shall exactly balance the first couple, and keep the body at rest, which may be done by using for the purpose the dish of a balance and sand, as in the former experiment. When the equilibrium is thus established, it will always be found that the weights w and w will bear to each other *the inverse proportion of the distance between the parallel cords,* that is to say, the weight w will be greater than the weight w in the exact proportion of the distance $p\ p'$ to the distance P P'; or, to express this in the usual manner by arithmetical symbols, we should have

$$\text{w} : w :: p\ p' : \text{P P'} ;$$

from which it follows that

$$\text{w} \times \text{P P'} = w \times p\ p'.$$

157. This conclusion involves the entire mechanical *theory of couples,* and may be enunciated as follows : —

Two equal and parallel forces acting in contrary directions on a body, have a tendency to make that body revolve round an axis perpendicular to a plane passing through the direction of such two parallel and opposite forces; and such tendency is proportional to the product obtained by multiplying the intensity of the forces by the distance between their directions ; and, consequently, all couples in which such products are equal and have their planes parallel are mechanically equivalent, provided that their tendency is to turn the body round in the same direction ; but if two such couples have a tendency to turn the body in contrary directions, then two such couples have equal and contrary mechanical effects, and would, if simultaneously applied to the same body, keep it in equilibrium.

158. Two forces not being parallel in their directions which are applied to different points of the same body, present two different cases, in one of which only they admit of a resultant.

1st. If the two forces applied at P and P′ (*fig.* 7.), not being parallel, are nevertheless in the same plane, their directions P M and P′ M′, if prolonged, will necessarily meet at some point such as o. In this case we may imagine the two forces to be applied at o, and their resultant will be represented in quantity and direction by the diagonal o c of a parallelogram whose sides represent, in quantity and direction, the two forces, according to the principle already explained.

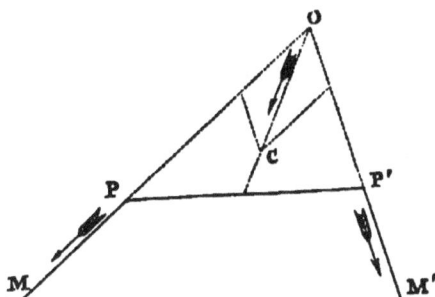

Fig. 7.

2nd. But if the forces applied at the points P and P′ (*fig.* 8.), not being parallel, are at the same time in different planes, then the directions, though indefinitely prolonged, will never intersect, and they will not have any single resultant; in other words, their mechanical effect cannot be represented by that of any single force.

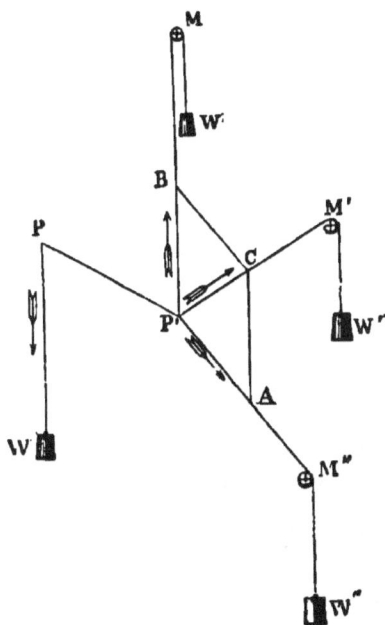

159. It can be demonstrated, however, that the mechanical effect of such a system of two forces as here described, whose directions lie in different planes, and which, though not parallel, can never intersect, will be mechanically equal to the combined action of *a couple* such as already described, and a *single force;* in other words, such a system will have a double effect on the body to which it is applied: 1st, a tendency to produce revolution; and, 2ndly, a tendency to produce a progressive motion; and if it were not held in equilibrio by some equal antagonist forces, the

Fig. 8.

body would at the same time move forward in some determinate direction, and revolve round some determinate axis.

To render this intelligible, let us imagine the forces to be represented by weights acting on strings, and passing over pulleys.

In *fig.* 8., let P and P′ be the two points of application of the two forces; let the force acting at P be vertical, and be represented by the weight w, suspended by a string at the point P.

Let the other force applied at P′ be horizontal, and in a direction P′ M′ perpendicular to P′ P, and let it be represented by the weight w′ suspended by a cord which passes over the pulley M′. Attached to the point P′, let another string be carried vertically upwards to M, and then passed over a pulley, and let a weight be suspended to it equal to the weight w. Now take upon the line P′ M′ a distance P′ c, consisting of as many inches as there are ounces in the weight w′, or, which is the same, in the force which stretches the cord P′ c. Take also upon the vertical line P′ M as many inches P′ B as there are ounces in the weight w, and draw the line B c. From c draw c A parallel to P′ B, and from the point P′ draw P′ A, parallel to B c. Carry a string from P′ along the line P′ A, and let it pass over a pulley M″, and suspend from it the weight w″, consisting of as many ounces as there are inches in the line P′ A.

Now, according to this statement, the weight w will consist of as many ounces as there are inches in P′ B, and the weight w″ will consist of as many ounces as there are inches in the line P′ A. It follows, therefore, from the principle of the composition of forces already established, that the combined effects of these two forces w and w″ acting in the line P′ B, and P′ A upon the point P′, will be the same as the single action of the weight w′ acting in the direction P′ c upon the same point P′, and that it may be consequently substituted for the latter without changing the effects upon the body. Let us then detach the weight w′ and relieve the point P′ from its action, leaving the weights w and w″ acting in its place. The point P′ and the body to which it belongs will then be affected in the same manner by the three weights w, w, and w″, as it was by the two original weights w and w′. It follows, therefore, that the effect of the two original weights w and w′ is mechanically equivalent to the effect of the weight w″ and the two equal weights w.

This is equivalent to stating that the two forces acting at the points P and P′, in the directions P w and P′ c, are equivalent in their effect to three forces, viz., a single force, represented in intensity by the line P′ A, and acting along that line, and a *couple* acting in a vertical plane passing through the line P P′; the distance between the two forces being equal to P P′, and their intensities being equal to w.

The total effect, therefore, would be equivalent to a single force acting in the direction F′ A, and a *couple* producing rotation round an axis perpendicular to a vertical plane through P P′.

160. **Suspension bridges.** — The principles upon which the

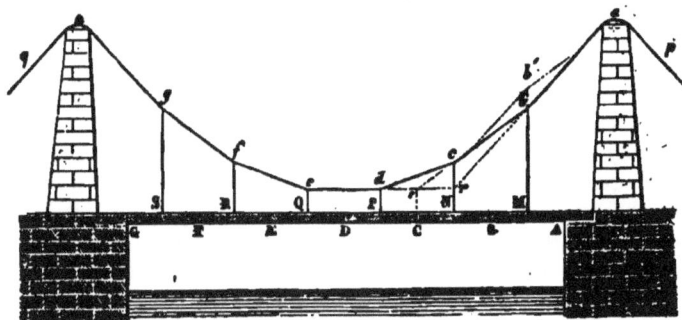

Fig. 9.

jointed rods and links by which the floors of these bridges are supported, admit of easy explanation upon the elementary theorems of the composition of force.

Supposing the floor of the bridge extending between the two quay walls to be represented by A B C D E F G (*fig.* 9.), and to be sustained by the parallel and equidistant vertical rods M, N, P, Q, R, S, it is necessary that these rods severally should so support the flooring that no one part of it shall have a tendency to rise above or sink below the adjacent part; for it is evident that if any such inequality existed in the sustaining force, one of two consequences would ensue, either of which would be incompatible with the necessary conditions of such a structure: if the flooring did not yield to these unequal forces, it would be subject to continual transverse strains, which would impair its strength; and if it did yield, its surface would not be level, but undulating. The condition, then, being admitted upon which the flooring is supported by the vertical rods in equilibrium in the horizontal position, we may imagine it divided at the middle points B, C, D, E, F of the successive intervals between the rods, as well as at the points A and G, where it meets the quay walls. In this way the several pieces A B, B C, C D, D E, E F, and F G are supported separately and severally by the rods M, N, P, Q, R, S. Although free to move up or down, being quite detached from each other, they will not do so, since such a derangement of their relative positions would be incompatible with the equilibrium of the forces supporting the flooring which has been supposed.

It remains now to determine the relative directions of the several links *a b*, *b c*, &c., which support the vertical rods. For this purpose we are to consider that the middle link *d e* is parallel to the flooring at a height above it altogether arbitrary, depending merely on the general height at which it may be found convenient to place the chain above the flooring. The half of the flooring extending from A to D, consisting of the three detached pieces A B, B C, and C D, may be considered as sustained exclusively by the links *a b* and *d e;* for if these last links were cut through, it is evident that the pieces suspended from the rods M, N, P, would fall. This being the case, the entire weight of the three pieces, acting vertically downwards at their centre of gravity N, will necessarily be in equilibrium with the tensions of the links *a b* and *d e*. It follows from this, by the principles of the composition of force, that if the line *e d* be continued to meet the vertical rod N at *r*, the direction of the link *a b* will be that of the line drawn from *r* to *a;* this, therefore, determines the direction of the first link *a b*. The direction of the second link *b c* is determined in precisely the same manner. We have only to consider the pieces B C and C D as being supported by the links *b c* and *d e;* and it follows, as before, that the weights of these pieces, B C and C D, acting vertically downwards from their centre of gravity C, will be in equilibrium with the tensions of the links *b c* and *d e*. If, therefore, the direction of *d e* be continued to meet the vertical through C at *s*, the line drawn from *s* to *b* will be the direction of the second link *b c*, that of the third link being necessarily the line *c d*.

The directions of the corresponding links at the other side of the centre will be the same.

The parts of the chain *p q* on the other side of the piers will be inclined to the vertical at the same angle with the links *a b* and *p q;* so that, the tensions being the same, the resultant of the two forces in each case will necessarily be vertical, and consequently the weight will be supported by the piers.

We have here considered only one chain. A bridge, however, must be at least supported by two, one at each side, and may be sustained even by four, two at each side. In such cases the force will be equally distributed among the chains, and the same principles will be applicable to determine the direction of their several links.

CHAP. II.

COMPOSITION AND RESOLUTION OF MOTION.

161. **Direction and velocity.** — Motion has two qualities, direction and velocity. If we would define in a precise and intelligible manner the state of a body which is in motion, we must therefore state, first, the direction in which it moves; and next, the rate at which it moves, or the speed which it has in such direction.

If the motion of a body be rectilinear, that is to say, if it move continually in the same straight line, then such straight line is its direction. But it is evident that a body may move in two opposite directions in the same straight line : thus, if the line of the motion be east and west, the body may move either from east to west, or from west to east, without departing from the line in question.

A term is wanting in our language for the convenient expression of this condition attending the motion of a body. In French, the word *direction* expresses the line in which the body moves, and the word *sens* expresses the direction of its motion in such line.

162. **Direction in a curve.** — If a body move in a curved path, such as A B (*fig.* 10.), the direction of its motion is continually changed, but at any point of the curve, such as P, it is considered to have the direction of a tangent P T at that point.

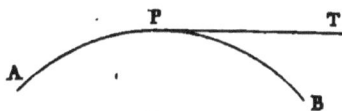

Fig. 10.

163. The velocity of a moving body is expressed by stating the relation between any space through which the body moves, and the time in which such motion is performed. Thus, we say that the speed with which a man walks is four miles an hour, the speed with which a stage-coach travels is ten miles an hour, and the speed of a railway train is thirty miles an hour.

It is evident that the same speed may be differently expressed, according to the different units of time or distance which may be adopted. We may express the velocity by stating the space moved over in a given time, or the time taken to move over a given space. The given time may be an hour, a minute, or a second; and the given space a mile, a foot, or an inch. If we say that a railway train moves at thirty miles an hour, or that it moves at eight hundred and eighty yards a minute, or forty-four feet a second, we express exactly the same speed, the only difference being that different units of time and distance are adopted.

The selection of the units of time and distance for the expression of velocity is of course arbitrary. It is usual, however, to adopt such units that the velocity may be expressed by a number which is neither inconveniently great nor inconveniently small. If the motion of the body be extremely rapid, we express its velocity by adopting a large unit of space or a small unit of time; and if, on the other hand, the motion be very slow, we express the velocity by a small unit of space or a large unit of time.

As spaces or times which are extremely great or extremely small are more difficult to conceive than those which are of the orders of magnitude that most commonly fall under the observation of the senses, we shall convey a more clear idea of velocity, by selecting for its expression spaces and times of moderate rather than those of extreme length. The truth of this observation will be proved, if we consider how much more clear a notion we have of the velocity of a railway train, when we are told that it moves over fifteen yards per second, or between two beats of a common clock, than when we are told that it moves over thirty miles an hour. We have a vivid and distinct idea of the length of fifteen yards, but a comparatively obscure one of the length of thirty miles; and, in like manner, we have a much more clear and definite idea of the duration of a second than we have of the duration of an hour.

164. In the following table are collected examples of the velocities of various objects, which may be found useful for reference, and which will serve as standards in the memory for various classes of motion: —

Table showing the Velocities of certain moving Bodies.

Objects moving.	Miles per Hour.	Feet per Second.	Objects moving.	Miles per Hour.	Feet per Second.
Man walking - -	3	4½	Air rushing into vacuum - - -	850	1247
Horse trotting - -	7	10½	Common musket-ball	850	1247
Swiftest racehorse -	60	88	Rifle-ball - - -	1000	1467
Railway train, English	32	47	Cannon-ball - -	1091	1600
" " Americ.	18	26½	Bullet discharged from air-gun, air being compressed into a hundredth part of its volume -	477	700
" " Belgian	25	36½			
" " French	27	39½			
" " German	24	35½			
Swift English steamboats navigating the channels -	14	20½	Sound when atmosphere is at 32° Fahr.	743	1090
Swift steamers on the Hudson -	18	26½	Do. 60° Fahr. - -	765	1122
Fast sailing vessels -	12	17½	Earth moving round sun - - -	68182	100000
Current of slow rivers	3	4½	A point on earth's surface at equator by diurnal motion -	1040	1525
" rapid rivers	7	10½			
Moderate wind - -	7	10½			
Storm - - -	36	52½			
Hurricane - -	80	117½			

165. **Uniform velocity.** — The speed of a body in motion may be *uniform* or *varied.*

The speed is uniform when all equal spaces, great or small, are moved over in equal times.

It is possible to conceive a body in motion, moving over a mile per minute regularly, and yet the speed not to be uniform; for though each successive mile may be performed in a minute, the subdivisions of such mile may be performed in unequal times: thus, the first half of each successive mile might occupy forty seconds, and the other half twenty seconds. To constitute a uniform motion, therefore, it is necessary that all equal portions of space, no matter how small they are, shall be passed over in equal times.

166. **Composition of motion.** — As forces which produce pressure would, if the bodies on which they act were free to move, produce motion in the direction of such pressure, whose velocity would be proportional to the pressure, all the principles which have been established in the last chapter respecting the components and resultants of forces will be equally applicable to motion. Hence, without further demonstration, we may consider as established the following principles, called the composition and resolution of motion.

If a body A (*fig.* 11.) receive at the same time two impulses, in virtue of one of which it would move in the direction A Y, over the space A B, in one second; and in virtue of the other, it would move in the direction A X, over space A C in one second; then the two impulses, acting upon it simultaneously, will cause it to move over the diagonal A D of the parallelogram, whose sides are A B and A C, in one second.

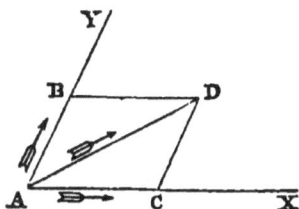

Fig. 11.

Conversely, also, a single motion A D may be considered as equal to the combination of two motions, A B and A C, along the sides of any parallelogram of which A D is a diagonal.

Now, as an infinite variety of parallelograms may have the same diagonal, it follows that any single force may be considered as equal to an infinite variety of combined forces or motions. In *fig.* 12., the line A D is the diagonal of the parallelograms A B D C, A B′ D C′, A B″ D C″, &c.

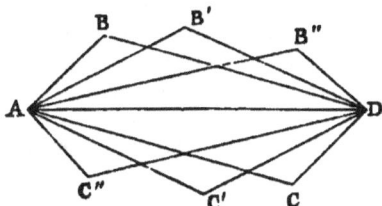

Fig. 12.

167. The effect of a force which acts upon a body in motion, must be either to change its velocity, or to change its direction, or to produce both these effects together.

If a body being in uniform motion in a certain straight line, such as A B (*fig.* 12.), receive at P the impulse of a force in the

A ⇒→ P B

Fig. 13.

direction in which it is moving, the effect of such impulse will be to augment its velocity; and such increase of velocity will be exactly equal to the velocity which the same force would impart to the body, if the body had been at rest. Thus, if the body had been moving towards B at the rate of ten feet per second, and that it received an impulse at P, in the direction of B, which would have imparted to it, being at rest, a velocity of five feet per second, then the velocity of the body after the impact will be fifteen feet per second. This is evident, because the previous motion which the body is supposed to have had in the direction of B cannot in any way impair the effect of a force tending to make it move in the same direction.

168. But if the impulse which it has received at P had been given in the direction P A, contrary to that of its motion, then such impulse would deprive the body of just so much velocity in the direction P B, as it would have imparted to it in the direction P A had it been at rest. In this case, therefore, the velocity after the impact will be diminished from ten feet per second to five feet per second.

These consequences are obviously analogous to the corresponding conclusions, which were explained in the last chapter, respecting the combined effects of forces acting upon a body in the same or any parallel directions.

169. If a body, being in motion from A to B (*fig.* 14.), receive at the point P an impact which, had it been at rest, would have caused it to move in the direction P C, then the body commencing from the point P will have a motion compounded of the motion

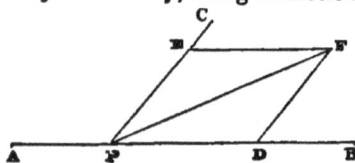

Fig. 14.

which it had before receiving the impact at P, and the motion which that impact would have given to it had it been at rest. The motion, therefore, which it would have on leaving P in this case, will be determined by the parallelogram of forces, according to the principles already established.

Thus, if we suppose that, by virtue of the motion which the

E 3

body had along the line A B, it would have moved from P to D in a second, and that the motion which it would, being at rest at P, have received from the impact would have carried it from P to B in one second, then the motion which the body will actually have in leaving P after receiving the impact, will be along the diagonal P F of the parallelogram whose sides are P D and P E.

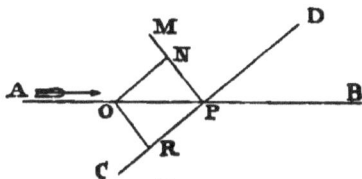

If a body moving with a certain determinate velocity in the direction A B (*fig.* 15.) encounter at P a fixed object that has a flat surface C D, its motion will not be destroyed, but will be modified by the resistance of this surface.

Fig. 15.

The effect of the surface, supposing it, as well as the body, to be perfectly hard and inelastic, will be to destroy so much of the motion of the body as is in a direction perpendicular to it.

To determine this, let us draw the line P M perpendicular to the surface C D, and, taking P O as the space which the moving body moves over in a second, from O draw O N parallel to P C, and O B parallel to P N, so that P N O B shall be a parallelogram, of which P O is the diagonal. The force, therefore, with which the body will strike the surface at P will be equal to the two forces, one in the direction of B P, and the other in the direction of N P, the sides of this parallelogram, inasmuch as these two forces are, by what has been already explained, equal to the single force represented by the diagonal O P. But the reaction of the surface C D will destroy the perpendicular force N P, and therefore the force B P alone will remain in the direction B P D. The body, therefore, after it strikes the surface, will move from P towards D with a velocity, in virtue of which it will describe spaces equal to B P in one second.

170. **Examples.** — The following examples will illustrate the principles of the composition and resolution of forces which have been explained in the present and the last chapter.

171. If a swimmer direct his course across a river in which there is a current, his body will be at the same time subject to two motions: first, that which he receives from his action in swimming, which, if there were no current, would carry him directly across the river in a direction perpendicular to its course in a certain time, as, for example, fifteen minutes; and the other in virtue of the current, by which, if he were to float on the water without swimming, he would be carried down the stream with a velocity equal to that of the stream, at a rate, for example, of five thousand feet in fifteen minutes.

Now, in actually passing over the river, the swimmer is affected at the same time by both these motions; and consequently his body will be carried in fifteen minutes along the diagonal of the parallelogram of which these motions are sides.

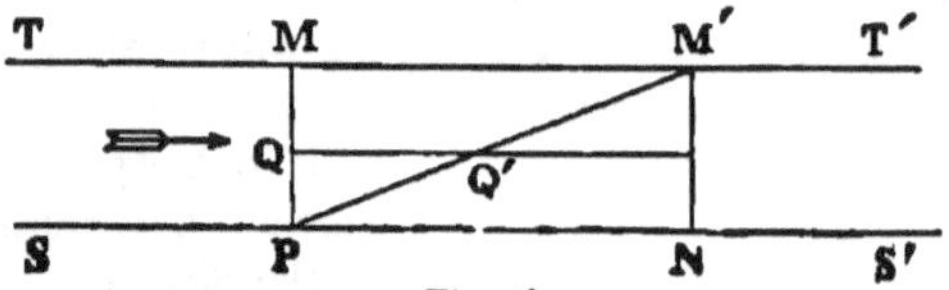

Fig. 16.

If s s' and T T' (*fig.* 16.) be the banks of the river, the direction of the stream being expressed by the arrow, and P be the point of departure of the swimmer, let P N be the distance down which the stream would carry the swimmer in the time which he would take, if there were no current, to cross the river from P to the opposite point M of the other bank. Then, as he crosses, he will be impelled by the two motions. By swimming, he will continually approach the bank T T'; at which he will arrive as soon as he would do if there were no current, because the current which crosses him laterally will not interfere with his course in swimming, which is perpendicular to it. Therefore, at the time that he would arrive at the middle point, Q, of the stream, if there were no current, he will still have arrived at the middle point, Q', of the stream; but in virtue of the current, that point will lie, not opposite the point from which he started, but lower down on the stream at Q'; and, in fine, he will arrive at the opposite bank at the point M', just so far below the point M as is equal to the space through which the current moves in the time which the swimmer takes to cross the river. The actual course followed by the swimmer, therefore, by the combined effects of his action in swimming and of the effect of the stream, will be the diagonal P M' of the parallelogram, one side of which, P N, represents the space through which the current moves in the time the swimmer takes to cross the river, and the other, P M, the space through which the swimmer would move in the same time if there were no current.

By a due attention to the principles of composition of motion, the swimmer may be enabled, notwithstanding the current, to cross the river in a direction perpendicular to its banks.

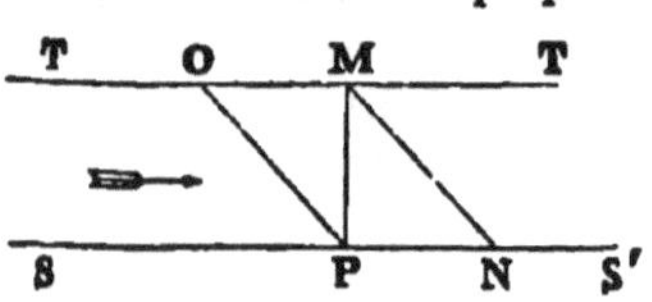

Fig. 17.

As before, let P (*fig.* 17.) be the point of the bank from which he starts, the current of the river being in the direction of the arrow.

Let P N be the distance through which the current would run in the time which the swimmer would

E 4

take to cross the river. From N draw the line N M to the point of
the bank directly opposite to P, and from P draw the line P O parallel to M N. Now, by the principles of the composition of motion
which have been already explained, a force in the direction of P M
will be equal to two forces simultaneously acting, one in the direction of P N, and the other in the direction of P O; consequently, if
the swimmer, leaving the point P, direct his course to the point O,
and use such action in swimming as would enable him to pass in
still water from P to O in the same time which the current requires
to run from P to N, then his actual motion, in consequence of the
combined effect of his action in swimming and the force of the
current, will be in the direction P M, directly across the stream; as
this direction will be the diagonal of the parallelogram, whose
sides represent the two forces to which his body is exposed.

But, under these circumstances, although he will pass directly
across the stream from P to M, instead of being carried diagonally
down it, as in the last example, along the line P M' (*fig.* 15.), he
will take a longer time and greater exertion to cross the river. In
the former example, as his motion was perpendicular to the stream,
the force of the current did not in any wise retard his course in
swimming, but merely changed the point of the opposite bank at
which he arrived; and although he passed over a greater space in
the water, yet the increased distance over which he moved was due,
not to his own exertion, but to the current. In the latter instance,
however, a part of his exertion was expended in stemming the
current, so as to prevent his descending the stream, and another
part in crossing the river.

172. A vessel impelled at the same time by wind and tide in
different directions presents another example of the composition
of motion. If we suppose the wind to impel the vessel in the direction of the keel, and at the same time the tide to act at right
angles to the keel, giving the vessel a lateral motion, the real
course of the vessel will be intermediate between the direction of
the keel and the direction of the tide at right angles to the keel.
The exact direction of this course may be determined by the composition of motion, provided the force of the tide and the effect of
the wind be known.

Let us suppose, for example, that the force of the tide is four
miles an hour, and that the effect of the wind is such that if there
were no tide the vessel would be carried in the direction of its keel
at the rate of seven miles an hour. The actual course of the vessel
will then be the diagonal of a parallelogram, one side of which is
in the direction of the vessel, measuring seven miles, and the other
at right angles to the keel, and in the direction of the tide, measuring four miles.

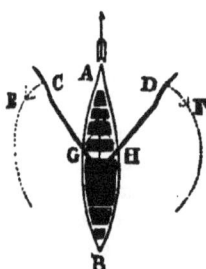

173. The action of the oars in impelling a boat is an example of the composition of forces. Let A (*fig.* 18.) be the head, and B the stern of the boat. The boatman presents his face towards B, and places the oars so that their blades press against the water in the directions C E, D F. The resistance of the water produces forces on the sides of the boat in the directions G L and H L, which by the composition of force are equivalent to the diagonal force K L, in the direction of the keel.

Fig. 18.

174. Similar observations will apply to almost every body impelled by instruments projecting from its sides, and acting against a fluid. The motion of fishes, the act of swimming, the flight of birds, are all instances of the same kind.

175. Numerous other examples may be derived from navigation, illustrative of the composition and resolution of forces and motion. The action of the wind on the sails of a vessel, and the effects produced by the reaction of the water, is one of these.

Let A B (*fig.* 19.) represent the position of the sail, and let the wind be supposed to blow in the direction C D oblique to the sail, its force being represented by the line C D. Let C D be considered as the diagonal of a parallelogram whose sides are E D and F D, the former perpendicular to the sail, and the latter in the direction of its surface. We may then consider the actual wind to be represented by two different winds, one of which will blow perpendicular to the surface of the sail, and the other will strike the sail edgewise, and will therefore produce no effect.

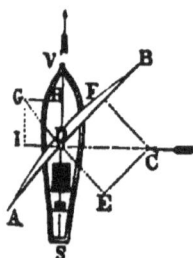

Fig. 19.

The effective part, therefore, of the wind C D will be represented by E D. Now, this force E D, thus acting perpendicular to the sail, will still act obliquely to the keel. Let the force therefore of the wind upon the sail be represented by D G, the diagonal of a parallelogram, one side of which, D H, is in the direction of the keel, and the other, D I, perpendicular to it. The component of the wind, therefore, expressed by D E, may be imagined to be replaced by two distinct winds, one, D H, in the direction of the keel, and the other, D I, at right angles to the keel.

The original force of the wind C D will thus be expressed by three distinct winds, one, D F, striking the sail edgewise, and therefore inefficient; another, D I, acting at right angles to the vessel, and therefore producing lee way; and the third, D H, acting in the direction of the keel from stern to stem, and therefore propelling the vessel.

In this case, it appears that a wind blowing at right angles to the course of the vessel, and having therefore in itself no tendency to propel the vessel, is nevertheless, by the resolution of forces, rendered efficient for propulsion.

It is easy to show that this principle may be pushed further, and that a wind which may blow in a direction nearly opposite to the course of a vessel may, by the proper application of the same principle of the resolution of force, be made to propel a vessel. In *fig.* 20. the wind is represented as blowing in the direction C D, forming an acute angle with the course of the vessel, and therefore to a proportional extent opposed to it. Let us suppose that the rigging admits of placing the sail so as to form a still more acute angle with the course of the vessel than does the wind.

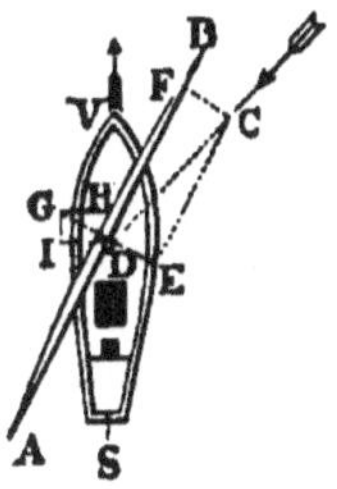

The sail will thus lie between the line C D, representing the direction of the wind, and the line D V, representing the direction of the keel. As before, C D, by two successive resolutions of forces, is shown to be equal to three different

Fig. 20.

winds : 1st, a wind represented by F D, blowing edgewise on the sail, and therefore inefficient; 2ndly, a wind, D I, blowing at right angles to the course of the vessel, and therefore producing lee way ; and 3rdly, a wind, D H, in the direction of the keel, from stern to stem, and therefore efficient for propulsion.

It is evident, however, that the more oblique the wind may be to the course of the vessel, the smaller will be the component of its force represented in the diagram by D H, which is efficient for propulsion; and the greater will be the component F D, acting edgewise on the sail, and therefore inefficient. This will be apparent from the mere inspection of the two diagrams.

The limit of the practical application of this principle in sailing is determined by the play given by the rigging to the sails. There is in each species of rigging a practical limit beyond which it is impossible to place the sails obliquely to the keel. A wind which blows upon the quarter near this limit cannot be rendered useful. It will be apparent, therefore, that different kinds of rigging supply different limits to the application of this principle; and we accordingly find that one form of vessel is capable of sailing nearer to the wind than another.

176. But even though the wind should blow in a direction

immediately opposed to the course of the vessel, we are enabled, by another application of the principle of the resolution of force, to press such an adverse wind into the service, and to render it available in carrying the vessel directly against its own force.

This object is attained by the expedient called *tacking*, which is nothing but a practical application of the resolution of force. Supposing the course which the vessel has to pursue to be due west, while the wind is blowing due east, then the course of the vessel due west is to be regarded as the diagonal of a parallelogram whose sides are directed alternately to some points north and south of west.

Let A W (*fig.* 21.) be the course of the vessel departing from A; it sails from A to D, some points north of west; from D it sails to D', some points south of

Fig. 21.

west; from D' it sails to D'', parallel to A D; and from D'' to D''', parallel to D D'; and so on. Thus, at first, instead of sailing direct from A to B', it reaches B' by sailing over two sides A D and D B' of a parallelogram whose diagonal is A B'. Again, instead of sailing directly from B' to B'', it sails from B' to D', and from D' to B'', over two sides of a parallelogram whose diagonal is B' B''; and so on. The vessel is thus conducted over the sides of a series of parallelograms, the succession of whose diagonals forms its right course.

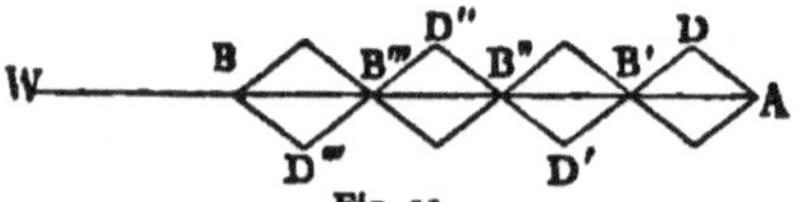

177. If a ball of lead be taken to the masthead of a vessel which is advancing under sail or steam, and be let drop downwards, it might at first be supposed that this ball would fall at a point vertically under that from which it was discharged, in which case it would necessarily strike the deck just so far behind the mast as would be equal to the space through which the vessel had advanced in its course during the time of the fall.

But we find, on the contrary, that the ball strikes the deck at the foot of the mast precisely in the place where it would have struck it had the vessel been at rest.

Fig. 22.

This fact is explained by the effect of the composition of motion.

Let A B (*fig.* 22.) be the course of the ship. Let B T be the position of its mast at the moment the ball is dropped; and let B' T' be the position of the mast at the moment the ball falls on the deck, the ship having advanced through the space B B' during the interval of the fall. The

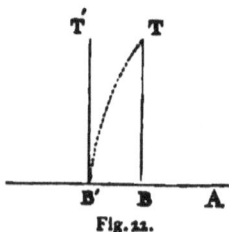

ball by being dropped from the top of the mast has in common with the ship its progressive motion; and if it had not been dropped, it would have moved through the space T T′, precisely equal to B B′, in the time of the fall.

This progressive motion during the fall is combined with the vertical motion, and if the vertical descent were made with a uniform velocity, the ball would, by reason of the combined effect of the two motions, move along the diagonal of the parallelogram, T B B′ T′. But the vertical descent of the ball not being uniform, but first moving more slowly, and then moving more rapidly, it will move, not along the diagonal of the parallelogram, but along a curve represented by the dotted line in the figure, which curve will be explained hereafter; but at the termination of the fall the ball will be found at the foot of the mast.

178. The same illustration is presented in a still more striking form by the movement of a carriage on a railway. Let us suppose that a carriage is moved along a line of railway, at the rate of 60 miles an hour, and a ball is dropped from it at the height of 16 feet as it moves. If this ball fell vertically it would strike the ground at a point 30 yards behind the point of the carriage from which it was dropped; for the time of falling 16 feet is one second, and in one second a carriage moving 60 miles an hour would move over 30 yards. The carriage would, therefore, be 30 yards in advance of the point at which the ball was let fall. But it is found that, as in the former case, the ball will meet the ground at a point vertically under the part of the carriage from which it was let fall, which renders it evident that during the fall the ball advances with the same progressive motion as the carriage.

Let T (*fig.* 23.) represent the point from which the ball is disengaged, and B the point of the rails vertically under it, the height T B being 16 feet. Let B B′ be 30 yards. In one second after the ball has been disengaged, the point T will have been carried for-

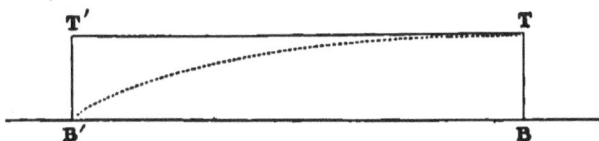

Fig. 23.

wards to T′ vertically above B′. At the moment the ball was disengaged from T, it participated in the progressive motion of the carriage; and consequently, having a motion in the direction T T′, which, if it had not been disengaged, would have carried it to T′ in one second, during the fall this motion is combined with that of the vertical descent, and the combination of the two motions will cause the ball to move over the curve represented by the dotted line, and to arrive at B′, the point vertically under T′, at the end of a second.

˙ These effects may be illustrated in a still more striking manner by supposing a vertical tube 16 feet high carried with the train, like the smoke-funnel of the engine. If a ball were held over the centre of the top of the mouth of this funnel, and let fall, it would, if the train were at rest, strike the bottom of its centre point, falling directly along the centre or axis of the tube. If the tube be carried forward with any velocity, such as 60 miles an hour, the side of the tube will not be carried against the ball by such progressive motion, as might be imagined, but the ball during its descent will still keep exactly in the centre of the tube, as it would have done had the tube been at rest.

179. The skill of the billiard player depends on his dextrous application of the composition and resolution of force. All the movements of the balls, in obedience to the cue, and whether reflected from each other or from the cushions, are determined by this principle. Let P (*fig. 24.*) be a billiard-ball driven in the direction P O by the cue. At O let it strike the cushion M N. The force of the impact may be represented by the dotted line O A, which is the diagonal of a parallelogram, one side of which is O B perpendicular to the cushion, and the other O C in the direction of the cushion. The effect, therefore, is the same as if, at the moment the ball struck the cushion, it were influenced by two independent forces represented by O C and O B. The force O B being perpendicular to the cushion is destroyed by its reaction; but the ball, being elastic, receives a rebound in the contrary direction O B′. The force O C being in the direction of the cushion is not destroyed, and being combined with that of the rebound O B′ will cause the ball to move along the diagonal O D of the parallelogram, of which these two lines O C and

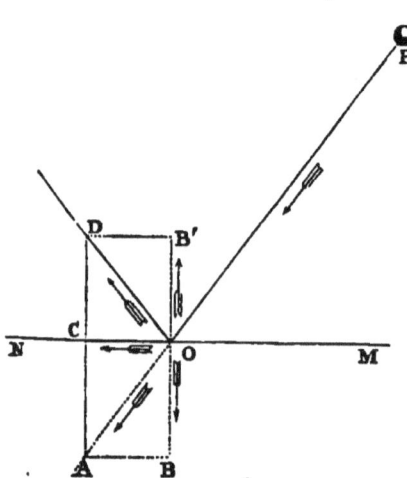

Fig. 24.

O B′ are the sides   If O, instead of being a point upon the cushion

had been a point upon the surface of another ball, the effects would be the same; only in this case the line M N would represent, not the cushion, but a tangent plane to the ball at the point of impact.

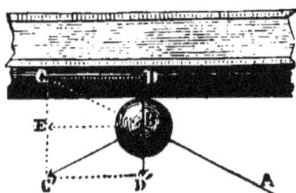

In *fig.* 25., the cushion of the table and the ball are shown, A B being the direction in which the ball strikes the cushion; B C' that in which it is reflected, and B D the force of the blow on the cushion.

Fig. 25.

The skill of the billiard player consists in a knowledge of the combination of such effects of the composition and resolution of force; but instead of deriving it from the physical principles, he obtains his knowledge by long-continued experience. He knows the effects, but cannot explain them.

In *fig.* 26. a stroke is represented in which a *cannon* is made, after successive reflections from each of the four cushions, at the

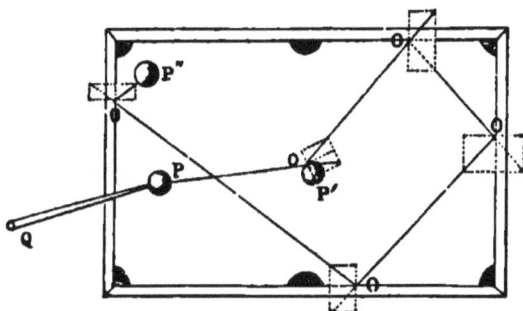

Fig. 26.

points marked o. The ball P is first directed in the line P o, upon the ball P', so that being reflected from it, it strikes the four cushions successively at the points marked o, and is finally reflected so as to strike the third ball P''. At each of the reflections from the ball P' and the four cushions, the same composition and resolution of force takes place as is represented in *fig.* 24.; and the diagrams showing such composition and resolution are given in *fig.* 26.

180. The principle of the composition and resolution of force has been ingeniously applied for the purpose of obtaining a direct demonstration of the diurnal motion of the earth.

If a high tower or steeple be erected on the surface of the earth, it is evident that, in consequence of the revolution of the globe upon its axis, the top of the tower will be moved in a greater diurnal circle than the base, being more distant from the common centre round which the entire world is moved. The top of the tower, therefore, and anything placed upon it, has a greater velocity from west to east, which is the direction of the earth's rotation, than has the bottom.

Now if we imagine a heavy ball to be let fall from the top of the tower towards the base, this ball will be affected by two motions : 1st, that which it has in common with the top of the tower from west to east, in virtue of the earth's diurnal motion ; and 2ndly, that vertical motion which it has in falling. The course it will follow will therefore depend on the combination of these two motions, and it will strike the ground at a point east of that which it occupied at the commencement of its fall, by a space equal to that through which the top of the tower is carried during the time of the fall. But during this same interval, the base of the tower is also moving eastward, but, as has been explained, through a less space.

Now as the ball is carried eastward through the space through which the top of the tower is carried, while the base of the tower is carried eastward through a less space, the ball, instead of falling at the base of the tower, which it would do, if there were no diurnal rotation of the earth, will fall just so much east of the base as is equal to the difference between the motion of the top and the motion of the bottom of the tower.

As the difference between its two motions must be an extremely minute quantity, it might be supposed that such an experiment, though beautiful in theory, would be impracticable, the quantity which would indicate the effect of rotation being smaller than could be accurately measured.

The experiment, nevertheless, has been made ; and the result has been, within the limits of error, such as would be produced by a diurnal revolution of the earth in twenty-four hours.

181. *Motion* is distinguished into *absolute* and *relative*.

182. If a man walk upon the deck of a ship from stem to stern, he has a motion relative to the deck, measured by the space upon it over which he walks in a given time ; but while he thus walks from stem to stern, the ship and its contents, including himself, are carried in an opposite direction.

If it should so happen that his own progressive motion from stem to stern is exactly equal to the progressive motion of the ship, then he will be at rest with regard to the surface of the sea, and will be vertically above the same point of the bottom ; for his own motion on the surface of the deck in one direction is, by this supposition, exactly equal to the motion of the ship in the other direction.

If he walk upon the deck at a slower rate than the progress of the ship, then he will have a motion relatively to the sea, in the direction of the ship's motion, equal to the difference between the rate of the ship and the rate of his own motion on the deck; but if he move from stem to stern on the deck at a more rapid rate than the ship advances, then he will have a motion relative to the sea contrary to that of the ship, and equal to the difference between the rate of the ship and the rate of his own motion on the deck.

But these motions are again compounded with that of the earth, in which the man who walks, the ship which sails, and the sea itself participate; and if the absolute motion of the man who walks upon deck is required to be determined, it would be necessary to combine, by the principles of the composition of motion, 1st, the motion of the man upon the deck; 2ndly, the motion of the ship through the water; 3rdly, the motion of the earth on its axis; and 4thly, the motion of the earth in its orbit.

183. Many of the feats exhibited in gymnastic and equestrian exhibitions are explained by the principles of the composition and resolution of motion.

As an example, let us take the case in which, the exhibitor standing on the saddle, a table or other elevated object is held before him, above the height of the horse, but below that of the rider, so that the horse may pass under the table, which would obstruct the progress of the rider. In this case the rider leaps over the table, and returns to the same point of the saddle which he left when the horse had passed under the table. Now, this feat demands from the rider a muscular exertion, extremely different from that which might be expected.

Let us suppose the course of the horse to be represented by

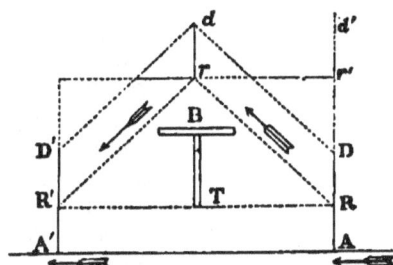

Fig. 27.

A A′ (*fig.* 27.), the table by T B, and the position of the rider upon the saddle when the horse approaches the table, and when the leap is about to be effected, by B D. At this point the rider leaps, not with a force which would project him over the table, but with one which would project him *vertically upwards* to the position represented by r′ d′. This motion, combined with the progressive motion which the rider has in common with the horse, and which is represented by ʀ т, will cause the rider to move, not directly upwards, in immediate obedience to his exertion, but in the diagonal direction ʀ r, so that at the moment the horse comes directly under the table, the rider is directly over it at r d. The upward force of the leap being here expended, the body of the rider begins to fall, and, if not urged by a progressive motion, would fall on the table. But retaining the progressive motion, the descending tendency of the fall is combined with such motion, and the rider accordingly descends in the diagonal direction r ʀ', and arrives at the point ʀ' precisely at the moment that the saddle borne by the horse arrives at the same point, so that the rider returns to his position on the saddle at ʀ' which he left at ʀ.

Strictly speaking, in this example, the motion of the rider does not take place in the right lines represented by the diagonals in the figure, but in lines slightly curved (*fig.* 28.). This, however, makes no difference in the principle involved in the case.

Fig. 28.

184. The flying of a kite presents an example of the principle of the composition of forces. Let ᴋ (*fig.* 29.) be the point on which the force of the wind may be considered as concentrated, and let the direction and quantity of this force be represented by the horizontal line ᴋ w. This line ᴋ w may be taken as the diagonal of a parallelogram, one side of which, ᴋ ʟ, is in the direction of the

Fig. 29.

surface of the kite, and the other, ᴋ м, perpendicular to it. The former component, ᴋ ʟ, being in the direction of the surface of the kite, glides off without effect; the latter, ᴋ м, being perpendicular to the kite, is effective. If the string of the kite be in

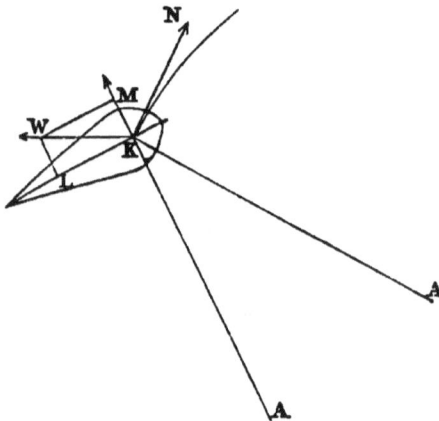

F

the direction κ ʌ, directly opposed to κ м, then the tension of the string will balance the force κ м, and the kite will remain suspended in the air, without rising higher; but if the direction of the string be κ ʌ', making an angle with κ м, then the tension of the string acting in the direction κ ʌ', and the component of the wind acting in the direction of κ м, will produce a resultant action in some intermediate direction such as κ ɴ, and this resultant will cause the kite to rise to the point ɴ, describing a circle round the centre ʌ', and the kite will ascend in this circle until the string, κ ʌ, takes the direction of the component of the wind to which it is opposed.

185. That a mass of matter moving in any manner exercises a certain force against any object which may lie in its way, is the physical law which, of all others, is the most early, the most frequent, and the most universal result of observation and experience. The child has hardly emerged from the nurse's arms, before it becomes conscious of the force with which its body would strike the ground if it fell.

186. **Momentum.** — The moving mass of a hammer-head will exercise a force upon a nail sufficient to make it penetrate wood; an effect which no common pressure could produce. The force of the hammer-head of the sledge-hammer will compress and vary the form of a mass of metal under it. By the force of a descending mass of matter, the die impresses upon a piece of metal the image and characters of the coin with more fidelity and effect than the pressure of the hand could produce them, if similarly applied to wax.

The force of a moving mass will cause a punch to penetrate a thick plate of iron, or a shears to cut the same plate, with as much facility as a needle would penetrate, or a common scissors cut this paper.

The moving masses of the spear, the javelin, and the arrow, and of the bullet and the cannon-ball, have been used as destructive projectiles in war, ancient and modern.

187. This quality equally appertains to matter in the liquid form.

The force of the torrent, when a river overflows its banks, has carried away buildings and levelled towns. The force of the stream is used to turn the mill and discharge various mechanical functions. They who have stood at the foot of Niagara, have been conscious of the frightful energy of the mechanical power developed by the motion of the mass of waters forming that cataract.

188. The same quality belongs to matter even in the attenuated form of air. The force of air in motion carries the ship over the sea, and, acting upon the diverging arms, impels the mill. The

tempest agitates the deep, and flings the largest vessel, with destructive force, upon the rock. The force of the moving atmosphere in the hurricane devastates countries, overturning buildings, and tearing up by the roots the largest trees.

It will therefore be understood how important it is to investigate the laws which govern a quality of matter which developes such effects.

189. It is not enough to know that matter in motion will exercise this force on any object which it encounters; we must be able to express, with arithmetical precision, the conditions on which the energy of this force in each case depends; we must be able to determine, when two different masses of matter are moved under known conditions, what is the proportion between the forces which they would respectively exert on any object which they might strike.

190. If a leaden ball, of a certain magnitude, move with a certain velocity, it will strike an object with a certain determinate force. If another leaden ball, of the same magnitude, moving beside it with the same velocity, strike the same object at the same time, it will evidently strike it with the same force, so that the force exerted by the two balls will be precisely double the force which would be exerted by either of them.

191. But if we suppose the two balls moulded into one, so as to make a ball of double magnitude, the force of the impact will still be the same. It results, therefore, in general, that if two bodies be moved with the same velocity, one having twice the quantity of matter of the other, the force of the latter will be twice that of the former.

By the same mode of reasoning, it may be shown generally that, in whatever proportion the mass of the body in motion be increased, the force with which it would strike an object in its way will be increased in the same proportion. This force, exerted by a mass of matter in motion, is called in mechanics by the term *momentum*, and sometimes by the phrase *moving force*.

When the velocity is the same, therefore, the momentum or moving force of bodies is directly proportionate to their mass or quantity of matter.

It is found that any force which would impel a ball with a given velocity must be doubled, if the ball require to be impelled with double the velocity, and increased in a threefold proportion if the ball be required to be impelled with three times the velocity, and so on. It is evident, therefore, that the moving force of a body will be augmented in the exact proportion in which its velocity is increased, its mass or quantity of matter remaining the same. Thus the force with which a ball weighing an ounce and moving

F 2

at ten feet per second will strike any object, will be exactly ten times the force with which the same ball, moving at one foot per second, would strike such an object.

These fundamental principles are so obviously consistent with universal experience, that they can scarcely be said to require proof.

192. **Examples.** — If we project a stone from the hand, we give it but a slight impulse, if our purpose be to impart to it a slow motion; but if we desire to project it with greater speed, we exercise a greater force of projection. Now, as the stone receives all the force communicated by the hand, it is evident that the increase of force in this case is exhibited by the increase of velocity of the motion of the projectile, the body projected being the same.

If we find a certain muscular exertion sufficient to project a ball 10 lbs. weight with a certain velocity, we shall find it necessary to use double the force if we desire to project a ball 20 lbs. weight with the same velocity.

193. A man's force exerted upon oars can propel a skiff weighing 100 lbs. through the water at the rate of ten feet per second; but the same force applied in the same manner to a vessel weighing a hundred tons would not move it faster than a sixteenth of an inch per second; for since the same force would be then applied to a body more than two thousand times heavier, the speed of the motion would be more than two thousand times less.

194. To express momentum arithmetically, let M express the mass of the body, and v its velocity; then M × v will express its moving force: and if M′ express the mass of another body, and v′ express its velocity, M′ × v′ will express its moving force.

Nothing can be more simple or easy than the use of such symbols, provided it be observed that the same units are used in each case for matter, and space, and time. Thus, for example, if in the first case the mass M be expressed by its weight in pounds, and the velocity v by the number of feet moved over per second, then it is necessary, in the second case, that the mass M′ should also be expressed in pounds, and the velocity v′ also in feet per second.

For example, if the mass M be 10 lbs., and the velocity v 5 feet per second, then the product M × v = 50; the meaning of which is, that the moving force of the body M having the velocity v, will be the same as the moving force of a body weighing 50 lbs., moved at one foot per second.

In like manner, if the mass M′ be 5 lbs. and its velocity v′ 10 feet per second, then its momentum will be M′ × v′ = 50, showing that in this case also the momentum is the same as that of a body weighing 50 lbs. moving through one foot per second.

In these two cases the momenta are equal, although the mass of one body be double that of the other; but it will be observed that the lighter mass obtains as much force by its superior velocity as the other has by its superior quantity.

195. **General condition of the equality of moving forces.**— These examples being generalized, we obtain the following theorem, which is of considerable use in mechanics : — *When the momenta of two bodies are equal, their velocities will be in the inverse proportion of their quantities of matter.* — The meaning of which is, that the velocity of the lesser body will be just so much greater than the velocity of the greater, as the quantity of matter in the latter is greater than the quantity of matter in the former.

A ball of cork which strikes upon a plank of wood, even with the greatest velocity, will scarcely produce an indentation of its surface. A ball of lead striking on the same plank with the same velocity, will penetrate it. The force of the latter will be greater than the former, in the same proportion as lead is heavier than cork.

CHAP. III.

THE COMMUNICATION OF MOMENTUM BETWEEN BODY AND BODY.

196. **Effects of collision.** — When a body in motion encounters another body, certain changes ensue in the motion and in the moving force of both bodies. These changes are in general of a complicated kind, depending on the degree of elasticity of the bodies, their form, weight, and other physical circumstances.

197. To simplify the question, however, at present, we shall consider the bodies completely devoid of elasticity, and so constituted that after the one impinges on the other they shall coalesce and move as one body in some determinate direction and with some determinate speed.

Although these conditions be not strictly fulfilled in practice, they lead to conclusions of the greatest practical utility, and which approximate more or less to the results of experience, which results, however, are modified by various physical and mechanical conditions, the consideration of which is here omitted.

198. If we suppose a mass of matter, M (*fig.* 30.), moving in a certain direction, A B, with a velocity V, to encounter another mass of matter, M', at rest, and that after the impact the two masses shall coalesce and move together; let it be required to ascertain

F 3

what will be the direction, velocity, and force of the united mass
after such impact.

The mass M', being supposed to be previously at rest, can have

Fig. 30

no motion save what it may receive from the mass M; and consequently, after the impact, it must move in the same direction
A B as the mass M moved in before the impact. Whatever moving
force M' may obtain by its coalition with M must be lost by the
mass M. This is an immediate and very important consequence of
the quality of inactivity or inertia, which has been already fully
explained.

Bodies cannot generate force in themselves, nor can they destroy
force which they have independent of any external agency. Whatever force, therefore, the mass M' may acquire after the coalition
of the two bodies, must be lost by the mass M; and consequently
the total moving force of the united masses, after the impact, must
be exactly equal to the moving force of the mass M before impact.
Now, according to what has been already explained, the moving
force of the mass M before impact was M × V. If v be taken to
express the velocity of the united mass after such coalition, then
since the quantity of matter in the combined mass is M + M', its
moving force after coalition will be expressed by (M + M') × v.
But this moving force cannot be greater or less than the moving
force of M before impact; for to suppose it greater would be to
assume that the bodies have a power of creating force in themselves, with which they were not endued by any external agency;
and to suppose it less would be to suppose them capable of destroying a force which they had independently of any external
agency.

Both of these suppositions, however, are equally incompatible
with the quality of inactivity or inertia, which implies the absence
of all power to create or to destroy any moving force. It follows,
therefore, that the momentum of the mass M, before impact, is
equal to the momentum of the united mass after impact, which is
thus expressed in arithmetical symbols : —

$$\text{M} \times \text{V} = (\text{M} + \text{M}') \times v.$$

That is to say, the mass M, multiplied by its previous velocity, is
equal to the united masses multiplied by their subsequent velocity.

To find, therefore, the velocity with which the united mass will move, it is only necessary to divide the moving force of the mass M before impact, by the sum of the united masses M and M'.

199. **Examples.**—As an example of this, let us suppose a boat, connected with a ship by a tow-rope, the latter being slack, to be propelled by rowers, at the rate of ten miles an hour, the boat weighing 5 cwt., and the ship weighing 250 tons, being 1000 times the weight of the boat. The moment the rope becomes tight, or, as seamen call it, *taut*, the moving force of the boat is divided between the mass of itself and of the ship, and the common velocity of the two will be determined by the principles above explained. To find it, therefore, it will be necessary to diminish the previous velocity of the boat in the proportion of the weight of the united masses of the ship and boat to that of the boat, that is, in the ratio of 1001 to 1.

The velocity of the boat having been ten miles an hour, or about $14\frac{1}{2}$ feet per second, the velocity with which the ship would be towed will be, therefore, the $_{\overline{1001}}$th part of a foot per second. Thus, the force which was sufficient to propel the towing-boat alone through 145 feet in ten seconds, will only tow the ship through little more than one-sixth of an inch in the same time.

200. **Collision of two bodies moving in same direction.**— We have here taken the case in which the body M' is at rest. Let us now suppose that it has a motion in the same direction AB, as that of the mass M, but with a velocity V' less than V the velocity of M, so that the body M shall overtake the body M', and that after coalition they shall move together with a common speed; what will this common speed be? It follows, as a consequence of the quality of inertia, that, after their coalition, the united mass cannot have a moving force either greater or less than they had before their union; consequently, the moving force of the coalesced bodies will be found by adding together the two moving forces which they had before their coalition. These two forces, according to what has been explained, being expressed by $M \times V$ and $M' \times V'$, the moving forces of the coalesced mass will be $M \times V + M' \times V'$. But if v express the velocity of the united mass, its moving force must also be expressed by this velocity v, multiplied by the total mass, that is, by $M + M'$; and, consequently, we have

$$M \times V + M' \times V' = (M + M') \times v.$$

It follows, therefore, from this, that if we divide the sum of the moving forces of the two bodies, before their union, by the sum of their masses, the quotient will be the velocity of the united mass after their union; the moving force of the united mass being equivalent to the sum of their moving forces before their union.

It is evident that this common velocity, which the combined bodies will have after their union, will be less than the speed of the body M, which overtakes the other, and greater than that of the body M′, which is overtaken. The one gains and the other loses velocity; and, consequently, it is evident that the one must gain and the other lose momentum. Now, this gain and loss, on the one side and on the other, is always precisely equal. Whatever momentum the body struck receives from the body striking, the latter loses, and neither more nor less. There is, therefore, a precisely equal change of momentum.

201. **Action and reaction.** — This general principle, which, as we have seen, is a direct consequence of the quality of inertia, and which, if not established, would lead to the conclusion that mere matter has a power in itself to produce or destroy its motion, is usually announced under the form of the following mechanical dogma or maxim: —

ACTION AND REACTION ARE EQUAL AND CONTRARY.

The term *action*, applied to the case above explained, is the power which the striking body has to give increased momentum to the body struck; and *reaction* expresses the correlative power which the body struck has of depriving the striking body of an exactly equal quantity of momentum.

This celebrated law of the equality of action and reaction, therefore, means nothing more than the equal interchange of momentum, in contrary directions, between two bodies which come into collision, and, as such, is an immediate consequence of the quality of inertia.

202. If a boat weighing a hundred-weight, rowed at the rate of fifteen feet per second, be suddenly connected by a towing-line with another boat which is rowed in the same direction at the rate of ten feet per second, and which weighs two hundred-weight, then, the momentum of the first being expressed by 15, and that of the second by 20, their combined momenta will be 35, which, being divided by their united weights 3, gives the quotient 11⅔, or 11 ft. 8 in., the space per second through which they will be rowed together.

203. Finally, let us consider the case in which the two bodies M and M′ (*fig.* 31.) are moving, not in the same, but in contrary directions. Let the body M be supposed to be moving from A towards C, and the body M′ from B towards C; and let C be the point at which they will coalesce, the momentum of M being supposed to be greater than that of M′. On their coalition, the mo-

mentum of M' will destroy just so much of the momentum of M as is equal to its own amount; for it is evident that equal and con-

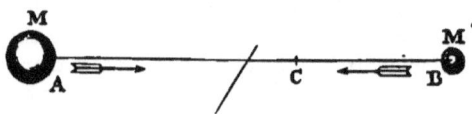

Fig. 31.

trary forces must destroy each other. After the union of the two bodies, the momentum, therefore, that will remain undestroyed, will be the excess of the moving force of M over the moving force of M'; and this excess will be in the direction c b of the progressive motion of the body M. But, as we have seen, the moving force of the body M is expressed by $M \times v$; and the moving force of the body M' is expressed by $M' \times v'$. It therefore follows that the moving force which the united mass will have in the direction a b will be $M \times v - M' \times v'$. But it is also apparent that the moving force of the united mass will be expressed by the quantity of the united mass multiplied by its velocity. If we express, as before, this velocity by v, then we shall have

$$M \times v - M' \times v' = (M + M') \times v.$$

This statement implies nothing more than that the momentum of the masses, after their coalition in the direction of the greater force, will be equal to the difference between their momenta in contrary directions before coalition.

It follows, also, from this, that the velocity of the united masses in the direction of that which has the greater force before their coalition, will be found by dividing the difference between their momenta by the sum of their masses.

204. This case affords another example of the application of the maxim, that action and reaction are equal and contrary. The effect which takes place when the bodies coalesce may be thus explained.

When the two bodies M and M' meet at c (*fig.* 31.), the action of the body M upon the body M' destroys the entire momentum of the latter in the direction b c, and, in addition to that, imparts to it a certain momentum in the contrary direction c b, which will be expressed by the mass M', multiplied by the velocity which that mass has in common with M, in the direction c b, after their coalition. This velocity is expressed by v; and, consequently, the momentum imparted to M', in the direction c b, after impact by the action of the body M, will be $M' \times v$. The total action, therefore, of the body M upon the body M' will be made up of the momentum of M', which is destroyed in the direction b c, and which is expressed by $\text{M}' \times \text{v}'$, and the momentum which is imparted to it in the contrary direction c b,. and which is expressed by $\text{M}' \times \text{v}$. Therefore, the entire action of m upon m' will be expressed by

$$\text{M}' \times \text{v}' + \text{M}' \times \text{v}.$$

Such, then, being the effect of the action of m upon m', let us now consider what will be the effect of the reaction of m' upon m. Upon the collision at c, when m' loses its entire momentum in the direction c a, it will destroy in the body m an exactly equal amount of momentum; that is to say, the body m will lose, in the direction a c, a momentum equal to $\text{M}' \times \text{v}'$, which is lost by the body m'. But m also imparts, by its action to m', a momentum in the direction of c b, which is expressed by $\text{M}' \times \text{v}$. Now, the reaction of m' upon m, in receiving this momentum in the direction of c b, must deprive m of exactly the same momentum in that direction; that is, it must deprive it of a momentum in the direction c b expressed by $\text{M}' \times \text{v}$. Hence it follows that the total amount of momentum lost by reason of the reaction of m' upon m is expressed by

$$\text{M}' \times \text{v}' + \text{M}' \times \text{v},$$

which is precisely equal to the momentum gained by m'. It appears, then, that in this case the law of equal action and reaction is still fulfilled; the action of m upon m' being precisely equal to the reaction of m' upon m.

205. When two equal bodies meet, moving with equal velocities in opposite directions, their shock will immediately destroy each other's momentum; for in this case, the momenta, being equal and contrary, will be mutually destroyed. The force of the shock produced by the two bodies in this case will be equal to the force which either, being at rest, would sustain, if struck by the other moving with double the velocity; for the action and reaction being equal, each of the two will sustain as much shock from reaction as from action.

206. **Examples.**—If two railway trains, moving in contrary directions at twenty miles an hour, sustain a collision, the shock will be the same as if one of them, being at rest, were struck by the other moving at forty miles an hour.

If two steam-boats, of equal weight, approach each other, one moving at twelve miles an hour, and the other fifteen miles an hour, each will suffer a shock from the collision, the same as if it were struck by the other moving at twenty-seven miles an hour.

207. In the combats of pugilists, the most severe blows are those struck by fist against fist, for the force suffered by each in such case is equal to the sum of the forces exerted by either arm. Skil-

ful pugilists, therefore, avoid such collisions, since both suffer equally and more severely.

208. In what precedes we have limited our observations to the cases in which the bodies which coalesce are moving either in the same or opposite directions in the same straight line. Let us consider now the case in which they move in different straight lines before their coalition.

Let a body, M (*fig.* 32.), be supposed to move in the line A B, and from A towards B. At some intermediate point, C, suppose it to be struck by another body, M′, moving in the line A′ B′ from A′ towards B′, and suppose that at the moment M arrives at C, M′ also arrives at that point and coalesces with it, and that after their union the bodies move together. The question is, in what direction, with what force, and with what velocity they will be moved. This question is easily solved by the application of the principle of the parallelogram of forces already explained.

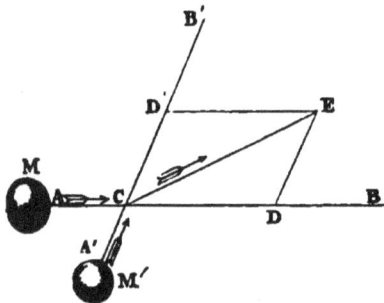

Fig. 32.

Let the velocity with which M moves in the direction A B be expressed by V, and let the velocity with which M′ moves in the direction A′ B′ be expressed by V′; then M × V will be the moving force of the body M directed from C towards B, and M′ × V′ will be the moving force of the body M′ directed from C towards B′. Let the distance C D represent the force M × V, and let C D′ represent the force M′ × V′, and complete the parallelogram C D E D′. Draw its diagonal C E. This diagonal will then represent the direction and the quantity of the momentum which the combined masses M and M′ will have after their coalition. If we would find the velocity, it will only be necessary to divide the number expressed by this diagonal C E by the number expressing the sum of the masses M and M′; the quotient will be the velocity with which the combined masses will move from C to E.

209. **Action and reaction modified by elasticity.**—When a body which strikes a hard surface is elastic, the effects of action and reaction are modified in a manner which it will be necessary to explain.

Let us suppose, for the simplicity of the explanation, that the form of the body is that of a sphere or globe. When it strikes a hard surface with any force, it will be momentarily flattened at the point of impact, and will take an oval form; the force of the impact will compress it in the direction of the blow, and it will be elongated in a direction at right angles to the body.

Thus, if its spherical form before the blow be represented in *fig.* 33., and if D C be the diameter which is in

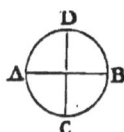
Fig. 33.

the direction of the impact, c being the point at which the impact takes place, and A B be the diameter at right angles to D C, then the body at the moment of receiving the force of impact will take an oval form represented in *fig.* 34., the diameter A B will be elongated, and D C contracted by the force of the impact. But the body, by force of its elasticity, will make an effort to recover its figure, and the point c will react upon

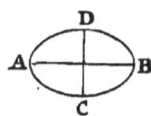
Fig. 34.

the surface which it strikes; and by that effort the body will recoil, because the diameter D C, in its effort to recover its original length, will press the matter of the body against the hard surface at c, and this pressure being resisted will cause the body to rebound.

210. **Perfect and imperfect elasticity.** — If the elasticity of the body be perfect, the force with which the spherical figure is restored will be equal to that with which it has been compressed into an ellipse, and this force being resisted by the surface, the body will rebound with the same force as that with which it struck it. But if the elasticity of the body be imperfect, then the restoring force will have less intensity than the compressing force, and the body will rebound with less force than that with which it struck the surface.

211. If an ivory ball, a substance which possesses elasticity in a high degree, be dropped upon any hard and smooth surface which is level, it will rise very nearly to the height from which it was dropped. It would rise exactly to that height, but for two causes: first, the want of perfect elasticity; and, secondly, the resistance of the air.

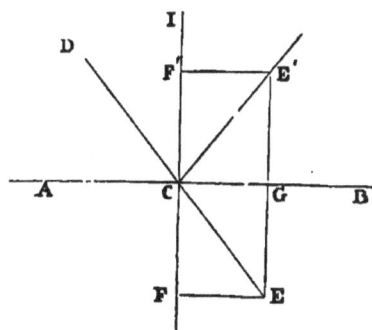
Fig. 35.

212. When an elastic body strikes a surface in a direction not perpendicular to that surface, it will be reflected in another direction, which will depend partly on the direction in which it strikes the surface, and partly on the degree of elasticity of the striking body. Let A B (*fig.* 35.)

be the surface; and let the body be supposed to strike it at c, having moved in the direction D c; and let us first suppose that the body is perfectly elastic. The force of the impact at c, being represented by c E, may be resolved into two forces, c F perpendicular to the surface A B, and c G parallel to it. We may therefore suppose the ball at. c to be affected by two such forces. Now, since the ball is supposed perfectly elastic, the component c F will cause a rebound in the direction of c F' equal to the force c F. If, therefore, we take c F' equal to c F, the body will, after the impact, be affected by two forces, c G in the direction of the surface, and c F' perpendicular to it; and it will accordingly move in the diagonal c E' of the parallelogram of which c F' and c G are sides; but this parallelogram being in every respect equal to the parallelogram c F E G, the angle formed by c E' with the surface c B will be equal to the angle formed by c E and c D with the same surface. Hence it appears that in this case the body will be reflected from the surface which it strikes at the same angle as that with which it strikes it; that is to say, the angle D c A will be equal to the angle E' c B.

213. This principle is usually announced thus: When a perfectly elastic body strikes a hard surface and rebounds from it, the angle of incidence will be equal to the angle of reflection, and these angles will be in the same plane.

By the angle of incidence is understood the angle which the direction, D c, of the original motion of the ball forms with the perpendicular, c I, to the surface struck; and by the angle of reflection is understood that which the direction, c E', in which the body recoils, forms with the same perpendicular.

214. But if the body which struck the hard surface be imperfectly elastic, then the recoil produced by the component c F (*fig.* 36.), perpendicular to the surface c B, will be less than the force with which the body strikes the surface at c. The line c F', therefore, which represents this force of recoil, will be less than c F, and the parallelogram c F' E' G, while it has the side c G in common with the parallelogram c F E G, has the side c F' less than c F. Consequently it is obvious that the angle which c E' makes with the perpendicular c I will be

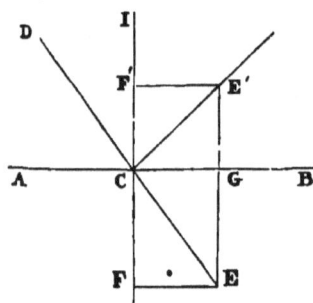

Fig. 36.

greater than the angle which D c makes with the same perpendicular

This principle is announced thus: When a body imperfectly elastic strikes a hard surface, the angle of incidence will be less than the angle of reflection, and the difference between these angles will be greater, the more imperfect is the elasticity of the body.

215. **The apparatus to illustrate collision experimentally,** which is generally found in collections for public instruction, shown in *figs.* 37, 38, 39., consists of balls, inelastic or elastic, suspended by diverging threads, the ends of which are attached to two parallel bars, and which are so adjusted that the balls, when quiescent, rest in a horizontal line and in mutual contact. Let us first suppose that two equal inelastic balls are suspended as shown in *fig.* 37. If one of them, being drawn aside, as in *fig.* 38., is allowed to fall against the other, it will strike it with a force proportionate to the distance, B A, from which it fell. After the collision, this force being divided equally between the balls, they will move together to a distance from their position of rest equal to half that from which the ball A fell. Thus, if the distance B A be 12 inches, the two balls, after collision, will move together to the distance of 6 inches, and will then swing together from side to side until they are brought to rest by the resistance of the air.

If three equal inelastic balls had been thus suspended, one

Fig. 37.

being, as before, drawn aside to the distance of 12 inches, the force, after collision, being equally distributed, all three would move together to the distance of 4 inches.

In general, if any number be suspended, as in *fig.* 39., the force of the single ball will be shared equally among all, after collision, and the whole will move to a distance, which ˉwill be found by dividing that from which the striking ball fell by the entire number of balls.

All these effects will be seen to be in complete accordance with the principles explained in the preceding paragraphs. Balls of putty or moist clay, are sufficiently inelastic for such an illustration.

Fig. 38.

If the balls be elastic, the effects are different and very remarkable. Let two equal ivory balls, for example, be suspended, as in *fig.* 37. When one, A, is drawn aside, as in *fig.* 38., and let fall against the other, B, the immediate effect of the collision, as in the case of inelastic balls, is, that A loses half its force, which it imparts to B. But B being compressed with this force recovers its figure with equal force, and reacts against A so as to deprive A of the other half of its force and to reduce it to rest. At the same time A being compressed by the reaction of B, recovers its figure with equal force, and reacting upon B imparts to it the other half of its force, so that B moves forward with the full force which A had at the moment of collision, while A remains at rest. The ball B, therefore, will move on to a distance equal to that from which A had descended. Thus, if A fall from 12 inches, it will remain at rest after the collision, and B will rise to 12 inches, from which it will descend and strike A, which in its turn will rise, B remaining at rest, and so on.

This alternate motion would continue indefinitely, but for the resistance of the air, by which the range of the vibration is gradually diminished.

If several ivory balls be suspended, as in *fig.* 39., and one be drawn aside and let fall, the effect of the collision will be transmitted through the series to the last ball, which will be affected exactly as if it were immediately acted upon by the first. Each of the intermediate balls in this case, being equally affected in opposite directions by the reaction of the contiguous balls in recovering their figures, will remain at rest, and the extreme balls alone will alternately rise and fall until reduced to rest by the resistance of the air.

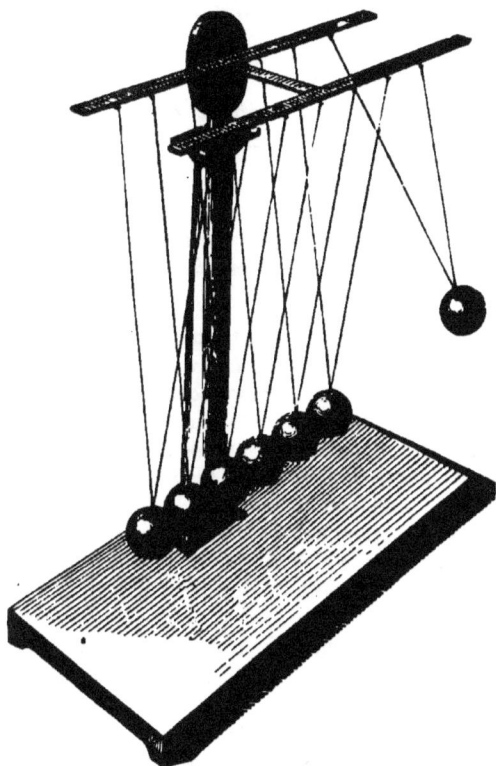

216. **The laws of motion.** — Before the true principles of inductive science were so well understood and so generally admitted as they are at present, the exposition of the property of inertia, and of its most important consequences, was embodied by Newton in three formularies, called by him *the laws of motion*, which have attained great celebrity in the history of mechanical science. Although these mechanical maxims have lost much of their importance by the more general diffusion of correct principles of physical science, they are, nevertheless, entitled to notice, and ought to be registered in the memory of all students, were it only for their illustrious origin, and to commemorate the difficulty which the true principles of induction had to struggle against, in the extermination of the errors of the old philosophy. These laws of motion are announced as follows : —

Fig. 39.

Every body must persevere in its state of rest, or of uniform motion in a straight line, unless it be compelled to change that state by a force impressed upon it.

Every change of motion must be proportional to the impressed force, and must be in the direction of that straight line in which the force is impressed.

Action must always be equal and contrary to reaction; or the actions of two bodies upon each other must be equal, and their directions must be opposite.

217. The first of these propositions is little more than a definition of the quality of inertia. The terms in which the second is expressed require qualification and explanation, without which they might be subject to erroneous interpretation.

If a body be at rest, it is true that every motion it receives must be proportional to and in the direction of the force impressed, that is, of the force which produces the motion; but if a body be already in motion in one direction, and receive a force in another direction, then the new direction which the body takes will not be in the direction of the force impressed, nor proportional to it, but will be in the direction of the diagonal of a parallelogram, one side of which represents the force with which the body previously moved, and the other the new force which is impressed upon it. This has been already fully explained.

In the third law, the word action means the moving force which one body receives when another acts upon it, and the word reaction means the moving force which the latter loses in consequence of communicating force to the former.

The equality of action and reaction is, therefore, subject to the same qualification as must be given to the terms of the second law of motion.

Both propositions are literally true only when the two motions in question are directed in the same straight line. When they are directed in different straight lines, then the propositions must be interpreted by the principles of the composition of forces, as already explained.

218. The consequence of the equality of action and reaction, combined with what has been explained respecting the indestructibility of matter in a former chapter, have led some philosophers to the adoption of the startling physical dogma, that there has been, and must be always, the same quantity of matter and the

G

same quantity of motion in the world; in other words, that nothing short of divine agency can create or destroy either the smallest portion of matter or the smallest moving force. So far as this applies to matter, it has been already explained; but a few words may be useful to elucidate the import of this maxim as applied to moving force or momentum, as it may naturally be objected, that since all creatures endowed with animal life have the power of spontaneous motion, how can it with truth be said that there is always the same quantity of moving force in the world?

219. Must not an animal, who by the act of its will puts its body in progressive motion from one place to another, and after a time, by another act of its will, causes this motion to cease, first create a new motion, and then destroy it? Is there not in such a power manifest contradiction to the maxim which states that there is always the same quantity of moving force in the world; for while the animal is in motion, is there not the momentum of the mass of its body in existence which did not exist before it began to move? and is not this momentum destroyed when it ceases to move?

220. Let us examine what takes place in such cases. If an animal commence to move its own body on the surface of the earth in any given direction, it obtains its progressive motion by the action of its feet upon the ground.

Between the mass of the body of the animal and the earth on which it treads, there is therefore an action and reaction, which are equal and opposite. Whatever moving force the body of the animal acquires in one direction, the earth loses in the other; and therefore the animal may be considered as robbing the earth, so to speak, of the moving force which its body gains.

But when the animal, after moving to the desired point, brings his body to rest, there is another action upon the earth. His body is deprived of the momentum which it had acquired by the action of the feet upon the ground, so that the momentum with which the mass of its body moved is now imparted to the earth, gradually, as his motion is retarded, and altogether when he comes to rest. It therefore appears, that the animal takes from the earth his progressive momentum when he begins to move, and returns it to the earth when he comes to rest.

221. If a railway train, weighing 100 tons, be started from a state of rest by a locomotive engine, and attain a velocity of 50 miles an hour, it will have acquired a moving force of 100 tons, moving at 75 feet per second, which is equivalent to a mass of 7500 tons moving at the rate of one foot per second. Now this momentum is obtained by the action of the driving-wheels of the engine on the rails, produced by the force of the steam.

This action is attended with an equal and opposite reaction upon the rail, and through the rail upon the earth. The earth, therefore, by this reaction, loses as much momentum in the direction in which the train moves as the train gains; therefore the earth loses, in this case, a momentum equal to 7500 tons moved one foot per second, which momentum the train has acquired.

Now, when the same train is about to stop, the moving force which it possesses is imparted to the rails, as it must be by the resistance of the rails on the wheels that the train is brought to rest. The rails, therefore, in this case, and with them the earth, receive back, gradually, its moving force, as the train is gradually stopped; and when the train has been brought to actual rest upon the rails, its entire moving force, equal to 7500 tons moving one foot per second, is restored.

222. It appears, therefore, that the earth may be regarded as a great reservoir of momentum as it is a great reservoir of matter, and that every moving force, produced upon it by any action whatever, whether mechanical or vital, must be borrowed from, and will be restored to it, when such moving force ceases. The analogy, therefore, of matter and momentum is complete, 'and the maxim above mentioned must be accepted. As all apparently new bodies must be composed of materials derived from the earth, and as all bodies apparently destroyed are merely decomposed, and their atoms restored to the common stock of matter which constitutes the globe, so all momenta must be obtained from the common reservoir of forces in the earth and restored to it.

223. But perhaps another objection may be raised in the minds of some to this reasoning. It may be asked whether, if the body of a man or animal, endued with life, could be imagined to be suspended in space, out of contact with the earth, could not such man or animal, by the act of its will, put its body in progressive motion? We answer at once in the negative.

Any attempt to move the limbs would produce in the body of such animal an equal action and reaction. If any limb were projected forwards with any given force, a reaction would take place in other parts of the body, which would be projected backwards with the same force, and the general mass of the body would have no progressive motion.

The memorable declaration of Archimedes, that if he had a point of support he could move the world itself, admits of being converted; and the philosopher might have said with equal truth, that without a support he could not move forward even his own body.

CHAP. IV.

TERRESTRIAL GRAVITY.

224. **The plumb-line.**— A small weight suspended by a light and flexible thread from a fixed point forms a combination called a plumb-line, and so denominated because the weight usually attached to the string is a ball of lead.

If several plumb-lines be placed near each other, it will be found that the strings, when at rest, will be precisely parallel to each other.

225. This common direction which the threads of plumb-lines assume is called the *vertical direction*.

226. If a quantity of liquid contained in a vessel be at rest, its surface will have a position which is called *level*.

If several fluids, near each other, be at rest, their surfaces will be found to be parallel to each other. Thus, the surface of water in a basin, the surface of a pond, lake, or river, are all parallel to each other, and parallel to the surface of the sea itself when calm.

The surface of the land is unequal and undulating, being formed into hills and vallies, and rising occasionally into mountains of considerable elevation. If this land, however, could be rendered fluid, the mountains and hills would subside, the vallies would rise, and the entire surface would assume one uniform level; and would, in fact, coincide with the general surface of the sea, and would be parallel to the surface of fluids at rest.

When it is said, therefore, that the level of a fluid surface at rest is parallel to the surface of the earth, it must be understood that by the surface of the earth is meant the general direction of the surface of the land, or the exact direction which it would assume, if, being rendered fluid, all the parts were allowed to subside to a common level. The direction which is taken by a plumb-line at rest, is found to be exactly perpendicular to such a surface.

227. These facts, which are the result of universal experience and observation, are explained by the supposition that the earth exercises an attraction upon all bodies placed upon or near it, and that the direction of this attraction is perpendicular to its general surface, that is to say, perpendicular to the surface of the sea, or of a fluid at rest. Thus, the fact that several plumb-lines have their strings parallel is explained by stating, that the weights suspended from the strings, being all attracted in a direction perpendicular to the surface of the earth, must be parallel to each other.

Every particle of a fluid contained in a vessel being equally attracted in the same vertical direction, the surface must become level and perpendicular to that direction; for if any portion of it were more elevated than another, the weight of such more elevated part would force it downwards, and press upward the lower particles, until all should attain the same level. This supposition, which is adopted to explain these familiar effects, is verified and conclusively established by a vast body of evidence supplied by astronomical researches, from which it appears that all bodies in the universe exercise upon each other attractions depending on their mass and on their mutual distance, in a manner which will be explained hereafter.

228. The globe which we inhabit participates in this common property. It therefore exercises an attraction upon all bodies placed on or near its surface, which is proportional to their masses, and is directed towards the centre of the earth, and therefore in a direction perpendicular to its surface.

229. If a body suspended at any height be disengaged, this attraction of the earth will cause it to fall in a vertical direction, that is to say, in the direction of a plumb-line.

230. And here it may be asked whether such effects are not incompatible with that principle of equality of action and reaction which we so fully developed in the last chapter? If a heavy body, disengaged at any height, be precipitated to the surface of the earth, does it not exhibit a moving force which did not previously exist, and is not this an action without a reaction?

It is, however, established by proofs which will be explained in another part of this work, that the earth not only attracts the bodies around it, but is attracted by them; and that the moving force which is impressed on them is balanced by a moving force impressed by them upon it. In fact, in the *attraction of gravitation,* as this physical agent is called, the equality of action and reaction is verified as completely as it is in the mutual impact of bodies.

For the present, however, it will not be necessary to dwell on this point. What more immediately concerns us is the explanation of those phenomena which are developed in the effects of gravity acting on bodies at or near the surface of the earth.

231. **All bodies fall with the same velocity.**—Gravity acts equally and independently on all the particles composing a body, and therefore has a tendency to make all these particles move with equal velocities, and in parallel lines perpendicular to the surface. It is easily conceived that if two leaden balls of equal magnitude be placed side by side at the same height, they will fall together with the same velocity to the surface, and strike the earth at the same moment side by side. Now, if the matter of these two balls be moulded into a single ball, the effect will remain the

same, since their form cannot affect the operation of gravity upon them.

In the same manner, if ten or a hundred leaden balls of equal magnitude be disengaged together, they will fall together; and if they be moulded into one ball of great magnitude, it will still fall in the same manner. Hence it follows that masses of matter, however they may vary in magnitude and weight, will descend to the surface of the earth with the same velocity, and if they fall from the same height will arrive at the surface of the earth in the same time, provided they be affected by no other force but that of gravity.

232. There are some circumstances developed in the fall of bodies, and the effects of the resistance of the air upon them, which are apparently incompatible with what has been just stated. If a feather and a leaden ball suspended at the same height be disengaged, it is evident that they will not fall with the same velocity. The leaden ball will be propelled with a rapidity much greater than that which affects the feather. But in this case the operation of gravitation is modified by the resistance of the air, which is much greater upon the feather than upon the leaden ball. That two such bodies would descend with the same velocity if relieved from the interference of the air, may be shown by the experiment which is familiarly known as that of *the guinea and feather*.

233. Let a glass tube (*fig.* 40.), of wide bore, as, for example, three or four inches, and of five or six feet in length, be closed at one end, and supplied with an air-tight cap and stop-cock at the other end. The cap being unscrewed, let small pieces of metal, cork, paper, and feathers, be put into it, the cap screwed on, and the stop-cock closed. Let the tube be rapidly inverted, so as to let the objects included fall from end to end of it. It will be found that the heavier objects, such as the metal, will fall with greater, and the lighter with less speed, as might be expected. But that this difference of velocity in falling is due, not to any difference in the operation of gravity, but to the resistance of the air, is proved in the following manner. Let the stop-cock be screwed upon the plate of an air-pump (*fig.* 41.), the cock being open, and let the tube be exhausted. Let the cock then be closed, and unscrewed from the plate. On rapidly invert-

Fig. 40.

ing the tube, it will then be found that the feathers will be pre-cipitated from end to end as rapidly as the metal, and that, in short, all the objects will fall together with a common velocity.

234. Since the attraction of the earth acts equally on all the component parts of bodies, and since the aggregate forces produced by such attraction constitute what is called the *weight* of the body, it is clear that the weights of bodies must be in the exact proportion of the number of particles composing them, or of their quantity of matter.

Hence, in the common affairs of commerce, the quantities of bodies are estimated by their weights.

It will appear, hereafter, that the weight of a body, or the force with which it is attracted to the sur-face, is slightly different in different places upon the earth; but this is a point which need not be insisted on at present.

At the same place the weights are invariably and exactly proportional to the quantities of matter com-posing the bodies. If one body have double or triple the weight of another, it will have double or triple the quantity of matter in the other.

235. **Motion of a falling body.**— It is not enough for the purposes of science to know merely the direc-tion of the motion which gravity impresses upon bo-

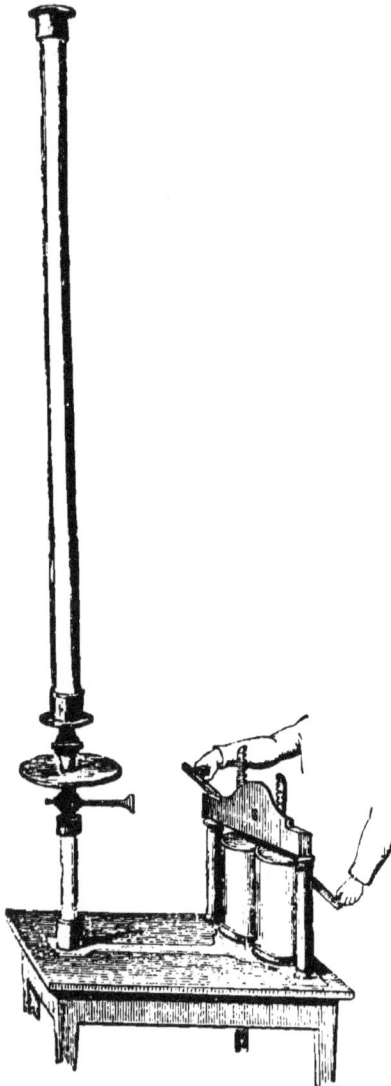

Fig. 41.

dies; we require to know whether the motion be one having a

uniform velocity; or, if not, in what manner does the velocity of
the falling body vary?

. If a man leap from a chair or table, he will strike the ground
without injury. If the same man leap from a house-top, he will
probably be destroyed by the fall. These, and innumerable similar
effects, indicate that the force with which a body strikes the ground
is augmented with the height from which it falls. Now, as this
force depends on the velocity of the body at the moment it touches
the ground, it follows that the velocity of the fall is augmented
with the height.

In short, when a body is disengaged and allowed to descend in
obedience to gravity, its velocity is gradually accelerated as it
descends. Meteoric stones which descend from the upper regions
of the atmosphere, strike the earth with such force that they are
often known to penetrate in it a considerable depth.

236. It might be naturally enough conjectured that the force
with which a body strikes the earth is proportional to the height
from which it has fallen, and such an illustration has been accordingly used by orators in speaking of the severity of censure proceeding from high quarters; but, like many other ornaments of
eloquence drawn from physical science, this is erroneous. The force of the fall is not, as we shall now
show, proportional to the height from which the
body has descended. A body falling from a double
height does not strike the ground with a double
force.

237. When a body, such, for example, as a leaden
ball, is disengaged at any height, and delivered to the
action of gravitation, the effect of this force upon it
is to impart to it a certain velocity. Now it is evident that the quantity of velocity which the attraction
of the earth gives to the ball in one second of time
must be equal to the force which it would give to it
in another second of time. Let us suppose, for example, that a moveable stage s (*fig.* 42.) is attached
to a wall or pillar, and is so adjusted that the ball
disengaged at B shall arrive upon the stage s precisely at the termination of one second. The body
will then strike the stage with a certain force.

Let another stage s' be placed at the same distance
from s as s is from B. If the ball, having been brought
to rest by the stage s, is again disengaged, it will
strike the stage s' at the end of another second, and
with the same force; and if the stage s'' be fixed at
an equal distance below s', the ball, having been brought to rest

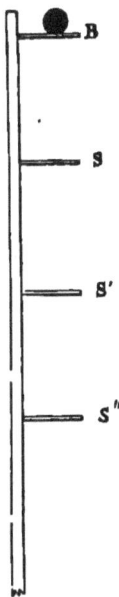

Fig. 42.

at s′, and then disengaged, will strike the stage s″ at the end of the third second, and with equal force.

In this case we have supposed that while the ball descends, the velocity it has acquired at the end of each successive second is destroyed by the resistance of the stages s, s′, and s″, &c. But suppose that on arriving at s, at the .end of the first second, the body was not deprived of the velocity it had acquired, but allowed to retain it in its descent, the retention of this velocity would not in the slightest degree prevent the action of gravity in imparting to it an equal quantity of velocity in the second second ; therefore at the end of the second second the body would have the velocity with which it struck the stage s, *in addition to* the velocity which it had acquired during the second second. In the second second, therefore, the body would descend through a much greater space than s s′, and at its termination would have a velocity double that which it had at the end of the first second. In like manner, if the .velocity acquired in the second second were not destroyed by the stage s′, the body would at the end of the third second possess this velocity, in addition to the velocity which would be imparted to it by the action of gravity in the third second.

In fine, it follows that the action of gravitation imparts to a descending body a certain velocity in every successive second of time during its action ; and consequently, the velocity which a falling body has at the end of ten or twenty seconds, is exactly ten or twenty times the velocity it had at the end of one second.

238. This principle, in virtue of which the velocity imparted by gravity to falling bodies accumulates in them, is expressed as follows : —

The velocity acquired by a body in descending by the force of gravity, increases in proportion to the time of the fall.

239. A motion, the velocity of which is thus augmented in proportion to the time counted from its commencement, is called *uniformly accelerated motion,* and the force which produces such a motion is called *uniformly accelerating force.*

Gravity, therefore, acting on bodies near the surface of the earth, is an uniformly accelerating force.

Since a body in falling moves with a velocity gradually and uniformly accelerated, its average or mean velocity will be that which it had precisely at the middle point of the interval which elapses during its fall. Thus, if a body fall during ten seconds, the average speed will be that which it had at the end of the fifth second. This is evident, inasmuch as, the speed imparted in each successive second being the same, the average of all the speeds at the end of each number of seconds, counted from the commencement, will necessarily be that which it had at the middle point of the time.

It follows from this also, that the final speed acquired by a body at the end of any time will be double the average speed counted from the commencement of its fall. This is evident, since, the velocity being proportioned to the time, the final speed is necessarily double that which is acquired in half the time, which is, as has been just shown, the average speed.

240. It follows from this also, that if a body were to move with its final velocity continued uniformly, it would, in a time equal to that of the fall, move over a space equal to double that through which it had fallen; for the final speed being double the average speed, the space described with the former will be double the space described with the latter in the same time.

To obtain a more exact estimate of the manner in which the descent of a heavy body is accelerated, it will be useful to investigate the spaces through which a body moves in its descent during every successive second of time.

241. Let us express by H the height through which a body falls from a state of rest in one second. At the end of such second, the body has acquired a velocity, in virtue of which it would, in another second, without the further action of gravity, move through a space 2 H; but during the next second gravity would cause the body to descend through another space equal to H, supposing it to move from a state of rest. Therefore, during the next second the body is moved through a space equal to three times H; that is to say, twice H in virtue of the velocity acquired at the end of the first second, and a space H in virtue of the action of gravity upon it during the next second.

Let us now consider the motion of the body during the third second. At the end of the second second, the body, having fallen through a height expressed by 4 H, has acquired a velocity in virtue of which, without any further action of gravity, it would move through a space equal to 8 times H in two seconds, and 4 times H in one second; but in addition to this, gravity also, in the third second, would move it through a space H; and from these two effects combined, the body in the third second would descend through a space expressed by 5 H. But we have seen that in the first two seconds it has fallen through a space expressed by 4 H; and therefore at the end of the third second it will have fallen through a height from the state of rest expressed by 9 H.

Pursuing its course further, we find that it begins its motion during the fourth second with a velocity such as would make it, in three seconds, without the further aid of gravity, move through a space equal to double that which it had fallen through from a state of rest, that is to say, 18 H; consequently, with this velocity, it would move in the fourth second through a space equal to 6 H; but, in addition to this, the action of gravity carries it in the fourth second through the space H, and by these combined effects it must move in this second through a space equal to 7 H. In the same manner, it may be shown that the space through which it moves in the fifth second is 9 H, while the space through which it moves in the first five seconds is 25 H; and the space through which it moves in the sixth second is 11 H, while the space through which it descends from a state of rest in the first six seconds is 36 H; and so on.

242. In the following table is expressed, in the first column, the number of seconds, or other equivalent intervals of time, counted from the commencement of the fall. In the second column is exhibited the space through which the falling body moves in each successive interval, the unit being understood to express the space through which a body falls in the first second of time. In the third column is expressed the velocity which the body has acquired at the end of each interval, counted from the commencement of the fall, and expressed by the space which, if such velocity continued uniformly, the body would describe in one second. In the fourth column is expressed the total heights from which the body falls from a state of rest to the end of the time expressed in the first column.

Tabular Analysis of the Motion of a falling Body.

Number of Seconds in the Fall, counted from a State of Rest.	Spaces fallen through in each successive Second.	Velocities acquired at the End of Number of Seconds expressed in First Column.	Total Height fallen through from Rest in the Number of Seconds expressed in First Column.
1	1	2	1
2	3	4	4
3	5	6	9
4	7	8	16
5	9	10	25
6	11	12	36
7	13	14	49
8	15	16	64
9	17	18	81
10	19	20	100

Although all the circumstances attending the descent of bodies falling freely are included with arithmetical precision in the above table, we may nevertheless render it more easy to obtain a clear conception of these important physical phenomena by the annexed diagram (*fig.* 43), in which the divided scale represents the vertical line along which the body is supposed to fall, o being the point from which it commences its descent. The points which it successively passes at the termination of 1, 2, 3, 4, 5, 6, and 7 seconds respectively are marked I, II, III, IV, V, VI, VII. The figures of the scale indicate the total heights through which the body has fallen at the end of each successive second, the unit being the height through which the body falls in the first second. The spaces included between brackets on the right of the diagram are those through which the body falls in each successive second. It will then be apparent, first, that the body is accelerated in its motion, inasmuch as the spaces through which it falls in each successive second are evidently increasing; secondly, that the space through which it falls in any number of seconds is expressed by the square of this number, the unit being the space fallen through in the first second; thirdly, that the spaces fallen through in each successive second are expressed by the odd numbers with reference to the same unit.

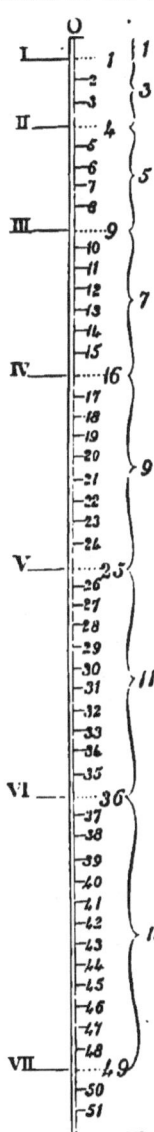

A direct experimental verification of the results exhibited in the preceding table and diagram, would be attended with several practical difficulties. The heights through which a body falls by gravity, acting freely in several seconds, are considerable, and a great velocity is soon acquired. The resistance of the air disturbs the result, and some difficulty would be found in observing, with sufficient precision, the points at which the falling body would be found at each successive second of time.

243. **Atwood's machine.** — This and other practical difficulties have, however, been surmounted by a beautiful and useful experimental apparatus, called from its inventor "Atwood's machine." By this apparatus, the intensity of the force of gravity can be diminished in any desired proportion without divesting it of any of its characters of an uniformly accelerating force. Thus we can make the falling body descend at so moderate a rate, that the effect of the atmospheric resistance becomes imperceptible, and the height, and all the circumstances attending the fall, can be observed with the greatest precision.

Fig. 43.   This contrivance consists of two equal cylindrical weights, w, w' (*fig.* 44.), connected by a fine silken thread, which passes in a groove over a nicely-constructed wheel B, turning on a horizontal axis, so as to be subject to an imperceptible friction. This wheel, and the stand which supports it, are placed upon a bracket A B, attached to a wall, or supported on a pillar, at a convenient height. Adjacent to the thread supporting one of the weights w, there is a divided vertical scale, by which the circumstances attending the descent of the weight can be noticed and measured. When the weight w is brought to the highest point of the scale, the weight w' will be near the ground; but the weight of the thread is so insignificant, that though unequal portions of it hang on each side of the wheel R, the difference of their weights produces no perceptible defect, and, accordingly, the two equal weights w and w' rest in equilibrio.

Now, if a small additional weight w be placed upon the top of the cylindrical weight w, it will cause the weight w to descend, and the weight w' to rise ; and this descent will have all the characters of a uniformly accelerating motion, for the force of gravity impresses on the preponderating weight w the same moving force which it would impress upon it if it were free ; but this moving force is, by the very condition of the apparatus, shared by w and the equal weights w and w', so that instead of imparting to w the velocity which such weight would have were it free, the velocity of the augmented moving mass, consisting of w, w, and w', will be diminished in precisely the same proportion as the mass moved is increased ; therefore, the weight w bearing upon it w, will fall with a velocity so much less than that with which w would fall, were it free, as the combined weights w', w, and w are greater than w alone. But the other circumstances attending the descending motion will be precisely similar to those which attend the descending motion of any falling body. The machine will, in fact, present a miniature representation of the phenomena of falling bodies ; the effects will be the same as though the attraction of the earth upon a falling body were diminished to such an extent that the velocity of the descent would be reduced to that with which the weight w falls.

Now, as we can adopt a preponderating weight w as small as may

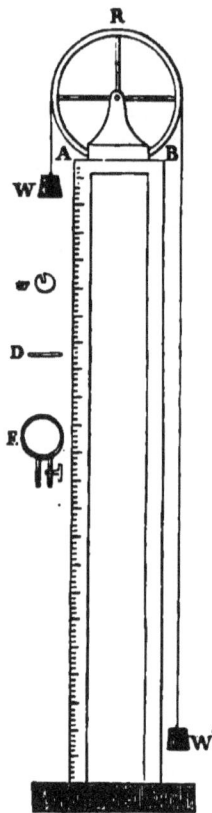

Fig. 44.

be desired, it is clear that we may reduce the velocity of the fall in so great a degree, that all the circumstances attending the motion during the descent can be deliberately and accurately observed.

Let us suppose, for example, that the weights w and w' are each twenty-four ounces, and that the preponderating weight w, placed upon w to produce its descent, is a quarter of an ounce. The total mass moved, therefore, by the action of gravity impressed upon the weight w, will be 193 times the weight w, for the weights w and w', taken together, are forty-eight ounces, that is to say, 192 quarters of an ounce ; and the weight w, which is one quarter of an ounce, being added to this, will make a total of 193 quarters of an ounce.

The attraction of gravity, therefore, instead of imparting velocity to one quarter of an ounce, has to move 193 quarters of an ounce, and, consequently, the velocity it imparts per second will be 193 times less.

We have here, with a view to simplify the explanation, avoided all reference to the motion imparted to the wheel over which the strings pass; but it will be evident that the force impressed by gravity on the preponderating weight w, must be shared with the matter of the wheel, as well as with the weights w and w'. If the matter of the wheel were all collected at its edge, it would then be moved with the same velocity as the weights w and w', and in this case it would be only necessary to consider the weight of the wheel as forming part of the masses w and w', and therefore to diminish the latter so that the total weight of w and w' and the wheel should make up forty-eight ounces.

But as the mass of the wheel is not all collected at its edge, it does not all receive the same velocity, but, on the contrary, its different parts are moved with less velocities the nearer they are to its centre. This difference of moving force imparted to different parts of the wheel, requires to be allowed for, by calculating how much matter collected at the edge of the wheel would have an equal moving force. Such a calculation, though presenting no difficulty, and subject to no inaccuracy or doubt, would involve mathematical principles and operations which cannot be conveniently introduced here ; and we may therefore assume that the momentum imparted to the wheel is represented by an equivalent portion of the forty-eight ounces assigned to the weights w and w', and that, in fact, the real weights of these must be a little less than those assigned to them, the difference being represented by the effect of the wheel.

Being provided with a pendulum beating seconds in an audible manner, and taking the thread which sustains the weight w' between the fingers, let the weight w be elevated until its upper

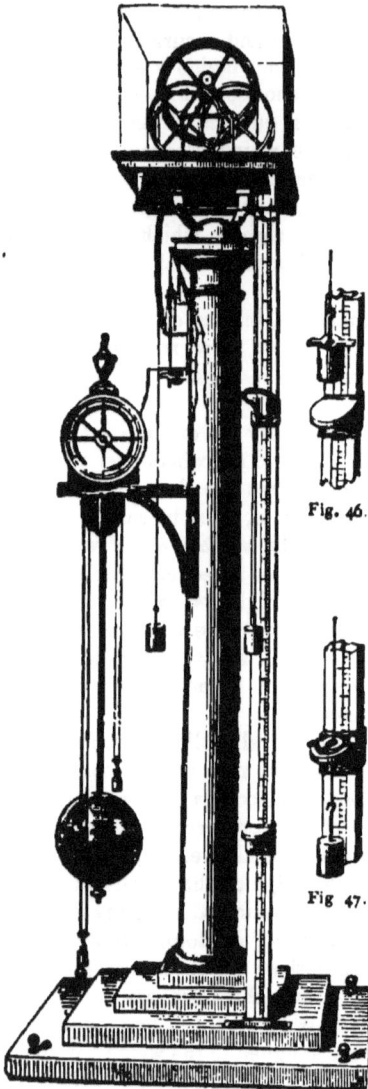

surface coincides with the zero of the scale. Listening attentively to the beats of the pendulum, let the thread be disengaged at the moment of any one beat

It will be found that, at the moment of the next beat, the weight w will have fallen precisely *one* inch. During the second beat it will have fallen through precisely *three* inches more; during the third beat it will have fallen through *five* inches; during the fourth beat it will have fallen through *seven* inches; during the fifth beat it will have fallen through *nine* inches; and so on. Now, if these distances be compared with those given in the second column of the preceding table, they will be found to correspond; the spaces through which the weight w descends in successive seconds being, as shown in this table, expressed by the odd numbers.

In the same manner, it will appear that the height through which the weight w falls during the first second, being one inch, the height through which it falls during the first two seconds will be four inches, the height through which it falls during the first three seconds will be nine inches, the height through which it falls during the first four seconds will be sixteen inches, the height through which it falls during the first five seconds will be twenty-five inches, and so forth.

Fig. 46.

Fig 47.

Fig 45.

These numbers correspond with and verify those given in the fourth column of the preceding table.

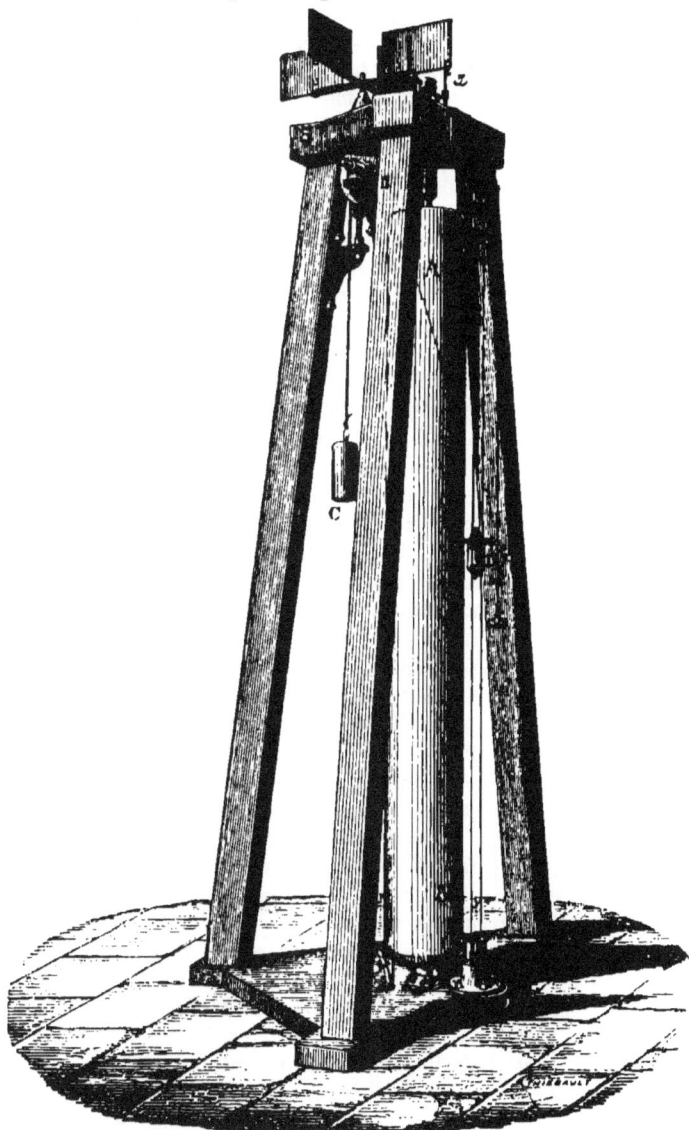

Fig. 48.

The principles of Atwood's machine being explained in *fig.* 44., its form, as actually constructed, with all its accessories, is shown in *figs.* 45, 46, 47.

244. **Morin's apparatus.** — Another apparatus for the experimental illustration of the laws which regulate the descent of bodies by gravity has recently been invented and constructed by M. Morin, the director of the *Conservatoire des Arts et Metiers* at Paris. It consists of a cylinder or drum A A, *fig.* 48., mounted on a vertical axis, on which it is moved uniformly by a train of clock-work B, impelled by a weight C, rendered uniform by a fly *a*.

A small cylindrical weight D, carries a pencil, the point of which is pressed gently against the surface of the cylinder. This weight is guided in its fall between two vertical wires, passing through holes in a plate fixed upon it, which projects from it at either side, so that in its fall it keeps constantly parallel to the cylinder and at the same distance from it. An apparatus for detaching this weight is fixed above the cylinder. By merely pulling a cord, like a bell-pull, shown in the figure beside the weight D, this weight can be let fall. Now, if the cylinder did not revolve, the pencil carried by the weight would evidently trace a vertical line upon its surface. If, on the other hand, the weight were stationary and the cylinder revolved, the pencil would trace a horizontal circle around it. But, if while the cylinder turns uniformly round its axis the weight falls, the pencil will trace around the surface of the cylinder a curved line, such as is shown in *fig.* 49. To facilitate the experiment, the surface of the cylinder is divided into equal parts vertically by the parallel lines *r r*, *s s*, *t t*, &c. &c.; and since the motion of the cylinder is uniform, the intervals between the moments at which these parallels pass under the pencil are equal. The vertical space through which the weight falls in the first interval will be *n'n*; in the first two intervals *p p*; in the first three *q'q*, and so on.

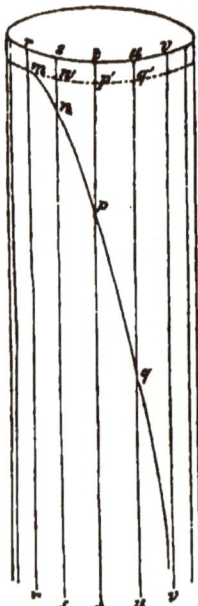

Fig. 49.

It is evident, therefore, that the circumstances of the descent will be indicated exactly by this means. In Atwood's machine it is not, properly speaking, a body falling freely, which is observed, but one, the rate of whose fall is diminished in a known proportion. Here, however, it is the actual descent which is exhibited and analyzed.

245. **Law of free descent.** — It is evident that the numbers, which express the heights through which the

bodies fall in any number of seconds, counted from the commencement of the motion, are the squares of the numbers of
seconds; and hence we have the following general principle: —
*When a body is moved by a uniformly accelerating force, such as
gravity, the spaces through which it moves, counted from the commencement of the motion, will be proportional to the squares of the
times, and the spaces through which it moves in equal intervals of
time will be as the odd numbers.*

These rules, which are of the highest importance, may be conveniently reduced to arithmetical symbols. Let us express by $\frac{1}{2}g$
the space through which a body, urged by an uniformly accelerating force from a state of rest, would move in one second, a
space which, in the case of gravity, is 16 ft. 1 in., or 193 inches.
Thus it is evident, from what has been stated, that we shall find
the space which the body would move through in any given number
of seconds, counted from the commencement of its motion, by
multiplying $\frac{1}{2}g$ by the square of this number of seconds.

246. Hence, in general, if T express the number of seconds
during which the body has been moving from a state of rest, $T^2 \times \frac{1}{2}g$
will express the entire space through which the body has moved
in the number of seconds expressed by T. If this space, then, be
expressed by H, we shall have

$$H = T^2 \times \tfrac{1}{2}g.$$

In like manner, since it has been established that the velocity
which is gained in falling during one second, is such, that in each
second the body would with that velocity move through a space
equal to twice that through which it had fallen, it follows that the
velocity acquired in one second is g; in other words, it is such,
that a body moving with that uniform velocity would move
through a space expressed by g in each second.

But it has also been shown that the velocity augments in proportion to the time, and that the velocity in two, three, four, and
five seconds is two, three, four, and five times the velocity in one
second. To find, therefore, the velocity acquired in any number
of seconds, we shall only have to multiply g by that number of
seconds. If, then, T express the number of seconds during which
the body has been falling, and v the velocity which it has gained
in the time T, we shall have

$$v = T \times g.$$

The two preceding formulæ include the whole theory of falling
bodies in vacuo. From these may be deduced the following for-

mula, by which the velocity which is acquired in falling through any given height is known : —

$$v^2 = 2 H \times g.$$

It remains now to show, that by Atwood's machine the numbers given in the third column of the preceding table may be verified; that is to say, to demonstrate, by direct experiment, that the velocity imparted to the body in its descent increases in the proportion of the time of the fall.

To accomplish this, the following arrangements are made. The preponderating weight used to produce the descent of w has the form of a long narrow bar D (*fig.* 44.), which is laid across the upper surface of the cylindrical weight w. A ring E, large enough to allow the weight w to pass through it, but not large enough to allow the bar resting on this weight to pass, is attached to the scale at the division marked 1. If the weight be now brought to such a position that its upper surface shall coincide with the zero of the scale, and if it be let fall at a moment corresponding with one beat of the pendulum, its upper surface will arrive at the ring E at the moment of the next beat, and the ring which allows the weight w to pass freely through will catch the bar, which will rest upon it. After the top of the weight, therefore, has passed the ring, the weight w being relieved from the bar, by whose preponderance its motion was accelerated, will continue to move downwards without further acceleration, with the velocity it had acquired at the end of the first second, such velocity being now continued uniform. If, then, the descent of this weight uniformly downwards be compared with the beats of the pendulum, it will be found to move uniformly at the rate of two inches per second.

Thus, we infer, first, that the velocity imparted at the end of the first second is such as to make the weight w move uniformly in one second through double the space through which it has fallen, and that such velocity is at the rate of two inches per second.

Let the ring be now moved to the fourth division of the scale, and the bar being put upon the weight w, let the experiment be repeated. It will be found that at the end of two seconds the bar will strike the ring and the weight will pass below it, moving with a uniform velocity; and by comparing its motion along the scale with the beats of the pendulum, it will be found that this velocity is at the rate of four inches per second.

Again, let the position of the ring be fixed at the ninth division of the scale, and, replacing the bar, let the experiment be once more repeated. It will be found that the bar will strike the ring at the end of the third second, and that the weight w, when disengaged from the bar, will continue to descend with the uniform

H 2

velocity of six inches per second. The same experiment may be repeated for as many seconds as the height of the scale may admit, and like results will be obtained. We may thus obtain a complete verification of the numbers contained in the third column of the table, p. 91.

247. From these experiments we are enabled to calculate the height through which a body would fall in one second of time by the effect of the force of gravity, and independently of any influence from the resistance of the air.

It appears from what has been stated, that when the magnitude of the weights w′, w, and w was so adjusted that the height of the descent was 193 times less than the height with which w would fall freely, the height through which it fell was one inch. It consequently follows, that if w were submitted to the unimpeded action of gravity, it would fall through 193 inches, or 16 ft. 1 in., in the first second.

248. The table at page 91. compared with this result, will show all the circumstances attending the descent of bodies falling freely, 16 ft. 1 in. being the unit of the table. Thus, if we desire to ascertain the height from which a body would fall in five seconds, we take the number in the fourth column of the table opposite 5 seconds, which is 25, and multiply it by 16 ft. 1 in., the product, which is 402 ft. 1 in., will be the height required.

In the same manner, if it be required to determine what space a falling body would descend through in the fifth second of its motion, we take, in the second column of the table, the number opposite 5 seconds, which is 9; we multiply 16 ft. 1 in. by this number, and find the product, which is 144 ft. 9 in., which is the space required.

In like manner, if it be required to determine with what velocity a body would strike the ground after falling during an interval of five seconds, we take the number in the third column of the table opposite 5 seconds, which we find to be 10, and we multiply 16 ft. 1 in. by this number. The product, which is 160 ft. 10 in., will be the velocity required; and we infer that the body thus falling would have, when it strikes the ground, a velocity of 160 ft. 10 in. per second.

It will be observed that the numbers in the first column of the table now referred to, and which express the time of the fall, are the square roots of the numbers in the fourth column, which express the height from which the body falls. We have therefore this general principle of uniformly accelerating motion:

When a body is moved by a uniformly accelerating force, the times required to move through any given space are proportional to the square roots of those spaces.

By the aid of this rule, and the results already obtained, we are enabled to ascertain the time which a body would take to fall from any given height. Thus, if a body be supposed to fall from a height of 10000 feet: Find the number of times which 16 ft. 1 in. are contained in 10000 feet, which is done by dividing 10000 by 16$\frac{1}{12}$. The quotient is 621·76.

This number is then the square of the number of seconds in the time of the fall. The square root of this obtained from a table of square roots being 24·9, we infer that the time a body would take to fall through the height of 10000 feet is 24·9 seconds.

In the same manner it follows, that since the velocity acquired by a body in its fall is proportional to the time of the fall, and since the time of the fall itself is proportional to the square root of the height, the velocity acquired must also be proportional to the square root of the height.

If we would, therefore, determine the velocity or force with which a body falling from a given height would strike the ground, independently of the effect of the resistance of the air, we are enabled to do so by these principles.

Thus, let it be required to determine the force with which a body falling from the height of 10000 feet would strike the ground. It has been just shown that the time of the fall would be 24·9 seconds, and it has been already demonstrated that the velocity acquired by the body would move it uniformly over a space equal to double the height through which it falls, and in the same time. Therefore, the velocity in this case would be such that in 24·9 seconds the body would move through 20000 feet; and consequently, by dividing 20000 by 24·9, we shall obtain the velocity in feet per second, which appears, therefore, to be 803 feet per second.

249. It appears, therefore, that the velocity or force with which a falling body strikes the ground increases in a much less proportion than the height from which it falls. If the height be augmented in a fourfold proportion, the force of the fall will only be augmented in a twofold proportion; and if the height be augmented in a ninefold proportion, the force of the fall will only be augmented in a threefold proportion; and so on.

250. This explains a fact of not unfrequent occurrence, and which sometimes produces surprise. Persons sometimes fall or leap from such heights as would seem to render their destruction inevitable, yet they are frequently found to escape without considerable injury. This is explained by the fact that the momentum, or shock produced by the fall, increases in a proportion so very much less than the height. A man can leap from a height of five feet with perfect impunity; if, however, he leap from a height of

H 3

ten feet, the force with which he will strike the ground, instead of being doubled, will be increased in a proportion less than one half; and if he leap from a height of twenty feet, the force with which he strikes the ground will be only doubled.

251. A further mitigation of the shock produced by a fall arises from the resistance of the air, which further diminishes the velocity acquired. A case occurred some years ago, in which a boy, dressed in a smock frock, accidentally fell down the shaft of a coal-pit having a depth of nearly 100 feet. It was expected that he would have been found dead at the bottom. On searching, however, he was found there almost uninjured. It is probable, that in this case, the frock he wore afforded a resistance to the air, somewhat resembling that of a parachute, which, combined with the principle already explained, that the velocity augments in a very much less proportion than the height, explained his safety.

252. **Retarded motion of bodies projected upwards.**—All the circumstances attending the accelerating descent of falling bodies, which have been explained in the present chapter, are exhibited in a reversed order when a body is projected upwards. Gravity then acts as a uniformly retarding, instead of uniformly accelerating force, depriving the body so projected of equal quantities of velocity in equal times; and further, it is apparent that the velocities which the force of gravity thus destroys in a body projected upwards in any given time are exactly equal to those which it would impart to a body in the same time when falling freely.

Thus, if a body be projected vertically upwards with the velocity which it would acquire in falling freely during one second, the body so projected will rise exactly to the height from which it would have fallen in one second, and at that point of its ascent it will have the velocity which it would have at the same point if it had descended.

To determine, therefore, the height to which a body will rise projected upwards with a given velocity, it is only necessary to determine the height from which it would fall to acquire the same velocity.

In like manner, to determine the time which a body would take to rise to a certain height when projected upwards, it is only necessary to determine the time which it would take to fall freely from the same height.

253. **Motion down an inclined plane.**—A plane and hard surface, which is neither in the vertical nor horizontal position, is called an inclined plane. In *fig.* 50., if the line w o be vertical, then w m will represent an inclined plane.

Bodies which descend upon inclined planes move with a uniformly accelerating force similar to that of gravity, omitting, as usual, the consideration of friction, and the resistance of the air.

Let w be a body placed upon the inclined plane. The force of gravity acts upon it in the vertical direction w o. Let this line w o, so representing the force of gravity, be considered as the diagonal of a parallelogram, of which w N and w M are sides, the side w N being perpendicular to w M. The entire force of gravity, therefore, represented by w o, and acting on the body w, will, by the principle of composition of forces, be equal to the two forces represented by the sides of the parallelogram w M and w N. But w N, being perpendicular to the plane, is counteracted by it, and exhibits itself merely in pressure upon it. The component w M, however, being in the direction of the plane and downwards, will cause the body to move down the plane.

Fig. 50.

The proportion of this accelerating force down the plane to that of gravity acting freely in the vertical direction, will, therefore, be that of the lines w M to w o. If w o be the height through which a body would fall vertically in one second, then w M will be the distance through which the body would fall in the first second on the inclined plane. It is evident, therefore, that by taking w o equal to 193 inches, the distance w M will be actually that down which the body w, independently of friction, &c., would fall in the first second.

If it be desired to ascertain the force with which the body w presses on the inclined plane, let w o be taken so as to consist of as many inches as there are pounds in the weight w. Then w N will consist of as many inches as there are pounds in the pressure which w exerts on the plane.

The motion down an inclined plane, therefore, being uniformly accelerated, like gravity, but only mitigated in its intensity in a certain ratio, depending on the inclination of the plane, all the circumstances which have been already explained in reference to the accelerated motion of bodies falling freely, will be similarly exhibited in the motion down an inclined plane.

H 4

Let w M (*fig.* 51.) be an inclined plane, and w o the vertical line, and let us suppose two bodies dismissed at the same moment from w, one falling down the vertical line w o, and the other down the line w M. Let I, II, III, IV, V, be the points upon the vertical line o, at which the body is found at the end of one, two, three, four, and five seconds.

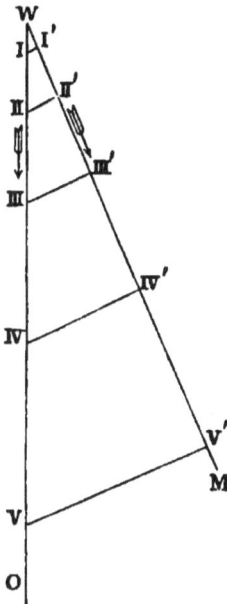

If from this point lines be drawn perpendicular to w M, the points I′, II′, III′, IV′, V′, where these perpendiculars will meet the inclined plane, will be those at which the body falling down such inclined plane will be found at the same epochs; that is to say, at the end of the first second the one body will be found at I and the other at I′, at the end of the second second the bodies will be found respectively at II and II′, at the end of the third second at III and III′, and so on.

The force down the inclined plane is just so much less in intensity than the force of gravity, as the spaces w I′, w II′, w III′, &c. are respectively less than w I, w II, w III, &c. Consequently, it is evident that these spaces, being in the proportion of the forces, will be described in the same time, as, indeed, has been already proved.

Fig. 51

In this manner, therefore, the circumstances of the motion down an inclined plane may always be determined with reference to the circumstances of the motion down a vertical line.

If it be desired to ascertain the points at which a body falling down an inclined plane will acquire the same velocities which it acquired in one or more seconds in falling freely in the vertical direction, it is only necessary to consider that the more feeble force down the plane requires a proportionally greater space to produce a given velocity. If, then, w M and w o (*fig.* 52.) represent, as before, an inclined plane and a vertical line, and if, as before, I, II, III, IV, V represent the points at which the body, falling vertically, would be found at the end of one, two, three, four, and five seconds, then the points on the plane where the same velocity would be attained as the body had at the points, I, II, III, IV, and V, will be determined by drawing lines from the points I, II, III, IV, and V respectively in the horizontal direction because, by these means, the line w V on the plane will be just so much longer than the line **w** 1 as the force of gravity, acting freely, is more intense than the force down the inclined plane; consequently, the velocity which will be acquired at v on the plane, will be the same as the velocity acquired at 1 in falling freely.

In the same manner, it will appear that the velocities acquired on the plane at the points v', v″, v‴, v⁗ will be the same as the velocities acquired in falling freely at the points II, III, IV, and V.

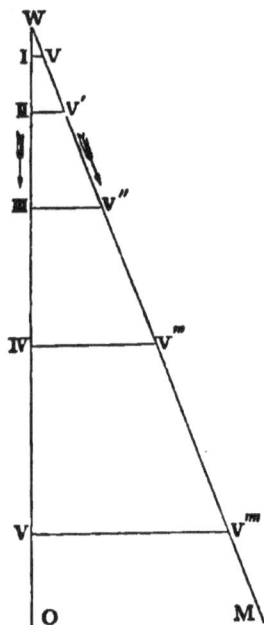

254. **Projectiles.**—We have considered the case where a body, acted on freely by the force of gravity, is either allowed to fall vertically downwards, or is projected vertically upwards. We shall now consider the other cases, in which a body is projected in any other direction, not vertical, and then abandoned to the action of gravity,—a problem which forms the foundation of the doctrine of projectiles. The solution of this problem follows immediately from the principles which determine the motion of a body falling freely, as explained in the present chapter, and the composition of motion.

255. Let us first take the case in which a body w (*fig.* 53.), as,

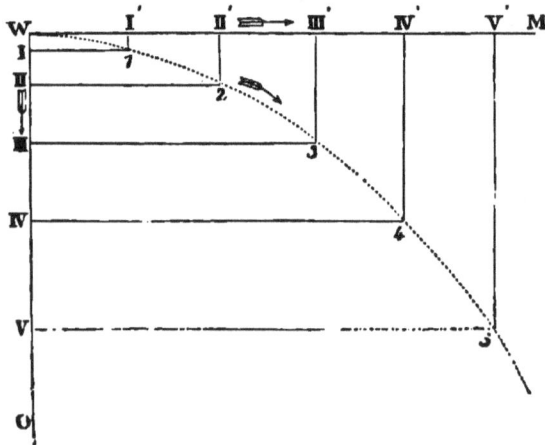

Fig. 52.

Fig. 53.

for instance, a ball shot from a cannon, is projected in the horizontal direction w M. If the force of gravity did not act on it, it would move forwards towards M with the velocity of projection continued uniform, and, in virtue of such motion, would pass over equal spaces in equal times. Thus, if, by the velocity of projection, the body would move from w to I' in the first second, it would move from I' to II' in the next second, from II' to III' in the following second, and so on, these successive spaces being equal.

But if, on the other hand, the body, without being projected at all, were disengaged at w, and left to the action of gravity alone, it would, as has been already explained, descend vertically, and would be found at the points I, II, III, IV, V at the end of the successive seconds, the distance being, as already explained, represented by the numbers 1, 4, 9, &c.

Now, the body leaving w, being submitted to both these forces simultaneously, will, by the composition of motion, be found at the end of each successive second at the extremities of the diagonals of a parallelogram whose sides represent these motions. Thus, at the end of the first second, the body will be found at the point I, being the extremity of the diagonal of a parallelogram whose sides are the space w I', through which the body would move in virtue of the velocity of progression, and w I the space through which it would fall freely in the same time by gravity. If the force of gravity would have made it move over w I with a uniform motion, then the body, in moving from w, would follow exactly the diagonal of the parallelogram. But the force of gravity imparting to the body not a uniform, but an accelerated motion, first very slow and then more rapid, the body will pass from w to I, not by a strict diagonal course, but by a curved line, as represented in the figure.

In the same manner, at the end of two seconds, the body will be found at 2. For it is actuated at the same time by two motions; first, the projectile motion, which, acting alone upon it, would carry it uniformly from w to II', and, secondly, the force of gravity, which, acting alone upon it, would cause it to fall from w to II. At the end of two seconds it will therefore be found at the point 2, being the extremity of the diagonal.

But, as before, the motion from w to II not being uniform but accelerated, first slow but afterwards more rapid, the body will pass from w to 2, not along the diagonal, but over the curved line represented in the figure.

The same explanation will be applicable to its remaining course, and it will follow that the body will pursue the curved course from w to 5 in five seconds, in consequence of the combination of the projectile velocity imparted to it, and represented by w v', combined with the descending motion imparted to it by gravity, and represented by **w v.**

256. In this case we have supposed, for simplicity, the body to be projected in the horizontal direction; but the same principles will explain its motion, if projected in an oblique direction, such as **w m** (*fig.* 54.).

As before, let the space which the body would move over in one second, in virtue of the projectile force alone, gravity being supposed not to act upon it, be **w r′.** It would move over the equal spaces terminating at r′, ir″, iii′, iv′, v′ in the successive seconds.

On the other hand, suppose the body to be acted on by gravity alone, independently of the projectile force. It would then, as before, moving in the vertical line **w o,** be found at the end of the successive seconds at the points i, ii, iii, iv, v.

Now, by the principle of the composition of motion, the body will actually be found, in consequence of the simultaneous effects of the two motions imparted to it by gravity, and by the projectile force, at the end of the successive seconds, at the points 1, 2, 3, 4, 5, which are the extremities of the diagonals of parallelograms, whose sides are respectively the spaces which the body would describe in virtue of the projectile force, and of gravity acting separately. The course of the body will be the curved line represented in the figure, and not the straight diagonal, for the reasons already explained.

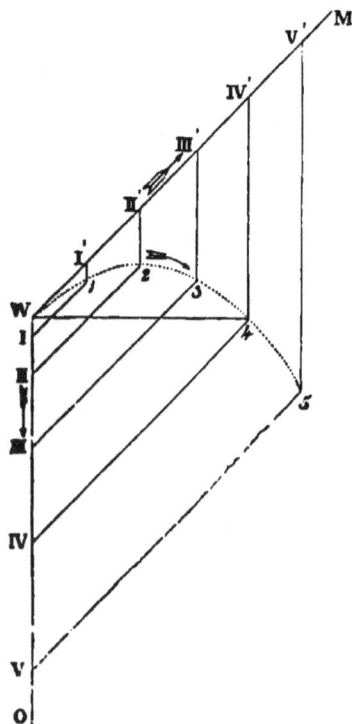

Fig. 54.

257. **Motion in parabolic curves.** — The path which the projectile follows in this case is a curve, known in geometry as the *parabola,* the property of which is, that the sides of the parallelogram whose diagonal determines its successive points, are related to each other as the successive whole numbers 1, 2, 3, 4, &c., and their squares 1, 4, 9, 16, &c.

258. Resistance of the air.—It must, however, we repeat, be remembered, that these conclusions rest upon the supposition that the body moves in a medium which offers no resistance to it, and which does not deprive it of any of the force imparted to it by projection or by gravity. In the actual case, however, the motion of all projectiles takes place through the atmosphere, which is a resisting medium, and, moreover, one of which the resistance varies, increasing in a certain high proportion with the velocity. The real path, therefore, of projectiles differs more or less from the parabola explained above. The deviation is not very considerable when the velocity of the moving body is not great; but when the projectile is driven with great velocity, as in the practice of gunnery, then the deviation from the parabolic path is so considerable, that the above theory becomes altogether inapplicable.

259. Application of projectiles in gunnery.—According to what has been explained above, a ball projected from any missile will not follow the direction of the axis of the barrel, but will proceed in a curved line, concave downwards, to which the direction of the barrel is a tangent; thus, for example, if the barrel be directed horizontally in the line w m, *fig.* 53, the ball proceeding along the dotted curve will fall below the line of aim; and if that line be directed to the object, the ball will miss it by passing too low; to hit an object, therefore, at any proposed distance, the gun must be aimed in a direction above that of the object, more or less, according to the distance of the object and the force of the charge; thus, for example, if the gun be discharged from w, *fig.* 54, and the object be at 4, the force of the charge being such as to cause the ball to move in the curve 1, 2, 3, 4, 5, it is evident that in order to hit the object at 4, the gun must be aimed in the direction w m. In practical gunnery, the force of the charge of each form and class of gun is known, and expedients are found by which the distance of objects aimed at may be determined with sufficient approximation for practical purposes. With these data the inclination at which aim is to be taken with each sort of weapon, in order to hit the object, is determined by simple rules, which can be, without difficulty, applied in the field.

Until recently the muskets placed in the hands of soldiers were usually aimed, so that the line of sight was at once parallel to the barrel, and directed to the object, as shown in *fig.* 55. But since the improvement which has recently taken place in musketry, and more especially by the introduction of the improved rifle, greater precision of aim has been attained. With an aim directed as in *fig.* 55., at the object, the ball must necessarily pass below it; so long as the range of a musket was of limited extent, and when great precision was not expected to be attained, this deviation was disregarded; but since, by the modern improvements, the range

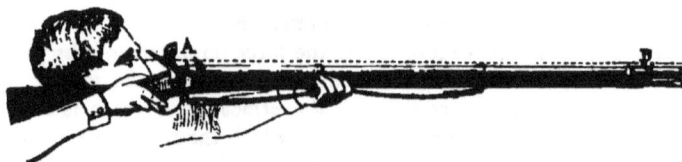

Fig. 55.

has been greatly augmented, the drop of the ball produced by the curvature of the projectile would become so considerable as to deprive the weapon of the necessary precision. On the modern guns, therefore, a double sight is provided, by which the elevation necessary to ensure point blank precision can always be given to the barrel; one of the sights B, *fig.* 56., is fixed in the usual manner on

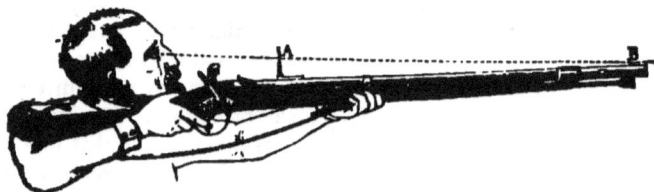

Fig. 56.

the extremity of the barrel, while the other, A, is one which is graduated, and sometimes provided with an adjustment, by which it can be adapted to objects at different distances, so as to hit them point blank.

260. **Effect of a hammer.**—When a nail is driven by the strokes of a hammer, the resistance which the moving force has to overcome, is the friction between the nail and the wood; the momentum of the hammer is, however, imparted to the wood, and if the latter have not sufficient resistance to prevent it from unduly yielding to the force, the nail will not

Fig. 57.

penetrate. In such a case, for example, as that which is represented in *fig.* 57., where the nail is to be driven into a board, having no support behind it, and not thick enough itself to offer the necessary resistance, the blows of the hammer, if strong enough, would break the board, but would not drive in the nail. The object is attained by applying behind the board *fig.* 58. a block of wood, or, still better, a lump of lead, against which the blows of the hammer will be directed. It is not, however, as might be supposed, by any increased resistance thus opposed to the blows

Fig. 58.

that in this case the object is attained. To comprehend the effect produced, it is necessary to consider that in each case the momentum of the hammer is equally imparted to the mass which it strikes; but in the one case (*fig.* 57.) this momentum is received by the board alone, which, having little weight, is driven by it through so great a space as to produce a considerable flexure, or even fracture; but in the second case, the same momentum, being shared between the board and the block of wood or metal applied behind it, will produce a flexure of the former, less in the same proportion exactly as the weight of the board and block applied to it together is greater than the weight of the board alone.

The same principle serves to explain a trick, or *tour de force*, sometimes shown by public exhibitors. The exhibitor, extending his body horizontally, his legs and shoulders being supported, causes a heavy anvil to be placed upon his chest and abdomen; men employed for the purpose then give successive blows of heavy sledge-hammers upon the anvil, without injury to the exhibitor; blows which would speedily put an end to his exhibitions if they were received without the interposition of the anvil.

To explain this feat, it is only necessary to consider that the whole momentum of the sledge, being imparted to the anvil, will give the latter a downward motion, just as much less than the motion of the sledge as the mass of the sledge is less than the mass of the anvil. Thus, if we suppose the weight of the anvil to be 100 times less than that of the sledge, its downward motion upon the body of the exhibitor will be also 100 times less than the motion with which the sledge strikes it, and the body of the exhibitor easily yielding to so slight a displacement, he passes through the ordeal with complete impunity.

Influence of time upon the effect of force. — Some extraordinary and apparently unaccountable mechanical effects are explained by the fact that the effect of forces are greatly modified by the continuance of their action ; a resistance which will be sufficient to counteract a force whose action continues only for a few seconds, will often yield to it if it act for as many minutes.

It is a fact familiar to all skaters, that they may pass with impunity over thin ice, which would break under their weight if they moved over it more slowly, and still more if they stood still upon it.

The effect of forces acting upon solid bodies is transmitted from molecule to molecule through their dimensions, before it can affect the aggregate of their mass; and if, from the nature of the force, the entire duration of its action be less than that which is necessary to propagate through the mass its effect, that effect can only be produced upon the part of the mass on which it immediately acts.

Let us suppose, for example, that a musket ball be pressed with the hand against a pane of glass, it will break the glass to pieces — the fractures extending to its very edges. If the same ball be flung from the hand with some velocity against the same pane, it will pass through it, making a hole considerably larger than the ball, surrounded by diverging cracks. But if the ball be propelled from a gun against the same pane of glass, it will pierce it with a hole whose diameter is nearly equal to that of the ball, and which shall have clean edges as if the glass were bored through with an augur. Now these effects are easily explained : when the ball is merely pressed against the glass, the force has such continuance of action, that its effects have time to be propagated to the limits of the pane, which is accordingly broken in pieces ; when the ball is projected against the glass with the hand, the action is more sudden and of shorter continuance, and consequently the fractures round the hole are much less extensive ; but when, in fine, it is discharged from a gun, its action being instantaneous, there is no time for the propagation of the force, and it merely drives before it the portion of the glass which lies in its way.

If a cannon ball be flung with the hand against the pannel of a door, suspended on hinges, it will cause the door to turn on its hinges, yielding to its force ; but if the same cannon ball be projected against the door by the cannon, it will pass clean through the pannel, making a hole equal in diameter to the ball, and without imparting any motion to the door.

In the former case, the action is slow enough to allow the force to be propagated to all parts of the door, and therefore to move it; but in the latter case, being instantaneous, it only affects the parts of the wood which lie immediately in its way.

CHAP. V.

CENTRE OF GRAVITY.

261. **Weight.**—If a body be prevented from moving in obedience to the force of gravity by a fixed axis passing through it, a fixed point from which it is suspended, or a surface placed beneath it, the effect of gravity upon it will be manifested by a pressure produced upon such axis, point of suspension, or surface. This pressure is called the *weight* of the body.

As gravity acts separately upon all the component particles of a body, the weight of such body is composed of the aggregate of the weights of all its particles. This, which is manifest from what has been already explained, may be rendered still more clear, from considering that if a body be divided into parts, no matter how minute and numerous, each of these parts will have a certain weight, and the aggregate amount of their several weights will be exactly equal to the weight of the body of which they are the fragments.

Such a division may be carried to the most extreme practical limit of comminution by pounding, grinding, filing, and other processes known in the arts, and the weight will still be divided as the matter is divided; nor is it possible, even in imagination, to conceive any degree of comminution so great that the same principle will not prevail; and it may, therefore, be considered as established that every individual atom which composes a body has weight, and that the weight of the mass is the sum of the weights of all its constituent atoms or molecules.

262. If the particles composing a body had no mutual coherence or other mechanical connection having a tendency to retain them in juxtaposition, each particle would obey the force of its gravity independently of the others, and they would fall asunder like a

mass of sand. But if they be so connected by their mutual cohesion, as they are in fact in all solid bodies, this cohesion will
resist the tendency of their weights to separate them; they will
maintain their juxtaposition, the body will retain its form, and the
several forces with which gravity affects them will become compounded, so as to produce a single force or pressure, which is the
resultant of all the separate forces impressed upon the particles.

263. As this resultant enters as a condition into every mechanical question affecting bodies, it is of the greatest importance to
investigate the conditions by which in every case its intensity and
the line of its direction may be determined.

It has been already shown that the weights of all the particles
composing a body act in directions parallel to a plumb-line, or
perpendicular to a level surface. But it has been also demonstrated (153.) that when any number of forces act in the same
direction in parallel lines, their resultant is a force acting in a line
parallel to them, and in the same direction in this line, and that its
intensity or quantity is equal to the sum of these forces. The
resultant, therefore, of the forces of gravity affecting all the particles of any mass of matter is a single force acting vertically
downwards, which is equal to the sum of all the forces affecting
the particles severally, and therefore equal to the weight of the
mass.

If, for example, A B, *fig.* 59., represent a mass of matter, and
the small arrows pointing vertically downwards
represent the direction of the gravitating forces of
the particles composing such mass, then it follows,
from what has been explained, that the resultant
of all these forces, or a single force equal to them,
will have a direction parallel to them, such as
D E, and will, in its intensity, be equal to their
sum.

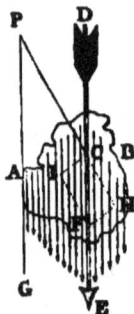

But this is not yet sufficient to indicate this
resultant in a definite manner. We as yet only
know that its direction is parallel to the common
direction of the gravity of the particles; but innumerable lines may be imagined passing through
the body vertically downwards, and the question
still remains to be determined which of these lines is the direction
of the resultant.

When the body in question has a determinate form and a uniform density, or even a density varying according to some known
conditions, the principles of mathematical science supply methods
by which the line of direction of the resultant may be determined,

I

but we shall here adopt a more simple and generally intelligible method of explanation.

If we suppose the line represented by the great arrow D E (*fig.* 59.) to be that of the resultant, then it is evident that if any point such as c in that line be supported, the body will remain at rest, because the resultant of all the forces acting upon the body having the direction D E will be expended in pressure on the fixed point c. The effect, therefore, will be that the whole weight of the body will press upon c, and the body will remain at rest.

The same would be true for any point whatever in the direction of the great arrow. If, for example, D were a pin from which a thread was suspended, and that this thread were attached to the body at any point in the line D c, then the body would still remain at rest, the whole weight being expended in pressure upon the pin at D; for, as before, the resultant of all the forces of gravity acting upon the component particles of the body would have the direction D E, and would therefore be supported by the fixed pin at D.

But if a point of support be selected which is not in the direction of the resultant D E, such as P, and a string be carried from P to any point of the body, such as c, then the body, although it will not be permitted to descend vertically, in obedience to gravity, will not nevertheless remain at rest.

If we suppose the weight of the body to be expressed by the line c F, let this line be taken as the diagonal of a parallelogram whose sides are c H and c I, one in the direction of the cord, and the other at right angles with it, — that portion of the weight which is represented by c H, and which is in the direction of the string, will act upon the fixed point P, and produce pressure upon it. The portion of the weight which acts in the direction c I will move the body towards the vertical line P G, which passes directly downwards from the point of suspension. The body will therefore begin to move towards that vertical line. If the body had been on the other side of the vertical line P G, it would still have moved towards it, and therefore in a direction contrary to its present motion.

It follows, therefore, that if a body be supported by a fixed point, it cannot remain at rest, unless the resultant D E of all the parallel forces which gravity impresses upon its particles pass through that point.

264. We are thus supplied with a practical means of ascertaining the direction of the resultant of the weights of all the component parts of a body with reference to any given point taken upon it, since we have only to suspend the body by a string attached to the given point, and allow it to settle itself to rest. When thus at rest, the resultant of the weights of all its particles will be in the direction of the string by which it is suspended.

If the same body be suspended by different points upon it, the parallel directions of the gravitating forces of its particles will differ in reference to the body, although they are the same in reference to the direction of the suspending string, being always parallel to it. Thus, for example, if an egg be suspended with its length vertical, the parallel forces which gravity impresses on its particles will be parallel to its length; but if it be suspended with its length horizontal, then the parallel directions of the gravity of its particles will be perpendicular to its length.

265. Since in each case the resultant of these parallel forces will coincide with the direction of the string, it must in the one case pass through the egg in the direction of its length, and in the other in a direction at right angles to its length. In like manner, the body being supported by any point whatever taken upon it, the direction of the string will be different for each such point; and consequently, there will be an infinite variety of resultants of the gravitating forces of the particles of the body, according to the different points by which it may be suspended.

Now, a question arises, whether there is any relation between this infinite variety of resultants; for if such be not the case, the determination of the resultant of the gravitating forces of a body, would be a problem which would present itself under an infinite diversity of forms and conditions for every individual body.

266. This question may be solved by a very simple experiment; and its solution is attended with a remarkable and important result.

Take a solid body of any form, regular or irregular, and composed of a material which is easily perforated without diminishing its mass, or considerably deranging its structure. Take, for example, a mass of putty of any form. Let this mass be suspended by a thread attached to a fixed point, which it may easily be if previously surrounded by a thread forming a loop. When at rest, the resultant of the forces of gravity, acting upon all its particles, will be a vertical line penetrating its dimensions in the direction of the suspending thread. Take a needle, and pierce the putty in this direction. The hole which is thus made through it will represent the direction of the resultant of the gravitation of its particles.

Let the mass be now detached from the thread of suspension, and let it be again suspended, but in a different position, which may be easily accomplished by the loops of thread surrounding it. The mass will again settle itself into a position of rest, and, as before, the direction of the resultant of all its gravitating particles will be a vertical line in the exact direction of the suspending thread. Let the putty, as before, be thoroughly pierced in this direction with a needle.

Let the same experiment be repeated in three or four other different positions of the mass, so that we shall obtain several holes pierced through the body by the needle, representing the direction of the resultant of the gravitating forces, in the several positions in which the body was suspended.

Now a curious relation will be found to exist between the several directions in which the needle has pierced through the putty. It will be found, in fact, that all these lines of perforation intersect, at a common point, within the dimensions of the body. This fact may be easily established.

Let a needle be inserted in any one of the perforations, and it will be found that another needle cannot pass through any of them, for its progress will be stopped by the needle already inserted. All the perforations, therefore, must intersect each other at a common point within the putty.

It appears from this experiment, that there is a certain point, within the dimensions of the body, through which the resultant of all the gravitating forces of the particles of the mass must pass, no matter in what position the body may be placed.

267. Centre of gravity.—This result, which is of high importance, may be further illustrated and verified in the following manner : —

Let a flat thin plate of metal, or a piece of card, of any form, however irregular, be pierced with small holes at several points, so that it may be suspended upon a horizontal pin, the plate itself being vertical. When so suspended, it can only remain at rest, provided the resultant of the gravitating forces of its particles pass through the pin ; for otherwise, as has been already explained, the body would move, in one direction or other, round the pin on which it is suspended.

If a plumb-line be suspended from this pin, it is evident that when the plate is at rest, the direction of the resultant of the gravitating forces must coincide with the direction of the plumb-line. Let a line then be traced upon the plate coinciding with the direction of the plumb-line. Let the body be then detached from the pin, and let the pin be inserted in another hole. The body will now hang in another position, the resultant of the gravitating forces of its particles again coinciding with the plumb-line. Let the direction of the plumb-line be traced upon the plate as before. In fine, let this experiment be repeated, with all the holes pierced in the plate, and it will be found that the lines traced upon the plate, indicating the various directions of the resultant of the gravitating forces of its particles, will intersect each other at one common point.

268. This common point, through which the resultants of the gravity of the particles of bodies pass, is called its *centre of gravity*. A line drawn in the vertical direction through the centre of gravity of a body, is called the *line of direction of the centre of gravity*.

269. If the centre of gravity of a body be supported on a point, or axis, and the body is free to turn round such axis, the body will, in that case, remain at rest in any position in which it may be placed; for, according to what has been already stated, the resultant of the gravitating forces of all its particles must be in the direction of a vertical line passing through the centre of gravity, and the whole weight of the body may be considered as acting in that line. But, if the centre of gravity be suspended by a pivot, or an axis, then the whole weight of the body will press upon such pivot or axis, no matter what be the position in which the body is placed. This may be easily verified by experiment.

Let the centre of gravity of any solid body be determined, by suspending it from different points, in the manner explained above, and let the body be placed upon a pivot or axis, passing through this point. It will be found to rest indifferently on such axis or pivot, in any position in which it may be placed. This experiment may be easily performed with a piece of card or pasteboard. The centre of gravity being determined, let a pin be passed through it, and it will be found that the card will rest in any position upon the pin.

270. **Case of regular figures.** — If a body, being of uniform density, have any regular figure, its centre of gravity will coincide with its centre of magnitude; for the matter composing the body will, in such case, be symmetrically arranged round that point, so that it is self-evident, that if this point be supported, the body will have no tendency to turn in any direction round it.

271. Thus, for example, it is evident, without experiment, that a ball or sphere of uniform density, such as a billiard-ball, has its centre of gravity at the centre of its magnitude. In like manner, a cube has its centre of gravity at the point where straight lines joining its opposite corners would intersect each other, that is to say, at its centre of magnitude.

272. If the figure of a body be such, that the matter composing it is uniformly distributed round any line passing through it, its centre of gravity must lie in that line, because, if it be suspended by a string in the direction of that line, it will remain at rest; since the gravity of its particles, acting equally on every side of such line, will have no tendency to move it, it will equilibrate. Thus, it is evident that the centre of gravity of a cone, being of uniform density, must be situate in its axis, that is, in a straight line drawn from the point of the cone to the middle of its base.

In the same manner it may be shown, that the centre of gravity of solids of an oval figure will be in the axis of the oval; the centre of gravity of a cylinder will be at the middle point of its axis; the centre of gravity of a straight rod of uniform thickness will be at the middle point of its length, and at the centre of its thickness.

It will be easy, in this and all similar cases, to verify the conclusions, by suspending the body in the manner already described.

273. **Imaginary centre of gravity.** — The centre of gravity of a body is not always placed within its dimensions. Thus, for example, the centre of gravity of a hoop is at its centre, an imaginary point, which does not constitute any part of the body in question. In like manner, in all hollow bodies the centre of gravity is an imaginary point. Thus it is in the centre of a hollow sphere. The centre of gravity of an empty box or cask is within it, at an imaginary point.

If a piece of wire, which when straight has its centre of gravity at its middle point, be bent into a curved form, its centre of gravity will be an imaginary point within the concave part of the curve. In like manner, if the wire be bent into the form of a v, the centre of gravity will be an imaginary point within the angle of the v.

These conclusions may be verified, and the centre of gravity in all such cases found, by suspending the body in different positions in the manner already explained.

274. Although the centre of gravity in such cases be not a material point, and not included within the dimensions of the body, it nevertheless still possesses those properties which it would possess were it actually included within the mass of the body.

To verify this by experiment, let us suppose a bar of metal, A B (*fig.* 60.), bent into a curved form. Let its centre of gravity be

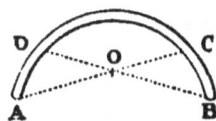

Fig. 60.

determined by suspension. When supported by the point A, let A C be the direction of the plumb-line, and when supported by the point B, let B D be the direction of the plumb-line. It follows, therefore, that the point o within the concavity of the circle where these two lines intersect, will be the centre of gravity. Let a light silk cord be attached to the points A and C, and stretched tight between them, and let another silk cord be stretched between the points B and D in the same manner. Now the point o, where these two cords cross each other, will be the centre of gravity.

Let a cord be tied to the junction of the strings at o, and let the upper extremity of this cord be attached to a fixed point, so that the wire may be thus suspended. It will be found that in this case the hoop of wire will rest in equilibrium in any position in which it may be placed. In this case, the weight of the silk string, being insignificant in comparison with the weight of the wire, does not disturb the position of the centre of gravity, which still remains at o.

275. If a body, without being absolutely fixed in its position so as to be immovable, be nevertheless partially restrained, so as to be capable of moving only under certain conditions or within certain limits, then the centre of gravity will have always a tendency to move into the lowest position which the conditions under which the body is placed will admit of; and in all cases it can never remain at rest unless its line of direction — that is to say, a vertical line passing through it — should pass through a point of support.

276. It may therefore be assumed as a principle of the highest generality, that in all cases in which a body is at rest, a vertical line passing through its centre of gravity must also pass through a point of support. If the point of support, therefore, through which this line passes be placed above the centre of gravity, the body is said to be suspended; if it be placed below, it is said to be supported.

277. If a body be suspended from a fixed point by a string, it will remain at rest, as has been already explained, provided its centre of gravity be placed in a vertical line under the point of support. But if the body be drawn out of that position, so that the centre of gravity will be on either side of such vertical line, then the body when disengaged will fall from such position to the vertical line, and in consequence of its inertia will continue its motion beyond the vertical line until it comes to rest; it will then return to the vertical line, and thus oscillate from side to side.

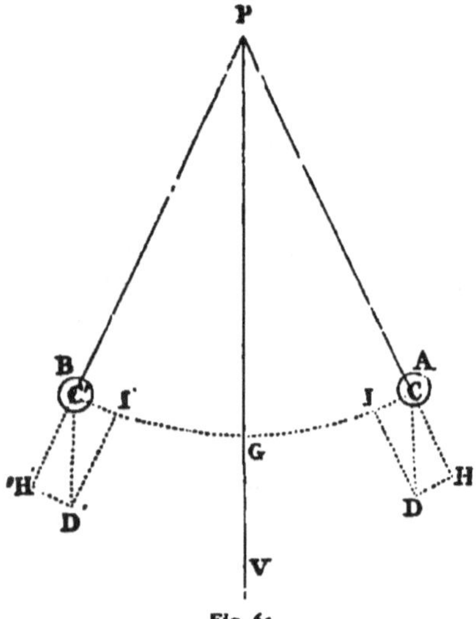

278. **Pendulum.**—Such a body constitutes what is called the *pendulum*.

Let P, *fig*.61., be the point of suspension. Let P A represent the string, and c the centre of gravity of the body. Let the weight of the body be represented by the vertical line C D. Let

Fig. 61.

this be taken as the diagonal of a parallelogram, one of whose sides

14

c H is in the direction of the string, and the other c I at right angles to it. The weight represented by the diagonal will thus, by the resolution of forces, be equal to two forces, one represented by c H and the other by c I. That which is represented by c H expends itself in pressure on the point of suspension; the other, represented by c I, will cause the body to move towards the vertical line P v, and in so moving the centre of gravity will describe the circular arc c G. When the centre of gravity arrives at G, it will be in the vertical line P v, passing through the point of suspension; and if the body were at rest it would remain there; but on arriving at G, the body has a certain velocity and moving force, which it will retain in virtue of its inertia, until deprived of it by some external agency. It will therefore continue to move to the left of G, and the centre of gravity will describe the circular arc G c'. In ascending this circular arc, the force of gravity has a tendency to destroy its velocity.

Let the weight of the body, as before, be represented by the vertical li-ie c' D': it will be equivalent to the two forces represented by c' H' and c' I'. The force c' H' is expended in pressure upon the point of suspension, P; the other, c' I', has a tendency to carry the centre of gravity, c', back to the point G, along the circular arc c' G. This component of gravity, while the body moves from G to c, gradually deprives it of its momentum, and if the momentum be entirely destroyed at the point c, then this same component of gravity, c' I', will cause the body to return along the circular arc to the point G. In this manner the body would oscillate continually from side to side of the vertical line P v, the centre of gravity describing alternately equal arcs, G c and G c'. But the resistance of the air and other impediments have a tendency continually to diminish the length of the arcs, by which it departs from the vertical line, until at length the body loses its vibration and settles itself in such a position that the centre of gravity c will be quiescent in the vertical line P v.

279. **Stability.**—The stability of a body resting in any position is estimated by the magnitude of the force required to disturb and overturn it, and therefore will depend on the position of its centre of gravity with respect to the base. If its position can be disturbed or deranged without raising its centre of gravity, then the slightest force will be sufficient to move it; but if its position cannot be changed without causing its centre of gravity to rise to a higher position, then a force will be necessary which would be sufficient to raise the entire weight of the body through the height to which its centre of gravity must be elevated; for, according to what has been already explained, the whole weight of the body may be considered concentrated at its centre of gravity.

280. Let в ʌ c, *fig.* 62., represent a pyramid, the centre of

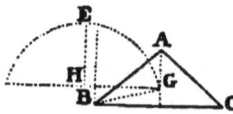 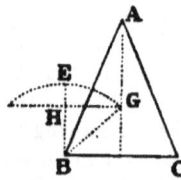 

Fig. 62.   Fig. 63.   Fig. 64.

gravity of which is ᴏ. To turn this over the edge ʙ, the centre of gravity must be carried over the arc ᴏ ᴇ, and must therefore be raised through the height ʜ ᴇ. If, however, the pyramid were taller relative to its base, as in *fig.* 63., the height ʜ ᴇ, through which the centre of gravity would have to be elevated, would be proportionally less; and if the base were still smaller in reference to the height, as in *fig.* 64., the height ʜ ᴇ would be still less, and so small, that a very slight force would throw the pyramid over the edge ʙ. It is evident, from examining these diagrams, that the principle may be generalized, and that it may be stated that the stability of any body depends, other things being the same, upon the distance of the line of direction of its centre of gravity from the edges of its base. The nearer this direction is to one edge of the base, the more easily will the body be turned over this edge.

281. **Line of direction.** — If the line of direction of the centre of gravity fall outside the edge, as in *fig.* 65., then the weight of the body concentrated at ᴏ will be unsupported, and the body will fall over its edge.

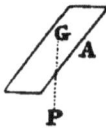

Fig. 65.

This will always take place, if the body be not attached to the ground at its base; but it happens, in some cases, that the body is so rooted to the ground at its base, that it will resist the tendency of its weight to make it fall, even though the line of direction of its centre of gravity should fall a little outside its base. Thus, we see trees not unfrequently leaning in such a position, that their centre of gravity obviously falls outside the limits of their trunk. Yet the trees nevertheless remain standing, the tenacity of the roots and their hold upon the soil being sufficient to resist the effect of their weight acting at the centre of gravity.

282. In the case of the celebrated leaning towers of Pisa and Bologna, although they are inclined considerably from the perpendicular, the lines of direction of their centres of gravity still fall within their bases. The tower of Pisa is 315 feet high; and it is inclined so that if a plumb-line hang from the side towards

which the inclination takes place, it will meet the ground at 12 ft. 4 in. from the base. The tower of Bologna is 134 feet high, and a plumb-line similarly suspended would fall at 9 ft. 2 in. from the base. Nevertheless, these structures have stood, and will probably stand, as permanently as if they were erected in the true perpendicular; for the line of direction of their centre of gravity falls sufficiently within the base to render their overthrow impossible by any common force to which they will be exposed.

283. If the line of direction of the centre of gravity, however, fall upon the edge, as in *fig.* 66., then the body will still stand; but it will be in a condition in which the slightest possible force will turn it over, as in this case it can be overturned without causing the centre of gravity to rise.

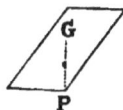

Fig. 66.

In *figs.* 67. 68. and 69. the line of direction falls within the base; and the position is stable.

Fig. 67.   Fig. 68.   Fig. 69.

In *fig.* 70. it falls within, and in *fig.* 71. outside the base. The

Fig. 70.   Fig. 71.

position is stable in the former case; and in the latter the body falls.

284. Hence appears the principle upon which the stability of loaded carriages or waggons depends. When the load is placed at a considerable elevation above the wheels, the centre of gravity is elevated, and the carriage becomes proportionally unstable. In coaches for the conveyance of passengers, the luggage is therefore very unsafely placed when collected on the roof, as is generally

done. It would be more secure to pack the heavier luggage in the lower parts of the coach, placing light parcels on the top ; for in such case the centre of gravity of the loaded vehicle would be in a lower position.

285. Drays for the carriage of heavy loads are often constructed in such a manner, that the load would be placed below the axle of the wheels. If a waggon or cart, loaded in such a manner that its centre of gravity shall be in an elevated position, pass over an inclined road, so that the line of direction of the centre of gravity would fall outside the wheels, it would cause the vehicle to be overturned.

The same waggon will have a greater stability when loaded with a heavy substance which occupies a small space, such as metal, than when it carries the same weight of a lighter substance, such as hay, because the centre of gravity in the latter will be much more elevated.

286. If a large table be placed upon a single leg in its centre, it will be impracticable to make it stand firm; but if the pillar on which it rests terminate in a tripod, it will have the same stability as if it had three legs attached to the points directly over the places where the feet of the tripod rest.

It will be still more stable if it be supported on three legs placed as in *fig.* 72.

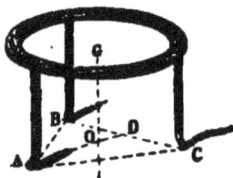

Fig. 72.

287. When a solid body is supported by more points than one, it is not necessary for its stability that the line of direction should fall on one of these points. If there be only two points of support, the line of direction must fall between them. The body is in this case supported as effectually as if it rested on an edge coinciding with a straight line drawn from one point of support to the other. If there be three points of support, which are not ranged in the same straight line, the body will be supported in the same manner as it would be by a base coinciding with the triangle formed by straight lines joining the three points of support. In the same manner, whatever be the number of points on which the body rests, its virtual base will be found by supposing straight lines drawn, joining the several points of support. When the line of direction falls within this base, the body will always stand firm ; and otherwise not. The degree of stability is determined in the same manner as if the base were a continued surface.

If there be four or more legs, the table may be unstable, even though the centre of gravity fall within the base, and will be so if the ends (A, B, C, D, *fig.* 73.) of the legs be not in the same plane ; for in that case they cannot all rest together on a level floor.

Fig. 73.

287. All the attitudes, gestures, and movements of animals are governed with reference to the centre of gravity of their bodies. When a man stands, the line of direction of his weight must fall within the base formed by his feet. If A B, C D (*fig.* 74.), be the feet, this base is the space A B D C. It is evident that the more his toes are turned outwards, the more contracted the base will be in the direction E F, and the

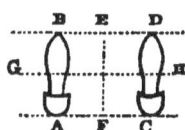

Fig. 74.

more liable he will be to fall backwards or forwards. Also, the closer his feet are toge- ther, the more contracted the base will be to the direction G H, and the more liable he will be to fall towards either side.

288. When a man walks, the legs are alter- nately lifted from the ground, and the centre of gravity is either unsupported or thrown from the one side or the other. The body is also thrown a little forward, in order that the tendency of the centre of gravity to fall in the direction of the toes may assist the muscular action in the propelling the body. This forward incli- nation of the body increases with the speed of the motion.

But for the flexibility of the knee-joints, the labour of walking would be much greater than it is, for the centre of gravity would be more elevated by each step. The line of motion of the centre of gravity in walking is represented by *fig.* 75., and deviates but

Fig. 75.

Fig. 76.

little from a regular horizontal line, so that the elevation of the centre of gravity is subject to very slight variation.

289. But if there were no knee-joint, as when a man has wooden legs, the centre of gravity would move as in *fig.* 76., so that at each step the weight of the body would be lifted through a more considerable height, and therefore the labour of walking would

be much increased. If a man stand on one leg, the line of direction of his weight must fall within the space on which his foot treads. The smallness of this space, compared with the height of the centre of gravity, accounts for the difficulty of this feat.

290. The position of the centre of gravity of the body changes with the posture and position of the limbs. If the arm be extended from one side, the centre of gravity is brought nearer to that side than it was when the arm hung perpendicularly. When dancers, standing on one leg, extend the other at right angles to it, they must incline the body in the direction opposite to that in which the leg is extended, in order to bring the centre of gravity over the foot which supports them.

291. When a porter carries a load, his position must be regulated by the centre of gravity of his body and the load taken together. If he bore the load on his back, the line of direction would pass beyond his heels, and he would fall backwards. To bring the centre of gravity over his feet, he accordingly leans forward (*fig.* 77.). If a nurse carry a child in her arms, she leans back for a like reason. When a load is carried on the head, the bearer stands upright, that the centre of gravity may be over his feet.

Fig. 77.

292. In ascending a hill we appear to incline forward, and in descending to lean backward; but, in truth, we are standing upright with respect to a level plane. This is necessary to keep the line of direction between the feet, as is evident from *fig.* 78.

Fig. 78.

293. A person sitting on a chair cannot rise from it without either stooping forward to bring the centre of gravity over the feet, or drawing back the feet to bring them under the centre of gravity.

If a person stand with his side close against a wall, his feet being close together, he will find it impracticable to raise the outside foot; for if he did, the line of direction of the centre of gravity of his body would fall outside the inner foot, and he would be unsupported.

294. When a quadruped stands with his four feet on the ground, the centre of gravity of his body is over a point found by drawing the two diagonals of the quadrilateral formed by his feet; that is to say, if a line be drawn joining his right fore foot with his left hind foot, and another joining his left fore foot with his right

hind foot, then the centre of gravity of his body will be very nearly over the point where these lines cross each other. Strictly speaking, it will generally be a little nearer to his fore feet than this point. It will, however, still be very nearly on the centre of the quadrilateral base formed by his four feet, and, therefore, in a position to give complete stability to the animal.

When a quadruped walks, he raises his right fore and left hind foot (the former leaving the ground a little before the latter), the diagonal line joining his left fore foot, and right hind foot supporting his weight. The centre of gravity of his body is a little in advance of this line, and his gravity, therefore, assists his forward motion. The left fore foot is raised a moment before the left hind foot is brought to the ground, and, in like manner, the right hind foot is raised immediately after the right fore foot comes to the ground. The effect of these motions is, that the weight of the animal is thrown alternately upon the two diagonal lines joining the right fore and left hind foot, and the left fore and right hind foot. .

When a quadruped trots, he also raises his legs from the ground, alternately, by pairs, placed diagonally; but in this case the two feet leave the ground and return to it precisely together, and each pair springs from the ground a moment before the other pair returns to it; so that there are short intervals between the successive returns of the feet, by pairs, to the ground, during which the entire body is unsupported. The weight is, in these intervals, projected upwards by the spring of the legs, so that the centre of gravity of the body describes a succession of arcs concave towards the ground. It is this motion of the body which produces the action sustained by the rider of a horse in trotting.

When a quadruped gallops, he raises simultaneously his two fore legs, and by the muscular action of his hind legs he projects his weight forwards. During the spring the centre of gravity is unsupported, but is thrown forward, describing a circular arc concave towards the ground. When this arc has been completed, the fore legs reach the ground, and immediately afterwards the hind legs; and the centre of gravity is again momentarily supported, and the animal is in an attitude to repeat the same action.

295. If a cylindrical body of uniform density be placed upon a horizontal plane, A B, *fig.* 79., its centre of gravity being its centre of magnitude, c, the line of direction c P will necessarily meet the plane at the point where the cylinder touches it, and the body will consequently remain at rest. If the cylinder be rolled upon the plane,

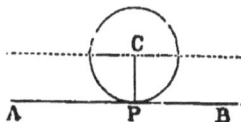

Fig. 79.

the centre of gravity will be carried in a horizontal line, parallel to the plane represented by the dotted line in the figure. Since, therefore, in this motion the centre of gravity does not rise, any force applied to the body, however slight, will cause it to move, since no elevation of its weight is required. But, on the other hand, the body will have of itself no tendency to change its position, because the centre of gravity is not only supported, but because by no change of the body can it assume a lower position.

296. If a board of uniform density be cut into the form of an ellipse, A B, *fig.* 80., and be placed upon a level surface, D E, with the longer axis A B of the ellipse parallel to the surface D E, the centre of gravity, C, will then be vertically over the point P, at which the board touches the surface, and the body will be supported at rest. If the body be disturbed slightly from this position, the end A being depressed, and the end B elevated, then the centre of gravity, C, will be elevated towards the point O; and if, on the other hand, the end B be depressed, and the end A elevated, then the centre of gravity will be raised towards the point O'. In either case, this elevation of the centre of gravity will require the application of such a force as would be sufficient to raise the body through that height, whatever it be, through which the centre of gravity has been elevated; and if, after such elevation, the body be disengaged, and left to the free action of gravity, the centre of gravity will descend to the lowest possible position, that is to say, to the position represented in the figure, and will oscillate from side to side of this position until the vibrating motion be destroyed by the resistance of the air and by friction. The centre of gravity will then rest in the position represented in the figure.

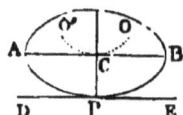

Fig. 80.

Let us now suppose that the same board is placed on the horizontal plane D E, with its longer axis vertical, as represented in *fig.* 81. The line of direction, C P, of the centre of gravity will now pass through the point of support of the body, and consequently the board will be supported. But if in this case the body be slightly turned from its position to the right or to the left, the centre of gravity will descend towards O or towards O', and cannot resume the original position at C until a force be applied to it which would be sufficient to raise the weight of the body through the height to which the centre of gravity has fallen. It is evident, therefore, that in this case, if the position of equilibrium, C P, be disturbed in the slightest degree by inclining the body a little either to the right or the left, the centre of gravity will move

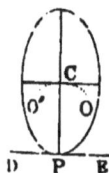

Fig. 81.

128   OF FORCE AND MOTION.

downwards, and the body will fall until it take the position represented in *fig.* 80., with its long axis horizontal.

297. **Stable, unstable, and neutral equilibrium.** — Now, it
will be observed, that in each of the three cases represented in
figs. 79., 80., and 81. there is equilibrium; but this equilibrium
is characterized in each case by particular conditions.

298. In the case represented in *fig.* 80., the equilibrium is
called *stable*, because, if it be deranged either to the right or to the
left, the body will of itself return to it, and settle definitively into
it, after some oscillation, the centre of gravity resuming its position
over the point P. This position of stable equilibrium is determined by the condition that no change of position can take place
in the body without causing an elevation in the centre of gravity ;
or, what is the same, it is that position in which the centre of
gravity is at the lowest point it is capable of assuming consistently
with the conditions under which the body is placed.

The state of equilibrium represented in *fig.* 81. is called *unstable*,
or tottering equilibrium. It is such a state of equilibrium, that if
the slightest derangement takes place in the position of the centre
of gravity, it will not return to the same point, but the body will
assume another position, in which the centre of gravity will be in
a state of stable equilibrium, as represented in *fig.* 80.

299. Unstable equilibrium, then, is characterized by the quality
that the centre of gravity is at 'the highest point which it can
assume compatibly with the conditions in which the body is
placed; and although it is vertically over the point of support, it
is nevertheless in such a condition that the slightest derangement
will cause it to descend, and the body to be overturned.

300. The case represented in *fig.* 79. is an intermediate condition between these two extremes, and is called the state of
neutral equilibrium.. It is neither stable nor unstable. It is not
stable, because the slightest force applied to the body will permanently change its position. It is not unstable, because the body
will not be overturned by the action of its own weight, however
its position may be changed.

Another case of neutral equilibrium is shown in *fig.* 82.

301. **Examples.** — The effects of
a variety of children's toys are explained by this principle. The centre
of gravity is, by loading one extremity in a manner not perceptible
to the eye, moved to a considerable
distance from the centre of magnitude. The object, therefore, will
only stand when the point which is
the real centre of gravity is in the lowest position.

Fig. 82.

Thus, a figure (*fig.* 83.) made of some light substance, such as elder pith or cork, has a piece of lead attached to one of its extremities. If it be placed on the other extremity, the end being rounded, it will apparently, by a spontaneous movement invert its position, and a sort of tumbler will be formed.

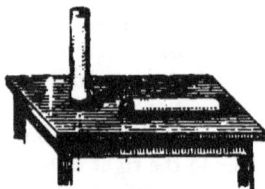

Fig. 83.

In *fig.* 84. a toy is represented, the principle of which will be easily understood from what has been explained. The figure rests steadily on the toe, because, by the weights at the lower ends of the rods, the centre of gravity is brought below the point of support.

302. Many of the feats exhibited by sleight-of-hand performers are explained by the principles of stable and unstable equilibrium. If any object, such as a sword, be supported on its point, it will be in unstable equilibrium so long as its centre of gravity is directly over its point; but as it cannot be maintained precisely so, the finger or other support of the point is moved slightly in one direction or another, so as to keep nearly under the centre of gravity, and to check the tendency of the sword to fall on the one side or on the other.

Fig. 84.

But these and similar feats are greatly facilitated, if the object thus balanced is made to spin upon its point; for in that case the centre of gravity, though not vertically over the point of support, is continually revolving round a vertical line passing through the point of support, and the tendency which it has at one moment to make the body fall on one side is instantly checked by a contrary tendency when it passes to the opposite side.

303. It is in this manner that the common effect of a spinning-top is explained. It would be stable, if placed as in *fig.* 85.; but it would be quite impracticable to make it stand on its point, as in *fig.* 86., if it did not revolve, or if it revolved very slowly; but if it have a very rapid motion of gyration, then it will stand steadily on its point.

It may be asked how the rapidity of the gyration affects the

K

question. This is easily explained. If the centre of gravity revolve round the line so slowly that the time taken in half a revolution is so considerable as to allow it to fall to any considerable

Fig. 85.

Fig. 86.

depth, then it cannot recover itself when it passes to the other side; but if the revolution be so rapid that half the time of one gyration is so small that the centre of gravity cannot fall through any sensible height, the top will maintain its position.

304. Public exhibitors place a circular plate on the point of a sword, the point being placed as near the centre of the plate as possible. But, however near the centre it may be placed, it is not always possible to ensure its coincidence with the centre of gravity of the plate. If in this case the plate were at rest on the point of the sword, it would not be balanced, but would incline and fall on that side at which the centre of gravity would lie. The exhibitor, therefore, prevents this effect, by giving to the plate a rapid rotation on the point of the sword. The centre of gravity of the plate rapidly moves in a small circle round the point of support, and its tendency at one moment to fall down at one side is checked the next moment by a contrary tendency to fall down at the other side, and the plate accordingly remains balanced on the point.

305. In some cases the centre of gravity of a body apparently ascends; but this is always deceptive, and its real motion is invariably a descending one. Let a cylinder of wood, A B, *fig.* 87., be pierced by a hole, O, near its surface, B, and let a cylinder of lead be inserted in this hole. The centre of gravity of the mass will then be, not at its centre of magnitude, but between that point and the centre of the cylinder of lead which fills the hole O, and will not be far removed from the centre of the lead, in consequence of the great comparative weight of that substance.

Fig. 87.

If such a cylinder as this be placed upon an inclined plane, M N, *fig.* 88., in such a position that the line of direction, O B, of the centre of gravity shall fall above the point of contact, P, of the cylinder with the plane, the cylinder will roll up the plane, because

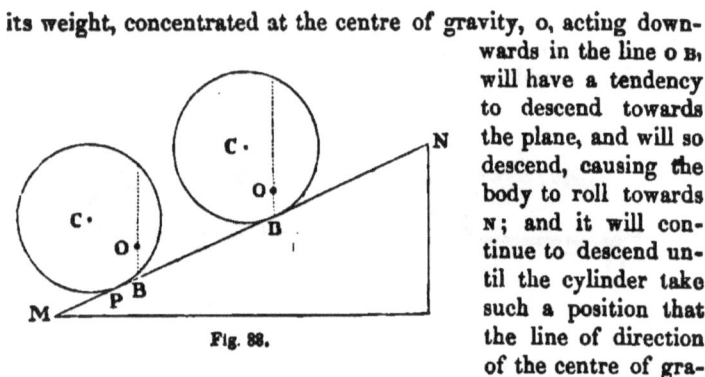

its weight, concentrated at the centre of gravity, o, acting downwards in the line o B,
will have a tendency
to descend towards
the plane, and will so
descend, causing the
body to roll towards
N; and it will con-
tinue to descend un-
til the cylinder take
such a position that
the line of direction
of the centre of gra-
vity, o B, shall pass through the point of contact of the cylinder
with the plane.

Fig. 88.

In these and similar cases, although the general mass of the body
rises, the particular part occupied by its centre of gravity falls;
that point is, in effect, simultaneously affected by two motions, one
produced by the progressive motion of the cylinder up the plane
from M to N, and the other by the motion of revolution of the
cylinder round its centre, c. In virtue of the former, the centre
of gravity would rise; and in virtue of the latter, it would fall.
The effect of the latter predominates until the centre of gravity
comes into the vertical line, passing through the point of contact
of the cylinder with the plane.

306. A case of the ascent of the centre of gravity is sometimes
produced by public exhibitors, the explanation of which may here
be found instructive. The exhibitor places a sphere of wood
upon an inclined plane, and standing upon it, he places his feet on
that side of the centre which is towards the elevation of the plane.
Immediately the globe begins to roll up the plane, and the exhibitor, by moving his feet so as to keep them near the highest
point of the globe, but still on the side next the elevation of the
plane, the globe continues to roll up the plane, the exhibitor
dexterously maintaining his position as here described.

In this case there is a real ascent of the common centre of
gravity of the globe and the body of the exhibitor. Now the
question is, What force in this case produces this ascent? for it is
evident that in the time during which it rolls to the top of the
plane, the entire weight of the globe and the body of the exhibitor
has been elevated through a perpendicular space equal to the
height of the plane.

The force which accomplishes this is the muscular action of the
feet of the exhibitor upon the surface of the globe. As the globe
rolls up the plane, if the feet of the exhibitor pressed upon the

K 2

same point of its surface, they would descend; and in that case
the common centre of gravity of the globe and the body of the
exhibitor, instead of ascending, would in fact descend, until the
feet of the exhibitor would sink down to the surface of the plane;
but this is prevented by the feet of the exhibitor continually
stepping backwards upon the surface of the globe, so as to stand
near the top; and thus, by moving his feet on the globe backwards continually towards the top of it, the exhibitor elevates the
centre of gravity of his body, while the action of his feet upon the
globe causing it to roll up the plane at the same time, raises the
centre of gravity of the globe.

307. **Magic clock.**—Time-pieces, with a transparent glass dial
and without any apparent works, thus called,
often excite the wonder of the curious. The
principle of their motion depends on the
tendency of the centre of gravity to take the
lowest possible position. In the centre of the
dial is a small circular gilt plate of metal,
which conceals a very short arm, A, of the hand
A C B, *fig.* 89. This short arm carries a
hollow ring, in the interior of which a small
weight is moved by clockwork, through
the intervention of an arbor which passes
through the centre of the dial, the works

Fig. 89.

which turn the rod being at a distance behind the dial, so as not
to be visible through it. This weight A, is thus made to move
uniformly round the ring, and to change continually the position
of the centre of gravity of the hand, so as to make it describe
uniformly round the centre of the dial the small dotted circle c g
in twelve hours, so that the successive hours are indicated by its
extremity B. The ring and weight A being concealed, the hand
seems to move without any apparent cause of impulsion; and what
adds to the illusion is, that the hand can be freely turned by applying the finger to it, but when left to itself it always returns to the
point from which it was turned, so as still to indicate the time, and
to be incapable of being set to a wrong hour.

308. **Centre of gravity of fluids.** — In all that we have stated
respecting the centre of gravity, we have supposed the body to be
solid; but this quality also plays an important part in the phenomena of fluids. The centre of gravity of a fluid mass is determined by the same conditions as if it were solid. It is that point
which would have the properties already defined, if the fluid mass
were supposed to be congealed.

Thus the centre of gravity of the water forming a lake is that
point which would have the properties already explained, if the

water of the lake were converted into a mass of ice. It will appear hereafter, however, to possess, in reference to fluid bodies, many important characters.

309. The centre of gravity of two separate and independent bodies is that point between them which would possess the characters already defined, if the two bodies were united by a straight rod which is itself devoid of weight. This point may be determined by a very simple mathematical process. Let the centres of gravity of the two bodies in question be conceived to be connected by a straight line, and let a point be found upon this straight line which shall divide it into two parts, which shall be in the inverse proportion of the weights of the two bodies. Then this point will be their common centre of gravity.

Thus let A and B, *fig.* 90., be the two bodies, and let a, b be their centres of gravity. Draw the line $a\,b$, and take upon it a point c, so that b c shall bear to a c the same proportion as the weight of A bears to the weight of B. In this case, c will be the centre of gravity of the two bodies.

Fig. 90.

Now if the line $a\,b$ were a rigid rod, devoid of gravity, the point c would have all the properties which have been already explained as belonging to the centre of gravity; thus, the bodies would balance themselves on c in any position.

CHAP. VI.

CENTRIFUGAL FORCE.

310. If a ball of metal or other heavy substance, placed upon a smooth and level surface, be attached to the extremity of a string, the other extremity of which is fastened to a fixed point upon the surface, and then whirled round in a circle, it is known, by universal and constant experience, that the string will be stretched with a certain force, which will be augmented as the velocity of the whirling motion is increased, or as the string is lengthened. That such force is not produced by gravity is evident, inasmuch as the level surface upon which the body moves supports its weight.

311. This force, which always attends matter that is moved

round a centre, in what manner and under what form soever the
motion be produced, is called *centrifugal force*, because it is manifested by a tendency of the matter which revolves to recede from
the centre of revolution; this tendency in the case just mentioned
being manifested by the force with which the string connecting the
body with the fixed point is stretched. This tension resists the
tendency of the ball to fly from the centre, and is therefore the
measure of its centrifugal force.

312. That centrifugal force is a mere effect of the inertia of
matter, may be easily shown. It has been already explained, that
in virtue of its inertia, a body, if in motion, can only move uniformly in a straight line. If, therefore, it be deflected from one
straight line into another straight line, it must be by the action of
some force impressed upon it at the moment of deflection; and if
a body be continually deflected from a straight direction, which it
must be if it move in a curve, then such body must be under the
operation of a force continually acting upon it, producing such
incessant change of direction.

Let P, *fig.* 91., be the fixed point to which the string is attached. Let A be the ball, and let A c F be the circle in which the
ball is whirled round. Let A c be a
small arc of this circle moved over in a
given interval of time. Starting from
A, the motion of the ball has the direction of the tangent A D to the circle,
and it would move from A to D in the
given interval of time, if it were not
deflected from the rectilinear course;
but it is deflected into the diagonal
A c, and this diagonal, by the composition of forces, is equivalent to two
forces represented by the sides A D,
A B. But the motion A D is that which
the body would have in virtue of its inertia; and therefore the
force A B, directed towards the fixed point P, is that which is impressed upon it by the tension of the string, and which, combined
with the motion A D, causes it to move in the diagonal A c.

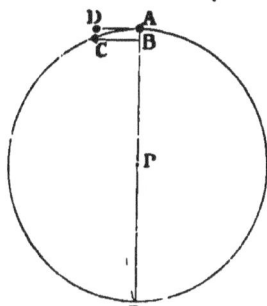

The tension of the string, therefore, is in fact a force directed
to the centre, P, which continually deflects the body from the
tangent to the circle in which it has a constant tendency to move
in virtue of its inertia.

313. It follows from the elementary principles of geometry, that
the space A B, which is that which represents the force of the string
upon the ball, or the centrifugal force, and which is in fact the
space through which the body is moved in a given small interval

of time by the tension of the string which measures the centrifugal force, is found by dividing the square of the number representing A c by the number representing the diameter A r of the circle, or twice the length of the string A P. But A c, being the space described in a given time by the revolving body, is its velocity.

If we would then compare the centrifugal force of the body with its weight, we have only to compare the space which the body would be moved through by the centrifugal force acting alone upon it, with the space which gravity would move the same body through acting equally alone upon it.

Let us then express the physical quantities involved in this question as follows :

$w =$ the weight of the revolving body.
$c =$ its centrifugal force.
$v =$ its velocity in feet per second.
$R =$ length of string in feet.
$\frac{1}{2}g = 16\frac{1}{12}$ feet, being the height through which w would fall freely in one second.

We shall have then the following proportion : —

$$w : c :: \tfrac{1}{2}g : \frac{v^2}{2R} ;$$

and therefore we have

$$c = w \times \frac{v^2}{R \times g}.$$

This formula, expressed in ordinary language, is as follows :

314. *The centrifugal force of a body revolving in a circle is found by multiplying its weight by the square of the number of feet which it moves through in a second, and dividing the product by the number of feet in the radius of the circle it described, multiplied by $32\frac{1}{4}$.*

But it is more convenient in practice to express the centrifugal force of a revolving body by reference to the number of revolutions it performs in a given time. Let us therefore express by N the number of revolutions, or fraction of a revolution, performed by the body in one second. The circumference of the circle which it describes, the length of the string being R, will be 6·2832 R.

If, then, this be multiplied by N, we shall obtain the space through which the body moves in one second, or its velocity; and since the square of 6·2832 is 39·4786, we shall have

$$v^2 = 39·4786 \; R^2 \times N^2 ;$$

and therefore we shall have the centrifugal force expressed by

$$c = 1·2273 \; w \times R \times N^2.$$

In this formula it must be understood, however, that the length of the string must be expressed in feet or fractions of a foot, and that N must express the number of revolutions, or fraction of a revolution, made by the body in one second.

This formula, expressed in ordinary language, is as follows:

315. *To find the centrifugal force of a revolving body, multiply its weight by the number 1·2273. Multiply this product by the number of feet in its distance from the centre round which it turns, and finally multiply this product by the square of the number of revolutions, or fraction of a revolution, which it makes round that centre in one second of time.*

Example.—Let it be required to find with what force a body weighing 2 lbs. would stretch a string 3 feet long, revolving four times per second. Multiply 2 lbs. by 1·2273, and we obtain 2·4546 lbs.; multiplying this by 3, the number of feet in the length of the string, we obtain 7·3638. In fine, multiply this by 16, which is the square of 4, the number of revolutions per second, and we have 117·8208. So that the centrifugal force with which the string is stretched would be 117-8/10 lbs. very nearly.

From the preceding conclusions it follows, that if two bodies of equal weights be whirled round their centres by strings or rods of the same length, their centrifugal forces will be in proportion to the squares of the number of revolutions which they perform in a given time. Thus, if one of them make three revolutions while the other makes two, the centrifugal force of the former will be to that of the latter as 9 to 4.

Again, if two bodies of equal weight are attached to centres by strings of different lengths, but perforn the same number of revolutions in a given time, their centrifugal forces will be in proportion to the lengths of the strings. Thus, if one be attached by a string of two feet, and the other by a string of three feet, the centrifugal force of the former will be to the centrifugal force of the latter as 2 to 3.

In general, if two bodies of equal weight be at different distances from the centres round which they revolve, and also make a different number of revolutions in the same time, then their centrifugal forces will be as the products found by multiplying their distances from the centre by the squares of the number of revolutions which they make in the same time.

316. **Whirling-table.** — These conclusions may be experimentally verified by an apparatus called a whirling-table, usually found in collections of philosophical apparatus.

A part of this instrument is represented in *fig.* 92., where c is a metallic bar, having two upright pieces *f f* at its ends, in which a polished metal rod is fixed parallel to c. On this rod a ball *g*

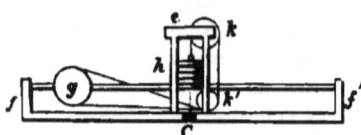
Fig. 92.

slides, being perforated by a hole corresponding to the rod. At the centre c is a vertical framework, which contains a number of thin circular weights h placed one above the other, and supported on a sliding stage, which is capable of rising and falling. At the centre of the top of this stage, carrying the weights, is a hook, to which two strings are attached, which are carried over grooves in the pulley k, and then pass over corresponding grooves in the lower pulley k', from which they are carried to the ball g to which they are attached. Now if the ball g be drawn along the rod ff' towards f with sufficient force, the weights h will be raised, and the force necessary to effect this may be increased or diminished by varying the number of weights at h. This apparatus has a square hole at the centre of its lower part c, by which it can be attached to a spindle, by which a regulated revolution can be imparted to it. The apparatus is provided with two such spindles, so that two instruments like that represented in *fig.* 92. can be fixed upon the table and put in rotation with any required velocities, the number of revolutions which they make in a given time having any desired ratio to each other. By this apparatus, the centrifugal force with which the ball g is affected can always be estimated by the weight which such centrifugal force is capable of raising at h. A rotatory motion is given to c, such that the centrifugal force of g is just sufficient to lift the weights h, but not to carry them to the top of the *frame*. When this takes place, the centrifugal force of g will be equal to the weight.

A variety of conditions affecting revolving bodies can be examined and determined by this apparatus. By varying the distance of the weights from the centres round which they revolve, the centrifugal forces in circles with different radii can be determined; and by varying the velocities of rotation, the effects of different angular motions, or of a different number of revolutions in a given time, can be ascertained. By this apparatus, therefore, the general principles which have been already established respecting centrifugal force can be verified.

Thus we can show :

1st. That when equal weights are at equal distances from the centres of revolution, their centrifugal forces will be proportional to the squares of the numbers of revolutions which they make in a given time.

2nd. When they revolve in the same time, then the centrifugal forces will be in the direct ratio of their distances from the centre.

3rd. When they are at different distances from the centre, and revolve in different times, then their centrifugal forces will be in the ratio of the products found by multiplying their distances from the centre by the squares of the numbers of revolutions which they make in a given time.

4th. But if the weights be unequal, then let the proportion of the centrifugal forces which they would have if they were equal be first found, and let the numbers expressing them be multiplied by the numbers expressing the weights; the product will then express the proportion of the centrifugal forces.

These propositions involve the whole theory of centrifugal force.

317. If two bodies revolve round a common centre in the same time, but at different distances from it, their centrifugal forces would be in the proportion of these distances if they were equal, the body at the greater distance having a greater proportional centrifugal force. But if the body at the lesser distance be increased in weight, so as to exceed the other in the same ratio as the distance of the other from the centre of rotation is greater, then the centrifugal forces will be equal; for what the centrifugal force of the lesser gains by distance, that of the greater gains by weight. Thus, if one of the bodies weigh three ounces, and the other five ounces, and if, further, the latter be at three inches from the centre, the other being at five inches from it, they will have the same centrifugal force, provided they revolve in the same time. This, which is an important proposition, may be experimentally proved by the whirling-table.

Let A and B, *fig.* 93., be two bodies connected by a wire, and

Fig. 93.

let a point, c, be taken upon this wire, in such a position that B c shall bear to A c the same proportion as the weight of A bears to the weight of B; then let the wire be attached to the spindle of the whirling-table at c, so that the balls shall be made to whirl round c. It will be found that the wire will maintain its position, although free to move in the direction of its own length; showing that the centrifugal force exerted by the greater ball, A, upon the wire in the direction c A, is equal to the centrifugal force exerted by the lesser ball, B, upon the wire in the direction c B. The lesser ball, B, therefore, gains as much centrifugal force by its greater radius, B c, as the greater, A, gains by its superior weight.

The point c, which divides the distance between the balls in the inverse ratio of their weights, is, as has already been shown, their common centre of gravity; and it therefore follows that if two bodies revolve round their common centre of gravity in the same time, they will exert equal centrifugal forces upon it.

The same important principle may be illustrated experimentally by the apparatus shown in *fig.* 94., which will be readily under-

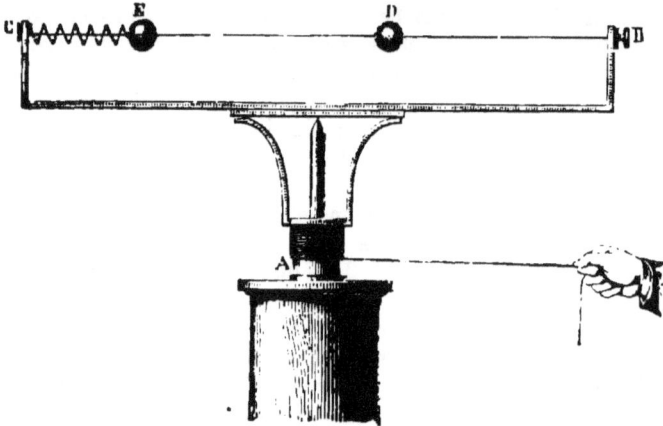

Fig. 94.

stood on inspection. The ball D is fixed, while E, sliding on the wire, re-acts on the spiral spring, C E. The centrifugal force causes E to compress the spring until it retires to such a distance from D that the centre of gravity coincides with that of rotation.

318. **Examples.** — Examples are presented of the effects of centrifugal force in almost all the motions which fall within our daily observation.

319. A horseman, or a pedestrian passing round a corner, moves in a curve, and consequently suffers a centrifugal force directed from the centre of the curve, which increases with his velocity, and which impresses on his body a force directed from the corner. He resists this force by inclining his body towards the corner. An animal made to move in a ring, as is customary in training horses, inclines his body towards the centre of the ring.

320. In all the equestrian feats exhibited in the circus, it will be observed, that not only the horse, but the rider, inclines his body towards the centre, *fig.* 95., and according as the speed of the horse round the ring is increased, this inclination becomes more considerable. When the horse walks slowly round a large ring, the inclination of his body is 'imperceptible; if he trot, there

is a visible inclination inwards; and if he gallop, he inclines still
more; and when urged to full speed, almost lies down upon his
side, his feet acting against the partition which separates the circus
from the adjacent parts of the theatre.

Fig. 95.

In all these cases, the facts are explained by considering that
the centrifugal force and the weight of the horse are compounded
together, and form a resultant which is directed upon the ground,
and is represented by the pressure of the horse's foot.

321. The actual amount of the centrifugal force, and the proportion which it bears to the weight of the animal or other body
which is moved in the circle, can be determined by the principles
already explained, if the radius of the circle and the velocity of
the body moving in it are known; and from these may be calculated the inclination which the body of the animal must assume in
order to be whirled round the circle without falling outwards by
the effect of the centrifugal force.

Let c (*fig. 96.*) be the centre of the ring round which the
animal moves. Let c ꜰ be
its radius, ꜰ being therefore
the point at which the feet of
the animal would act. Take
the line ꜰ ᴀ, perpendicular to
ꜰ c, and consisting of as many
inches as there are pounds
weight in the animal. Take
ᴀ ʙ parallel to ꜰ c, consisting
of as many inches as there
are pounds weight in the cen-

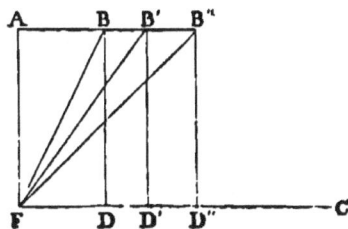

Fig. 96.

trifugal force. Then F B will represent the inclination which the animal must assume in order to prevent it from falling either outwards or inwards. A less inclination than this would cause him to fall outwards, and a greater inwards.

To demonstrate this, let the weight be conceived as acting at B, and to be represented by B D, which is equal to A F; the centrifugal force is represented by B A, and these two forces combined will produce a resultant represented by the diagonal B F. If the body of the animal be inclined according to this line B F, then the resultant will press upon its feet; but if it be inclined at a less angle, the resultant will cause it to fall outwards; and if at a greater angle, it would cause it to fall inwards.

If the centrifugal force be increased, as will be the case if the animal moves with increased speed, then it would be represented by A B'; and the resultant of it, and of the weight, would be represented by B' F. Again, if the centrifugal force be further increased, and represented by A B'', the weight being represented by B'' D'', the resultant of these will be represented by B'' F, which line must then be the inclination of the body of the animal. It is clear, then, that an increase of the centrifugal force, which arises from increased speed, will cause the body of the animal to incline more and more towards the centre of the circle.

For example, if a horse move in a ring of 60 feet diameter, with a speed of 15 feet per second, the ratio of the centrifugal force to his weight will be that of the square of 15, or 225, to 60 × 32½, which is equal to 1930; the ratio, therefore, of the centrifugal force to the weight is 1 to 8½ very nearly. We shall therefore find the inclination corresponding to this, by taking A F (*fig.* 96.) equal to 8½ inches, and A B equal to 1 inch: the line F B would represent the inclination of the horse.

322. A carriage not having voluntary motion cannot make this compensation for the disturbing force which is called into existence by the gradual change of direction of the motion in turning a corner; consequently it will, under certain circumstances, be overturned, falling, of course, outwards, or *from* the corner. If A B be the carriage, and C (*fig.* 97.) the place at which the weight is principally collected, this point C will be under the influence of two forces; the weight, which may be represented by the perpendicular C D; and the centrifugal force, which will be represented by a line C F, which shall have the same proportion to C D as the centrifugal force has to the weight. Now, the combined effect of these two forces will be the same as the effect of a single force represented by

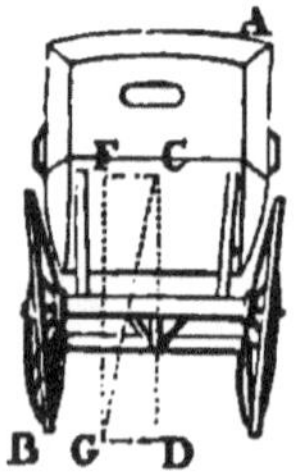

Fig. 97.

c g. Thus, the pressure of the carriage on the road is brought nearer to the outer wheel B. If the centrifugal force bear the same proportion to the weight as c F (or D B), *fig.* 98., bears to c D, the whole pressure is thrown upon the wheel B.

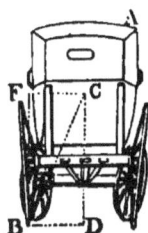

If the centrifugal force have to the weight a greater proportion than D B has to c D, then the line c F, which represents it (*fig.* 99.), will be greater than D B. The diagonal c G, which represents the combined effects of the weight and centrifugal force, will, in this case, pass outside the wheel B, and therefore this resultant will be

Fig. 98.

unresisted. To perceive how far it will tend to overthrow the carriage, let the force c G be resolved into two; one in the direction of c B, and the other, c K, perpendicular to c B. The former, c B, will be resisted by the road; but the latter, c K, will tend to lift the carriage over the external wheel. If the velocity and the curvature of the course be continued for a sufficient time to enable this force c K to elevate the weights so that the line of direction shall fall on B, the carriage will be overthrown.

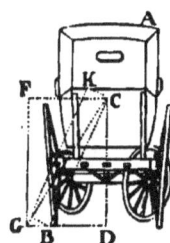

Fig. 99.

It is evident, from what has been now stated, that the chances of overthrow, under these circumstances, depend on the proportion of B D to c D, or, what is to the same purport, of half the distance between the wheels to the height of the principal seat of the load. It was shown in the last chapter, that there is a certain point, called the centre of gravity, at which the entire weight of the vehicle and its load may be conceived to be concentrated. This is the point which, in the present investigation, we have marked c. The security of the carriage, therefore, depends on the greatness of the distance between the wheels and the smallness of the elevation of the centre of gravity above the road; for either or both of these circumstances will increase the proportion of B D to c D.

323. If a stone or other weight be placed in a sling which is whirled round by the hand in a direction perpendicular to the ground, (*fig.* 100.) the stone will not fall out of the sling, even when it is at the top of its circuit, and consequently has no support beneath it. The centrifugal force in this case acting from the hand, which is the centre of rotation, is greater than the weight of the body, and therefore prevents its fall.

324. In like manner, a bucket of water may be whirled so rapidly,

that even when the mouth is presented downwards, the water will still be retained in it by the centrifugal force (*fig.* 101.).

Fig. 100.   Fig. 101.

The tendency of the water to recede from the centre of rotation may be also shown by the apparatus represented in *fig.* 102.

325. If a glass of water be fixed upon the whirling-table, and be made to revolve rapidly round its central line A B (*fig.* 103.), as an axis, the water will rise upon its sides and sink upon its centre, the surface assuming a concave form. This effect is due to the centrifugal force, which affects every particle of the water, and gives it a tendency to fly from the centre of revolution; it would be possible to impart to the glass a revolution so rapid as to cause the water to flow over the sides.

326. Centrifugal drying-machine for laundries. — The
agency of the centrifugal force has been applied with great ingenuity in a machine for drying linen in some of the larger class

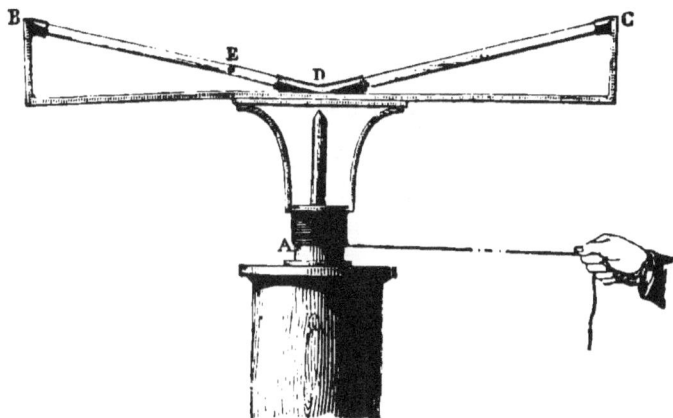

Fig. 102.

of laundries in France. This machine, which is represented in
fig. 104., consists of a large hollow drum, within which there is a
central space left empty, through which the axle passes. The

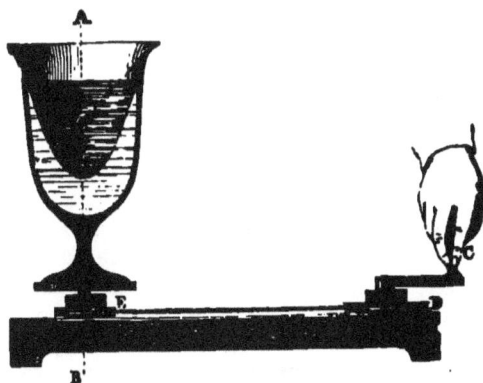

Fig. 103.

space A A, shown in section, surrounding this vacant central space,
in enclosed by two cylindrical surfaces, an exterior and an interior.
This circular cylindrical chamber is closed by covers, by opening
which the linen to be dried can be introduced. The bottom of
this chamber is pierced with holes like a sieve, through which the
water expressed from the linen can flow off; a rapid rotation

being given to this cylinder, the linen, by the effect of the centrifugal force, is urged against the exterior surface of the cylinder, and is there squeezed with a force which increases with the

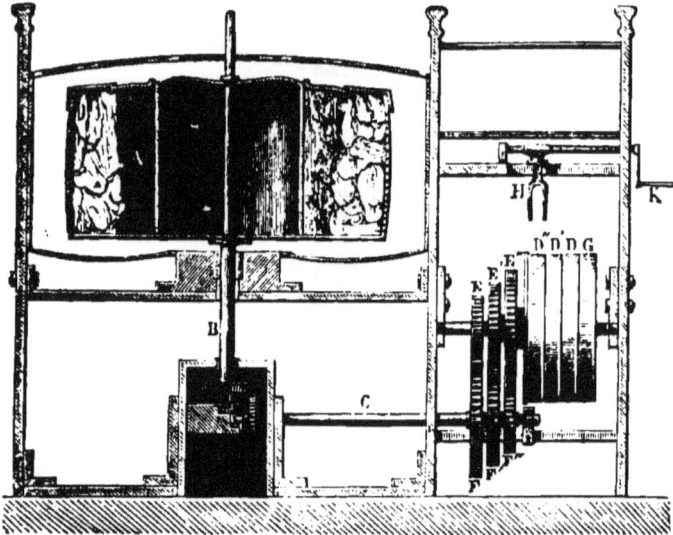

Fig. 104.

rapidity of the rotation, by the effect of which the water is pressed out of it, and escapes through the holes in the bottom. A rotation so rapid as to produce 25 turns per second, or 1500 per minute, is given to these drying cylinders, by which the linen, however moist it may be, is rendered so nearly dry that a few minutes' exposure in the air renders it perfectly so.

The mechanism by which this rapid rotation is produced consists of a horizontal shaft c, connected with the vertical shaft B by a pair of bevelled wheels; the shaft c receives its motion from the system of toothed wheels F E, F' E', F" E", which are themselves driven by a system of hoops D, D', D". These hoops receive their motion from a shaft driven by any moving power with which they are connected by an endless band.

As the very rapid rotation necessary to produce the desired effect could not be instantaneously or even very rapidly produced without injury to the mechanism, an ingenious expedient is adopted by which it is gradually imparted without any injurious shocks. The endless band, which imparts the motion, and which is shifted horizontally by the fork H, moved by the screw and winch x, is first placed upon the hoop D; this hoop is fixed upon

L

the axle of the toothed wheel E, which drives the wheel F, the axle c, and the cylinder A A, with a moderate velocity; the wheels F' F'', also fixed upon the axle c, are moved at the same rate. and they impart motion to the wheels E' and E'', which are respectively fixed upon hollow axles, which pass one within the other, and which carry respectively the hoops D' and D'',

When the apparatus is thus put in motion, with a certain moderate velocity, the endless band is transferred by the apparatus from the hoop D to the hoop D', and the motion is then conveyed to the shaft c, through the wheels E' and F', instead of E and F; and the proportion of these wheels is so arranged, that the same velocity of the hoop D' will impart an increased velocity to the shaft c and the cylinder A A.

In the same manner exactly, by transferring the band from D' to D'', the velocity is still further increased, and the full effect produced.

When the motion of the machine is suspended, to remove the dried linen and recharge the cylinder, the band is transferred to the hoop G, which turns on the axle without being attached to it.

327. If a body B (*fig.* 105.) move down a curved surface G F, whose centre is at c, it may acquire such a velocity that the centrifugal force will cause it to leave the surface, and to be projected forwards to the ground. The conditions under which this would take place are easily explained.

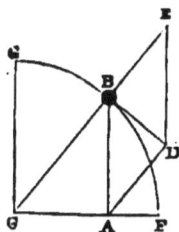

Fig. 105.

Let the weight of the body be expressed by B A, and its centrifugal force by B E, and let the parallelogram be completed. Then the diagonal B D will be the resultant of these two forces, and will be the line in which the body has a tendency to move. So long as this diagonal forms an acute angle with B c, the body will remain on the curved surface; but so soon as it forms a right angle with B c, then it will become a tangent to the surface, and the body will fall off.

328. **Body revolving on a fixed axis.** — The most important classes of problems in mechanical science in which the principles determining the centrifugal force are practically applied, are those which relate to solid bodies revolving on an axis. If a solid body be pierced by a straight and round hole in which a cylindrical rod is inserted, and the body be made to turn rapidly round this rod in an axis, each particle of matter composing the body will revolve in a circle round such axis; all these circles will be described in the same time, and consequently the centrifugal forces of the particles exerted upon the axis will be in proportion to their distances

from the axis, such distances being the radii of the circles which they describe respectively round it.

All the particles of the body which are at the same distance from the axis will therefore exert equal centrifugal forces upon it, and those which are at greater distances will exert centrifugal forces proportionally greater than those at less distances. The particles distributed round the axis will produce forces directed from the axis in the direction of perpendiculars connecting them with the axes respectively.

Now it is evident that as many different forces will thus be exerted upon the axis as there are different particles of matter composing the mass of the revolving body.

Three cases are presented which may arise in the combination of these forces.

329. They may be in equilibrium; that is to say, the centrifugal forces exerted by all the particles composing the body on the axis may neutralize each other. In this case the axis would suffer no strain in consequence of the centrifugal forces, and the body would spin round it without producing any effect upon it.

In such a case, if the axis were withdrawn from the hole in which it is inserted, the body would still continue to spin as before; because, since the axis suffered no pressure or strain from the revolving matter, its presence or absence can make no difference in the motion.

330. The centrifugal forces produced by the particles of the revolving mass may be such as to be represented by a single force applied at some point of the axis, and at right angles to it. In this case the axis will suffer a corresponding pressure or strain at this point. If this point could be fixed, the remainder of the axis might in that case be withdrawn; because, since no other point of it suffers any strain from the effect of the motion, its presence or absence can produce no difference in the motion.

331. The centrifugal forces may be such that their combined effect cannot be represented by a single force, but may be represented by two equal and parallel forces acting on two different points of the axis, and in contrary directions. This combination is what has been already called a *couple*, and its effect is to twist the axis round some point intermediate between the two contrary forces. In this case the axis could not be withdrawn unless two fixed points were provided, representing the points to which the opposite forces of the couple are applied.

332. The combination of forces produced by the revolving matter may be such as to be incapable of being represented either by a single force or by a couple. In this case it can be proved that their combined effects will be represented by a single force and a

couple taken together. The effect, in such a case, is a pressure upon the axis at right angles to its length at the point where the single force is placed, and a tension or twist produced at the same time by the couple.

These are all the possible cases which can be presented by a solid body revolving on a fixed axis.

333. **Examples.** — Their complete analysis and demonstration would require the use of the principles and formulæ of the higher parts of mathematical science, the introduction of which would not be suitable to the purpose of the present treatise. We shall therefore limit ourselves to some examples which will convey a sufficiently clear notion of the general effects produced by the rotation of solid bodies on fixed axes.

334. If a series of particles of matter placed in the circumference of a circle are made to revolve by a common motion round an axis, passing through such circle, and perpendicular to its plane, their centrifugal forces will be evidently in equilibrium, and no pressure on the axis will be produced. A circular series of such particles is represented in *fig.* 106.; the radii represent the direction of the centrifugal forces, which are all equal, because the particles are equal and the distances from the centre are equal.

It is evident on inspection that these forces equilibrate round the centre, and that the central point, therefore, would

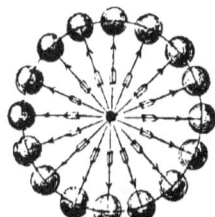

Fig. 106.

suffer no pressure in one direction rather than in another.

335. A flat circular plate of uniform thickness and density may be considered as consisting of a series of concentrical rings of such particles. If such a plate revolve round an axis passing through its centre, and perpendicular to its plane, the centrifugal forces of the particles will be in equilibrium, and no pressure will be produced on the axis.

336. A cylinder may be considered as composed of a number of such circular plates placed one upon the other, and the axis of the cylinder will be the line formed by the centres of these plates. If such a cylinder, therefore, revolve round its axis, the centrifugal force of its mass may be in equilibrium, because each separate plate being in equilibrium, the entire pile would necessarily also be in equilibrium.

337. There is an extensive and important class of solid bodies having a geometrical form, which gives to their axes this property. They are called *solids of revolution.*

Let A B D (*fig.* 107.) be a triangle with a right angle at A. If this be supposed to revolve round the side A B as an axis, it will by its revolution generate a solid called a cone; that is to say, the space through which it would pass as it revolves, and which it would include within its sides in revolving if filled by any solid matter, would form a cone.

If a right-angled parallelogram (*fig.* 108.) revolve round its side A B, it would generate in the same way a cylinder.

If a triangle A B D (*fig.* 109.) revolve round a side A B, not including a right angle, it would generate the double cone.

 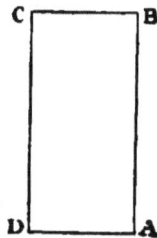 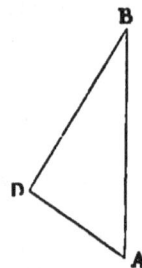

Fig. 107.   Fig. 108.   Fig. 109.

If a semicircle (*fig.* 110.) revolve round its diameter A B, it would generate a sphere or globe.

If a semi-ellipse (*fig.* 111.) revolve round its shorter axis A B, it would generate an oblate spheroid, being a figure resembling an orange or a turnip.

If a semi-ellipse (*fig.* 112.) revolve round its longer axis, it would generate a prolate spheroid, being a figure resembling an egg, only that its ends are similar.

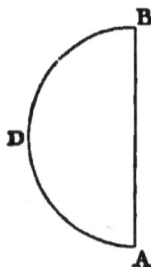 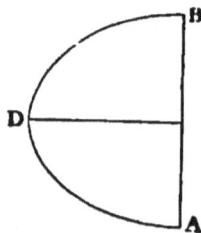 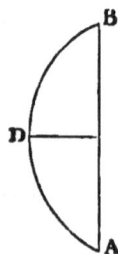

Fig. 110.   Fig. 111.   Fig. 112.

Now it is evident, that in these and all solid bodies whose figures are determined by the same principle, all sections made by planes perpendicular to their axes of revolution are circles through the

centre of which such axes pass.   Since the centrifugal forces of the
particles of matter composing each of such sections are in equilibrium, the centrifugal force of the entire mass of the body is
necessarily also in equilibrium, such mass being composed of these
several sections.   The body, in short, may be considered to be
made up of a number of circular plates laid one upon another,
varying in their diameter according to the form of the body.

    If a solid of revolution, therefore, be made to revolve upon its
geometrical axis, the centrifugal forces of its mass will be in equilibrium, and no pressure or strain whatever will take place upon its
axis.

    If a globe or sphere be composed of any materials capable of a
change of figure, it will be converted into an elliptical flattened
spheroid by the centrifugal force produced by its rotation.   This
fact is shown experimentally by an apparatus composed of elastic
hoops put in rotation by the whirling-table, as shown in *fig.* 113.

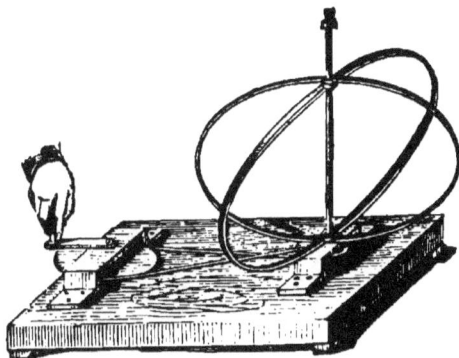

Fig. 113.

    338. The same reasoning will be applicable to all solids which
have an axis round which the particles of such section made by a
perpendicular plane are so arranged, that every particle of matter
at one side has a corresponding particle at an equal distance at the
other side; for in this case every pair of such equidistant particles will exert equal and opposite centrifugal forces on the axis.

    A great number of solid bodies, including the class of solids of
revolution described above, fulfil this condition.

    If the sections of a solid be all equal, and have any regular
geometrical figure having a point within it forming its geometrical
centre, and through which all lines drawn are bisected, then such
a solid participates in the above property, and the centrifugal

forces of its mass, when revolving round such an axis, will neutralize each other.

339. A column formed by laying one upon another equal flat plates of the same figure has this property. Such a column is called, in geometry, a rectangular prism.

340. A pyramid formed by laying upon each other similar plates not equal, but diminishing gradually in magnitude, will have the same property.

It follows, therefore, that the axes of the rectangular prisms and pyramids whose bases have a centre of magnitude enjoy the above property.

341. It will be observed that the axis round which the centrifugal force equilibrates in the examples given above will pass through the centre of gravity of the bodies in question. It may therefore be asked, whether such property belongs to all lines whatever passing through the centre of gravity of a body; that is to say, whether, if a body is made to revolve upon an axis passing through its centre of gravity, the centrifugal force will be in equilibrium, and whether such axis will be free from strain or pressure. It is easy to show that this will not be the case in general, and that it is only certain lines passing through the centre of gravity which have the property above mentioned.

If a rectangular plate having unequal sides, A B and B C (*fig.* 114.), be made to revolve round a line M N, passing through the

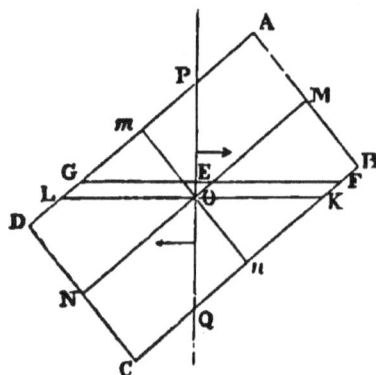

Fig. 114.

middle points of the opposite sides A B and D C, the centrifugal forces will be in equilibrium; because every point on the one side of M N has a corresponding point at an equal distance on the other side.

In like manner, the line *m n*, passing through the middle points of the sides A D and B C, would enjoy a like property. But if any other line, such as P Q, be drawn through the centre of gravity, o, and the plate be made to revolve round such line, then it will be evident that the centrifugal force will not be in equilibrium. For through a point such as E, let a line F G be drawn perpendicular to the axis P Q. This line F G will be divided into unequal parts at E, and the centrifugal force produced by the particles between E and F will be greater than the centrifugal forces produced by the particles between B and G. The same will be true of all lines drawn perpendicular to O P; and the combined effect of all the centrifugal forces acting in that part of the axis between O and P will be to produce a greater strain on the side of the angle A than on the side of the angle D, and the centrifugal force will have a resultant directed towards the side A. By the same reasoning it may be shown that the centrifugal forces of that part of the plate which is below K L will have a resultant directed to the side of the angle C, and this resultant will be equal to the resultant of the centrifugal forces above K L, directed to the side of the angle A.

It follows, therefore, that the axis P Q round which the plate revolves will be affected by forces having two resultants parallel to each other and in opposite directions, one perpendicular to O P, and directed towards the side A, and the other perpendicular to O Q, and directed towards the side C. These two forces form a couple, and have a tendency to turn the axis of revolution towards the position M N, as represented by the arrows in the diagram.

342. This property is general, although it cannot be demonstrated in all its universality without the aid of the language and principles of the higher analysis. It may be stated thus :— In all bodies whatever, there are three lines passing through the centre of gravity which are at right angles to each other, each of which is so placed, in reference to the mass of the body, that the centrifugal forces produced by the revolution of the body round them respectively will be in equilibrium; and such lines, when the body revolves round them, will suffer no strain or pressure.

But if the body be made to revolve round any other line passing through the centre of gravity, the centrifugal force will produce a strain, which will be represented by two equal opposite and parallel forces acting upon the axis at opposite sides of the centre of gravity, and having a tendency to turn the axis in the position of one or other of the three axes of equilibrium here mentioned.

343. **Principal axes.** — An axis round which the centrifugal force equilibrates is called a *principal axis;* and from what has been explained, it appears that there are three principal axes through the centre of gravity at right angles to each other.

344. There are some particular cases in which every line passing through the centre of gravity is a principal axis ; such, for example, is the case with a sphere or globe of uniform density. Such a solid, whatever diameter it may revolve round, will be a solid of revolution, and the sections perpendicular to such diameter will be circles. The same principle is true of all the regular solids. All lines, for example, passing through the centre of a cube are principal axes.

345. If a solid be made to revolve round an axis which does not pass through the centre of gravity, but which is parallel to one or other of the principal axes passing through that point, the centrifugal forces will not equilibrate round such axis, but they will be represented by a single force perpendicular to it, and passing through the centre of gravity.

346. Such an axis is called a principal axis, and in general there are three such axes corresponding with such points taken in the body, which are parallel respectively to the principal axes passing through the centre of gravity.

347. It may therefore be stated, in general, that if any point in a body be taken different from the centre of gravity, there are three lines passing through it at right angles to each other, round each of which, if the body is made to revolve, an effect will be produced by the centrifugal forces which can be represented by a single force, perpendicular to the circumference, and passing through the centre of gravity.

348. If the body be made to revolve round any line passing through a given point in it which is not a principal axis in the sense just referred to, then the centrifugal forces produced by such revolution cannot be represented either by a single force or by a pair of equal and opposite parallel forces, but will be represented by both of these together. This, therefore, is the character of all axes, round which a body would move, which do not pass through the centre of gravity, and are not parallel to either of the principal axes passing through that point.

349. **Experimental illustrations.** — Many of these important properties admit of being experimentally illustrated in a very striking manner by a simple apparatus.

Let a body be suspended by a thread, or, better still, by a skein composed of several threads, from a fixed point. Let this fixed point be so arranged, that a rapid rotatory motion may be given to it. This rotatory motion will soon be imparted to the body suspended to the threads by the twisting of the skein, and, after a time, the rotation of the body will become extremely rapid. It will first take place round the line, passing vertically through it in the direction of the skein, when it hangs quiescent. If this line happen to be one of the principal axes passing through its centre of gravity, it will continue to revolve round it ; but if it be not one of these principal axes, the centrifugal force, as has been already explained, will produce an effect upon it, represented by two equal, opposite, and parallel forces tending to twist it. This effect will be soon rendered manifest in a remarkable manner : the body, when it revolves rapidly, will not continue in the same position which it had when it was quiescent. The line which was in the direction of the string will begin to take another direction, the point where the string is attached to it will not remain in its first position, and the body will throw itself into a position at a greater or lesser angle to the string, and, in a word, will at length, after the revolving motion has become sufficiently rapid and continued, assume such a position, that a principal axis, passing through its centre of gravity, will become vertical, and the body will spin round it.

It is evident, therefore, that this effect takes place in spite of the opposing influence of the gravity of the body; for in the new position which the body assumes, the line of direction of its centre of gravity ceases to be represented by the skein. This experiment may be varied in a great variety of ways, exhibiting most instructive and amusing effects.

If a metallic ring be suspended by the skein by being attached to a point in its side, the ring, when quiescent, will hang with its plane vertical; but when the rotation becomes rapid, the ring will throw itself into a horizontal position, and will spin round a vertical axis through its centre, and perpendicular to its plane.

If the experiment be made with an oblate spheroid suspended by a point P (fig. 115.) in its equator PQ, its lesser axis A B

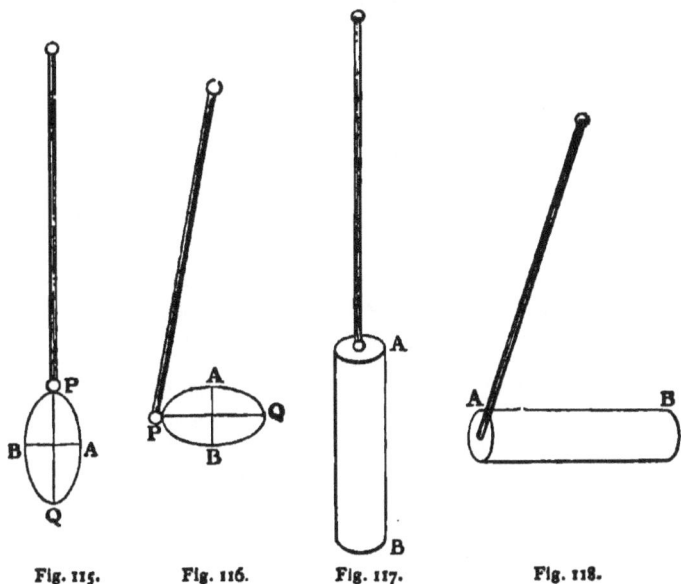

Fig. 115.   Fig. 116.   Fig. 117.   Fig. 118.

being horizontal, it will, when it acquires a rapid motion, take the position represented in fig. 116., in which the shorter axis A B is vertical, and the equator P Q horizontal, and it will spin in this position round the principal axis A B, although the centre of gravity is unsupported by the string.

If a cylindrical rod A B (*fig.* 117.) be suspended by a point at the centre of one of its ends, and a rapid revolution be imparted to it, it will not continue in this position, but will assume the position represented in *fig.* 118., in which its length will be horizontal, and it will revolve round an axis passing through its centre, and at right angles to the length.

If a metallic chain, the ends of which are united, as in a necklace or bracelet, be suspended from any point, and put in rapid revolution, the chain will at first make some irregular gyrations, but after a time it will gradually open, and settle itself into the form of a precise circle, the plane of which will be horizontal, and the string will then be attached to a point in this circle. In short, the chain will assume the form of a solid ring in a horizontal plane.

CHAP. VII.

MOLECULAR FORCES.

350. **Pores the region of molecular forces.** — It has been demonstrated in a former chapter, that the space included within the external surface of a body is not all occupied by the matter composing that body. We have seen that bodies, however dense or ponderous, are capable of being diminished in their volume by compression, or by diminution of temperature. It follows, therefore, that the component particles forming the mass of a body of a uniform density are uniformly distributed throughout its volume, each particle being separated from those around it by a space of greater or less extent unoccupied by matter.

These interstitial spaces are the regions which form the theatre of action of those important physical agents called molecular forces. A multitude of phenomena, familiar to all observers, show that between the particles which compose the mass of a body, there exist attractive or repulsive forces, the sphere of whose action is in general limited to distances imperceptible to the senses, and which only admit of being proved by indirect means.

351. **Cohesion.** — The qualities of solidity and hardness, and, in general, those properties by which a body resists fracture or flexure, or any other derangement of its form, arise from the

energy with which its component particles attract each other and resist any force which tends to separate them.

Molecular attraction manifested in this manner is called the *attraction of cohesion.*

352. **Adhesion.** — If the surfaces of two bodies be brought into very close contact, it is found that they cannot be separated without the exertion of some force of greater or less intensity, according to the circumstances of the contact.

Molecular force manifested in this manner is called the *attraction of adhesion.*

353. **Capillary attraction.** — Certain bodies being placed in contact with a fluid, the fluid will enter their dimensions and occupy their pores; as, for example, when a sponge or a lump of sugar is brought in contact with water. The fluid in these cases rises in opposition to its gravity, and fills all the interstices of the sponge or the sugar. Molecular force manifested in this manner is called *capillary attraction.** The effects of this force will be explained in a subsequent part of this work.

354. **Chemical affinity.** — When two bodies of different kinds are mixed together, their constituent particles will in certain cases unite, and form by their combination the constituent particles of a compound, differing in its sensible qualities from either of the components. For example, if two gases called oxygen and hydrogen be mixed together in a certain proportion, and a light be applied to the mixture, an explosion will take place; the atoms of the two gases will unite one with another, and the entire mass will be converted into water, the weight of the water being exactly equal to the sum of the weights of the two gases. In this case, each atom of the oxygen is attracted by an atom of hydrogen, and their combination forms an atom of water.

Molecular attraction manifested in this manner is called *chemical attraction,* or *chemical affinity.*

355. **The atomic attraction and repulsion.** — It has been shown that all bodies submitted to the action of mechanical forces of sufficient energy are capable of being compressed and diminished in their volume. By such means, therefore, their component particles are forced into closer proximity.

But all of these resist such compression with a certain force, and most of them have a tendency to recover the volume which they had before compression. This general fact indicates the existence of another force, contrary in its direction to the attraction of cohesion, the sphere of whose action is within that of the latter

---

* So called from *capilla,* a hair; the magnitude of the pores being in this case estimated at about the thickness of a hair.

attraction. To explain this phenomenon we are compelled to suppose that each atom composing a body is surrounded with a sphere of repulsion within which adjacent atoms cannot enter unless urged by a certain force. But outside this sphere of repulsion there exists the sphere of attraction, by which such atoms attract all the surrounding atoms, which gives the character of solidity and hardness to the mass.

356. The attraction of cohesion is manifested in solid bodies by the force which is necessary to derange their form by fracture or flexure, or by any other change of figure. The same force is manifested in liquids in their tendency to form into spherical drops, a globe being the greatest volume which can be contained within a given surface.

357. Thus particles of water falling in the atmosphere attract each other, and collect in spherules forming rain. If such spherules after their formation be exposed to cold, they harden and form hailstones. If a little mercury be let fall on a sheet of paper, it will collect in small silvery globules, notwithstanding the tendency of the gravity of its particles to make it spread over the paper in fine dust. Innumerable examples present themselves of this class of phenomena. The tear as it falls from the eye collects in a spherule upon the cheek; the dew forms a translucent globule on the leaves of plants.

358. The manufacture of shot presents one of the most striking examples of this phenomenon in the arts. The lead, in a state of fusion, is poured into a sieve, the meshes of which determine the magnitude of the shot, at the height of about two hundred feet from the ground. The shower of liquid metal, after passing through the sieve, forms, like rain in the atmosphere, spherules, which, before they reach the ground, are cooled and solidified.

These spherules form the common shot used in sporting, and the precision of their spherical form shows how regularly the liquid obeys the geometrical law, that a sphere contains the greatest volume within a given surface.

359. This disposition of fluids to affect the spherical form may be further elucidated in considering that any other figure which a body could take would necessarily place different parts of its surface at different distances from its central point, a circumstance which would be incompatible with the combined qualities of attraction of cohesion and fluidity. By fluidity, all the particles forming the mass are free to move amongst each other, and by the attraction of cohesion they are drawn round their common centre with the same force. To suppose that they could rest at different distances from their common centre, would necessarily involve the

supposition either that the attraction by which they are affected was unequal, or that the mass had not perfect fluidity.

This principle, which is so evident, may be inverted, and we may assume that in all cases where natural bodies are found in the spherical form, even though they be solid, they must have been, at the epoch of their formation, in a fluid state.

360. Hence it is inferred that the earth and the other bodies of the solar system were once fluid, and that our globe existed formerly in a liquid state.

361. The mutual repulsion of the atoms of gases is manifested by their indefinite expansibility and compressibility. Air included in a cylinder under an air-tight piston will expand so as to fill the increased volume as the piston is drawn up, and to this expansion there is no practical limit. This is explained by supposing that around each molecule of the air or gas there is a sphere of repulsion, so that each particle repels those around it. When the piston is raised to twice its former height, the air beneath it will expand into double its former volume.

In this case it must be concluded that the vacant spaces between the particles of air are twice as great as they were before the piston was raised. If the piston be again raised to double its present height, the same effect will take place. The air will again expand in virtue of the repulsive force prevailing among its particles, and the interstitial spaces separating the particles will be proportionally augmented.

There is no known limit to this expansive quality, and it consequently follows that the region through which the repulsive forces of gases act has a corresponding extent.

362. Methods will be explained in a succeeding part of this work by which several of the gases have been reduced to the liquid state, and analogy justifies the conclusion that all gaseous bodies are capable of this change. In the liquid state, the attraction of cohesion is rendered manifest, as has been already shown. But we have a still further evidence of the attraction of cohesion amongst the particles of gases, inasmuch as some of them have been reduced to the solid state; and by analogy we may conclude that all are capable of this change. It has been already shown that the solid state is only a consequence of the attraction of cohesion.

363. These and other phenomena lead to the conclusion that in this case of the gaseous bodies, there is beyond the sphere of cohesion a sphere of repulsion. When the particles, either by the application of cold or compression, or both of these agencies, are brought into such close contact as to be within the space of cohesion, then they become a liquid or solid, as the case may be.

364. The mutual repulsion found to prevail among the con-

stituent particles of bodies is by some attributed to the agency of heat, and it is certain that the energy of this repulsion is increased or diminished, according as heat is imparted to or subtracted from bodies. In a solid body, such as a mass of gold, in its ordinary state, the attraction of cohesion between its particles greatly predominates over the influence of the repulsion already mentioned ; but if heat be applied to this mass, the energy of the repulsion is gradually increased, until at length it becomes so nearly equal to that of cohesion, that the gravity of the particles overcomes that part of the cohesion not balanced by the repulsion, and the constituent parts of the mass no longer holding together in the solid form, the metal is converted into a liquid. If heat be still applied to this liquid, the temperature will rise and the liquid will expand ; but after a certain quantity of heat has been imparted to it, the repulsive force between the particles themselves is so great, that, in spite of their gravity and of the attraction of cohesion, they separate and disperse into a vapour which possesses the qualities of gas, being capable of expanding without limit.

365. Thus it appears that the same body may exist in the solid, liquid, and gaseous forms, according to the conditions under which it is placed in reference to heat.

366. **Adhesion of solids.** — If the surfaces of two pieces of metal, being rendered perfectly smooth, are brought into close contact by a strong pressure, they will adhere together with considerable force. That this adhesion is not due to atmospheric pressure can be demonstrated by showing that the adhesion will continue in a vacuum. In this case the superficial molecules of the two bodies are brought into contact so close as to be within the sphere of each other's attraction.

Innumerable examples of the adhesion of solid bodies are familiar to daily experience. We may write with chalk, or with a pencil, or charcoal on a wall or on a ceiling, although the effect of gravity would be to cause the particles abraded from the chalk, the lead, or the charcoal to fall from the wall or the ceiling. Dust floating in the air sticks to the wall or ceiling, in spite of the tendency of its gravity to fall from them.

The force of adhesion of solid surfaces one to another may be ascertained by placing the adhering surfaces in a horizontal position, the lower one being attached to a fixed point, and the upper one connected with the arm of a balance. The weight necessary to separate them is the measure of the adhesion. If we desire to ascertain the amount of adhesion per square inch of surface, it is only necessary to divide such weight by the magnitude of the adhering surface expressed in square inches.

367. It is on the adhesion between metallic surfaces when pressed

strongly together that the efficacy of a locomotive engine depends. The driving-wheels press with a great weight upon the rails, and are made to revolve round their own centres by the force of the engine. If there were no adhesion, or even insufficient adhesion between the tire of the wheel and the rail on which it is pressed, the wheel would turn without advancing; and this actually does happen in cases where the rails are greasy, and very frequently when they are covered with a hoar frost, the contact being then interrupted, and the matter between the wheel and the rail not offering the necessary adhesion.

368. **Effect of lubricants.** — On the other hand, when the force applied to break the adhesion is directed perpendicularly to the adhering surfaces, a fluid or unctuous matter smeared upon the surface often increases the adhesive force.

Thus two metallic plates will adhere with greater force together if they are smeared with oil than when they are clean. This may partly arise, however, from the fact that the film of oil which covers them excludes air more effectually than could be accomplished in the case of surfaces so considerable by mere pressure.

369. **The bite.** — The effect known amongst workers in metal as the *bite* is the adhesion of two metallic surfaces brought into extremely close contact. It may be doubted whether this adhesion would not be diminished if some fluid were introduced between the surfaces.

370. **Glues, solders, and like adherents.** — The adhesion of the surface of solids may be rendered more intense than even the cohesion of the particles of the solids themselves by interposing between them some substance in a liquefied form, which hardens by cold, and which when hard has a strength equal to or greater than that of the solids which it unites. Glues, cements, and solders supply remarkable examples of this. Two pieces of wood glued together will break anywhere rather than at their joint. The processes of gilding and plating also supply examples of the adhesion of metals to each other.

371. The process of silvering mirrors is an example of the adhesion of metal to glass; and that of mortar in building is an example of the adhesion of earthy matters to each other.

Two pieces of caoutchouc, if pressed together upon freshly cut surfaces, will be found to unite as completely as if they composed one independent piece.

BOOK THE THIRD.

THEORY OF MACHINERY.

~~~~~~~~~~

CHAPTER I.

GENERAL PRINCIPLES.

372. **What constitutes a machine.** — A machine is an instrument or apparatus by which a force applied at a certain point, and having a certain determinate intensity and direction, is made to exert a force at another point, more or less distant from the former, and generally different in intensity and direction. Thus, for example, a horse moving on a horizontal road in a circle is made to raise a weight vertically in the shaft of a mine, or water from the shaft of a well.

Men pulling at a rope in some direction more or less oblique are enabled to raise a mass of heavy matter from the hold of a ship, and to transfer it to an adjacent wharf.

373. The force which is applied to and transmitted by a machine is technically called the power; the point at which it is applied is called the point of application; its direction is the line in which the force has a tendency to make the point of application move; and its intensity is usually expressed by a weight which, acting at the same point of application, would produce a like effect upon it.

The moving powers applied to and transmitted by machinery are infinitely various. In the capstan of a ship, the moving power is human force applied to it; in a common pump, the same moving force is used; a horse is the moving power applied to vehicles of transport on common roads, and a steam-engine on railways; the wind is the moving power applied to a sailing-vessel, and to a windmill; the momentum of water acting against the float-boards of a wheel, or its weight acting in the buckets, is the moving power of a water-wheel; the elastic force of steam acting on the piston in the cylinder is the moving power of the steam-engine.

374. That part of a machine which is immediately applied to the resistance to be overcome is called the *working point*.

375. The resistance, whatever be its nature, to which the

M

working point is applied, is technically called the *weight* or *load*. In many cases weight is the actual resistance which machines are applied to overcome; as, for example, in raising water from a well, or from mines; also in raising ore. In some cases the resistance to be overcome is friction, used for the purpose of fracturing and pulverising material substances. This is the case in flour-mills.

In some cases the resistance to be overcome is the friction of surfaces, and the resistance of the air. This is the case when carriages are moved on level roads or railways.

Whatever be the nature of the resistance, a weight which would produce an equivalent force acting against the moving power may be assigned. Thus, for instance, if the traces of a carriage drawn by horses, or the chain connecting a locomotive with a railway train, be stretched by the resistance of the carriage or the train, a weight may be substituted, which, being suspended vertically from the traces or the chain, would produce the same tension. The resistance in such case is expressed by stating the amount of this weight.

376. In the exposition of the theory of machinery, it is expedient to omit, in the first instance, the consideration of many circumstances, of which, however, a strict account must be subsequently taken before any practically useful application of them can be made. A machine, such as we must for the present contemplate it, is a thing which can have no real or practical existence. Its various parts are considered to be free from friction. Thus, the surfaces composing it, which move in contact with each other, are assumed to be infinitely smooth and polished; the solid parts are all considered to be absolutely rigid and inflexible.

The weight and inertia of the matter composing the machine itself are wholly neglected, and we reason upon it as if it were divested of these qualities. Cords, ropes, and chains are supposed to have neither stiffness, thickness, nor weight; they are regarded as mathematical lines, infinitely flexible and infinitely strong. The machine, when it moves, is assumed to encounter no resistance from the air, and to be in all respects circumstanced as if it were in vacuo.

These suppositions being all false, it follows that none of the consequences immediately deduced from them can be true. Nevertheless, as it is the business of art to bring machines as near to this state of ideal perfection as possible, the conclusions which are thus obtained, though false in a strict sense, yet deviate from the truth in no considerable degree.

These conclusions may, in fact, be regarded as a first approximation to truth.

377. The various effects which have been previously neglected are afterwards taken into account. The roughness of surfaces, the imperfect rigidity of the solid parts, the imperfect flexibility of cords and chains, the resistance of the air and other fluids, and the effects of the weight and inertia of the machine itself, are afterwards severally examined, their properties explained, and the manner in which they modify the transmission of the power to the weight developed. These modifications and corrections being applied to the conclusions obtained, a second approximation to the truth is made, but still only an approximation; for, in investigating the laws which govern the several effects last mentioned, we are compelled to proceed upon a new group of false suppositions.

To determine the laws which regulate the friction of surfaces, it is necessary to assume that the surfaces are uniformly rough, and subject to uniform pressure; that the solid parts which are imperfectly rigid, and the cords and chains which are imperfectly flexible, are constituted throughout their entire dimensions of a uniform material, so that the imperfections do not prevail more in one part than in another. Thus, all irregularity is left out of account, and a general average of the effects taken. It is obvious, therefore, that, even in this second system of reasoning, we have still failed in obtaining a result exactly conformable to the real .state of things. But it is equally obvious that we have obtained one much more conformable to that state than had been previously accomplished; and, in fine, it is found that the conclusions thus obtained are sufficiently near the truth for practical purposes.

378. The imperfections in our process of investigation, manifested in this laborious system of successive approximation of the truth, is not peculiar to Natural Philosophy. It pervades all departments of natural science. In Astronomy, the motions of the celestial bodies, and their various changes and appearances as developed by theory, assisted by observation and experience, consist of a like series of approximations to the real motions and appearances which take place in nature.

It is the same in Art. The first labours of the artist produce from the rude block of marble a rough and rude resemblance of the human form. The next attempts remove the greater inequalities and protuberances, and reduce the form to a closer resemblance to the original. It is not, however, until after a long succession of operations, in which smaller and smaller portions of the stone are detached, that the last labour of the chisel of the master completes the resemblance.

379. We shall therefore for the present consider the machine, by which the effect of the power is transmitted to the working point, as divested of weight and inertia; we shall consider all the

pivots, axles, and surfaces which move in contact absolutely devoid of friction; we shall consider all cords, ropes, and chains to
be absolutely and perfectly flexible, and to be moved in contact
with the grooves and wheels without friction; and, in fine, we shall
consider the machine itself, as well as the agent exerting the power,
and the matter composing the weight, to move without resistance
from the air or any other fluid.

380. The exposition of the effects of machinery is often invested
with the appearance of paradox. Astonishment is excited at what
seems incompatible with the results of common experience, rather
than admiration of the genius and skill by which simple and obvious principles are so applied as to produce unexpected results.

Thus it is stated that, by means of a machine, a power of comparatively insignificant amount is capable of supporting or raising
a vast weight; as, for example, it is affirmed that the fingers of an
infant pulling a thread of fine silk, which a pound weight could
snap asunder, are capable by this or that machine of supporting or
raising several hundred weight.

Statements like these, if literally understood, are fallacious; if
rightly explained, they involve nothing which is not consistent with
our habitual experience.

381. In every machine there are some fixed points or props,
and the arrangement of the parts is always such that all that
portion of the weight not directly acting against the power is distributed among these props. If the weight, for example, amount
to 20 cwt., it is possible so to arrange it that any proportion of it,
however great, may be thrown upon the fixed points or props of
the machine: the remaining part only can properly be said to be
supported by the power, and this part so supported can never be
greater than the power. Considering the effect of a machine in
this manner, it appears that the power supports just so much of
the weight, and no more, as is equal to its own force, and that all
the remaining part of the weight is sustained by the machine.

The force of this observation will become more and more apparent when the conditions are explained, under which a power and
weight can maintain each other in equilibrium, through the intervention of a machine, whether simple or complex.

382. But if the power, instead of merely supporting the weight
at rest, be employed to raise this weight a given height, it may be
asked how it can be explained that a power indefinitely small can
lift a weight indefinitely great. The paradoxical character of this
statement arises, as is the case generally in such propositions, from
the omission of an important condition. It is quite true that a
feeble power is capable of raising a great weight, but it is necessary
to add that in doing this, the feeble power must act through a

space just so much greater than that through which the weight is raised, as the weight itself is greater than the power. Thus, if a weight of 1000 lbs. be raised one foot by a power whose force is only equal to 1 lb., then such power in raising the weight must move through 1000 feet.

Now when this condition is stated, the proposition is stripped altogether of its paradoxical character. There is nothing at all astonishing in the fact, that one thousand successive exertions of a force of one pound, each exertion being made through the space of one foot, should raise a 1000 lbs. weight through the height of one foot. There is nothing more surprising in such a fact than if the 1000 lbs. weight, being divided into 1000 equal parts, were raised by a thousand successive efforts of the power without the intervention of any machinery.

383. It will be necessary to consider the effect produced by means of a machine under three distinct relations between the power and weight, viz.,

    I. When the power equilibrates with the weight.

    II. When the power is greater than that which equilibrates with the weight.

    III. When the power is less than that which equilibrates with the weight.

384. The power and weight are said to be in equilibrium when they are so related to each other that when placed at rest they will remain so. It is a great, but very common error, to suppose that equilibrium, as applied to a machine, necessarily implies rest or the absence of motion. It is easy to show that if the power and weight, being in equilibrium, are put in uniform motion, they will continue that uniform motion exactly as a mass of matter would do in virtue of its inertia, if moving independently of any machine; for if we were to suppose that such motion would cease either suddenly or gradually, we must necessarily also suppose a definite force applied to the machine to stop its motion. Since the power and weight are in equilibrium, they cannot of themselves stop or retard the motion. It is true that the motion will in practical applications be gradually retarded, but that will be the effect of friction and atmospheric resistance, both of which are at present excluded from consideration.

We cannot, on the other hand, suppose the uniform motion imparted to the power and weight to be accelerated, without supposing the application of some adequate force to produce such acceleration, the power and weight being excluded by the very condition of their equilibrium.

Let us suppose the power and weight to be connected with the machine by cords, by which they are suspended from their respective points of application, both being, as usual, represented by equivalent weights. Now the cords by which they are suspended will be stretched with the same force whether the power and weight be at rest or in uniform rectilinear motion; and consequently the relation between them in both cases must be the same.

385. The most common state of machines which are under the operation of equilibrating forces, is that of uniform motion, and not that of rest, as commonly stated. If a wind or water-mill be in regular operation, its driving-wheel moving with a uniform speed, then the power of the wind or water will be in equilibrium with the resistance, whatever that may be. If a steam-engine be in regular operation, its piston will move at a uniform rate, and the force of the steam upon it will be in equilibrium with the resistance which it is applied to overcome. If a locomotive engine draw a railway train at a uniform speed, then the power exerted by the engine will be in equilibrium with the resistance opposed by the train.

386. Let us now consider the case in which the power is greater than that which equilibrates with the weight or resistance. In this case the motion imparted to the object moved will be accelerated; for so much of the power as would equilibrate with the weight or resistance would impart, as has been already shown, a uniform motion to the object moved. The surplus power above this amount, therefore, must be employed in accelerating the motion.

387. For example, if a locomotive engine exert a greater power than is equivalent to the resistance opposed by the train which it moves, then such surplusage of power can only act upon the inertia of the train, and will impart to it an equivalent amount of moving force. So long as this surplus power, therefore, acts, the mass of the train will receive from it a corresponding augmentation of its momentum, and consequently will receive a proportionate increase of speed. If the resistance, however, opposed by the train to the moving power augments with the speed, then it may at length become equal to the amount of the moving power; and when it does, their equilibrium is established, and the train is moved by the power at a uniform speed.

These conditions are by no means imaginary. They are realized in every case in which a train is started from a state of rest, and in general when any machine whatever is first put in motion.

388. The power in commencing its action must necessarily be greater than the resistance opposed by the load; for if it were not, it would only equilibrate with the resistance, and no motion would ensue. The surplus power is absorbed by the momentum acquired by the moving mass; and as the velocity augments, more and

more momentum is imparted. The velocity will at length become uniform, either because the energy of the power will be diminished, so as to become equal to the resistance, or because the resistance will be augmented, so as to become equal to the power; or, in fine, as most generally happens in practice, both of these effects are combined, the resistance increasing and the power diminishing. This is always the case, therefore, when a machine is impelled by a surplus power, and when, on the other hand, there is a less than ordinary resistance on the side of the machinery and of the load. When first starting, the velocity being inconsiderable, the resistance of the air and other agencies depending upon speed is less. As the velocity increases, these resistances augment. This augmentation of resistance, however, as the speed increases, is generally much less than the diminution of the moving power; in short, a considerable surplus power is generally necessary, at starting, to impart to the load, and to the moving parts of the machinery the necessary momentum. But after this momentum has once been imparted, then nothing remains for the power but to balance the resistance of the load, properly so called.

This excess above the equilibrating power and the accelerated motion are reciprocal consequences. Such excess necessarily infers the accelerated motion of the load, and the accelerated motion of the load indicates such excess.

389. Effects directly the reverse of these are developed when the power applied is inferior to that which would equilibrate with the weight. Let us suppose, in this case, the machine to have been in uniform motion, and therefore the power and weight to have been in equilibrium. Let the power then be diminished by any amount, however small: the moment this diminution takes place equilibrium is destroyed, the power becomes inferior to the resistance, and there is an action in a direction contrary to that of the power, and therefore contrary to that of the motion which the load had already acquired, equivalent in amount to the difference between the resistance and the power. This force will act against the momentum of the load, and will continually diminish it, until, at length, it brings the load to rest. From the moment, therefore, that the power becomes less than the resistance, the motion of the load will be gradually retarded. The inferiority of the power to the resistance, and a gradually retarded motion, are, therefore, reciprocal consequences of each other.

It may be useful to illustrate still further these effects, which are of considerable importance in practice. Let us suppose the resistance which a machine is employed to overcome to be represented by the weight A, *fig.* 119., and let the power which acts against such resistance through the intervention of the cord A B C be re-

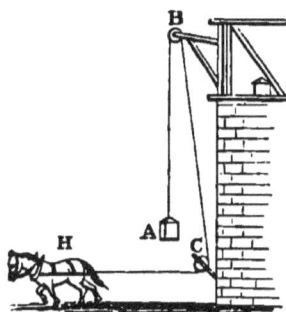

presented by the force of an animal H. When the animal is at rest before starting, the cord A B C H is stretched with a force exactly equal to that of the weight. When the power begins to move, a momentum is imparted to the weight through the intervention of the cord. The cord is therefore stretched with an additional force proportional to this momentum. The speed of the power gradually increases from the moment its motion commences until it attains that speed which is continued

Fig. 119.

uniform. During this increase of the speed of the power H, a corresponding and continual increase of momentum is imparted to the weight A, and consequently, during this interval, the tension of the cord is constantly greater than the weight. When, however, the speed of the power H becomes uniform, then no further momentum will be imparted to the weight, and the force exerted by the power will diminish so as to become exactly equal to the weight. During this uniform motion the tension of the cord will be the same as it would be if the power and weight were at rest.

When the weight approaches its point of destination, and is about to be brought to rest, the power slackens its exertion, and, at the moment that it becomes less than the weight, a moving force takes effect equal in intensity to the difference between the power and weight directed from B to A. But against this there is the momentum of the weight directed from A to B in virtue of the uniform velocity with which it had been moving. The moving force, therefore, from B to A, represented by the difference between the power H and the weight A, will act against this momentum, and will gradually diminish it.

Although the upward motion of the weight, therefore, will continue after the diminution of the power, it will be gradually retarded, and after a certain interval will be altogether exhausted, and the weight will come to rest.

These effects take place in all machines whatever when they are started and stopped, and the circumstances and mechanical laws which govern them are precisely the same as in the illustration here given.

390. **Proper functions of a machine.** — The use of a machine is to adapt the power to the resistance. If the intensity, direction, and velocity of the power were identical with the intensity and direction of the resistance, and the velocity required to be imparted

to it, then there would be no need of a machine ; the power might
be applied immediately to the resistance. But if a power of feeble
intensity is required to act against a great resistance, then a machine must be interposed which will augment the intensity of the
power. Or if a power moving in one direction be required to
impart motion to a resistance in another direction, then a machine
must be interposed which will transmit the effect of the power to a
new direction.   Or if a power having a certain velocity be required
to impart a greater or less velocity to the resistance, then a machine
must be interposed which will modify the velocity in the required
proportion.

But even these, though the principal, are only a few of the
infinite varieties of change and modification which machines are
required to effect in the transmission of the power to the resistance.   Independently of the directions, intensities, and velocities
of the moving power and resistance, the character of the respective
motions may differ in an infinite variety of ways : thus the moving
power may be one which acts with a reciprocating motion between
two points ; as, for example, that of the piston of a steam-engine ;
and this moving power may be required to produce a continuous
motion in a straight line, like the motion of a train along a railway.

The machine which connects such a power with such a weight
must therefore be so constructed as to convert the reciprocating
motion between two points into a continuous motion in a straight
line.

The moving power may act in a straight line, while the resistance requires a circular motion. Thus, the wind which acts upon
the arms of a windmill is a continuous rectilinear force. The mill-
stones to which that force is transmitted, revolve by a continual
circular motion round vertical axes. The machinery of the windmill must therefore be adapted to convert the rectilinear force of
the wind into the circular motion of the stones.

In every class of machines, and in every individual machine of
each class, the relation between the velocities and directions of the
power and weight, and the change produced on the character of
the motion of the power when transmitted to the weight, depends
solely on the structure of the machinery.

No variation in the magnitude of the power and weight can
alter this relation. Thus the ratio of the velocities of the power
and weight on a *lever* or an *inclined plane*, so long as their form and
proportions remain the same, will be unaltered, and, whatever
power or weight be applied to them, they will have this particular
ratio.

391. **No machine can really add to the mechanical
energy of the power.** — A machine being composed of inert

matter cannot generate force, and consequently the working point cannot exert more force than is transmitted to it from the point of application of the power.

It will, in fact, exert less, because friction and other sources of resistance must intercept a portion of the action of the power in its transmission from its point of application to its working point; but as, for the present, the consideration of this species of resistance is neglected, and machines are considered as exempt from them, we shall assume that the influence of the power is transmitted undiminished to the working point. But it is important on the other hand to remember, that *no more* moving force can be so transmitted.

392. Now the energy or momentum of the power is determined by multiplying the weight which is equivalent to it, by the space through which it is moved; and, on the other hand, the moving force imparted to the resistance is also estimated by multiplying the weight which is equivalent to this resistance by the space through which it is moved.

The moving force of the power is determined in the same manner as the moving force of a weight equivalent to it, and moving with the same velocity, would be determined. Thus, if we multiply the power, or its equivalent weight, by the space through which it moves in a given time, that is to say, by its velocity, we shall obtain a product which expresses its moving force or mechanical effect.

393. This product is called the *moment of the power.* Thus if P express the power and p the space through which it moves in one second, then $P \times p$ will be its *moment.* In like manner, the moving force imparted to the resistance at the working point will be expressed by multiplying the resistance, or the weight equivalent to it, by the space through which it is moved in a given time. Thus if w be the weight, and w be the space through which it is moved in one second, then $w \times w$ will be the moving force of the weight, and this product is called the *moment of the weight.*

394. The relation between the moments of the power and weight determines their mechanical state.

Three cases are here presented : —

I. When the moment of the power is equal to the moment of the weight; that is, when

$$P \times p = w \times w.$$

II. When the moment of the power is greater than the moment of the weight; that is, when

$$P \times p \text{ is greater than } w \times w.$$

III. When the moment of the power is less than the moment of the weight; that is, when

$$P \times p \text{ is less than } W \times w.$$

In the first case, it is manifest that the power and weight will be in equilibrium, and that they will be either at rest or in uniform motion. For, since the moment of the power is the expression of its moving force, and since this moving force is transmitted without increase or diminution by the machinery to the weight, and since, by the supposition we have made, it is equal to the moving force of the weight, these two forces must balance each other, and therefore be in equilibrium.

395. The condition therefore of equilibrium is, that the moment of the power is equal to the moment of the weight, or

$$P \times p = W \times w.$$

396. If the moment of the power be greater than the moment of the weight, then the moving force of the power, exceeding that of the weight, and being transmitted to the working point undiminished, will prevail over it, and the power and weight must either have an accelerated motion in the direction of the power, or a retarded motion in the direction of the weight.

397. If the moment of the power be less than the moment of the weight, then the moving force of the power being transmitted to the weight and being less than the moving force of the latter, the latter will prevail, and therefore the power and weight must have either a retarded motion in the direction of the power, or an accelerated motion in the direction of the weight.

398. If the moments of the power and weight be equal we may infer that the power will bear to the weight the same ratio as the velocity of the weight bears to the velocity of the power, or

$$P : W :: w : p;$$

or, as it is sometimes expressed, the power and weight will be to each other inversely as their velocities. This is another mode, then, of expressing the conditions under which the power and weight will be in equilibrium.

399. **Power always gained at the expense of time.** — It is this inverse proportion which is intended to be expressed, when it is said that power is never gained save at the expense of time; the meaning of which is, that if a small power work against a great resistance, the rate at which it moves the resistance will be just so much slower than that at which the power itself moves, as the resistance is greater than the power.

400. This condition of equilibrium, when rightly understood,

removes all paradox from the statement of the effects of machinery
A small power working through a large space, raising a great
weight through a small space, is merely an expedient by which a
feeble power is enabled to accomplish its task, by a long succession
of efforts, without dividing the weight.   To raise the weight of a
ton by a single effort one foot, would require a force equivalent to
the weight of a ton.   But if, by the intervention of a machine, a
power is enabled to accomplish this object by 2240 distinct efforts,
each effort working through one foot, then such power need not
be more than one pound, or 2240 efforts made through the space
of one foot, each effort exerting the force of one pound will be
mechanically equivalent to 2240 lbs., or one ton raised through
one foot, and the effect produced will be the same as if the weight
were actually divided into 2240 equal parts, and the power applied
successively to raise each of these parts one foot.

401. A very inadequate estimate would, however, be formed
of the objects and the utility of machinery, if we were to suppose
them only directed to this particular class of problems.   Cases
innumerable occur, on the contrary, where small resistances are
moved by great powers.   For example, in a locomotive engine,
while the piston in the cylinder moves once backwards and forwards, the train, which is the resistance overcome, is moved
through a space equal to the circumference of the driving-wheel.
Now, if we suppose the length of the cylinder, as frequently
happens, to be one foot, and the circumference of the driving-
wheel to be fifteen feet, then the velocity of the piston, or the
power, will be to the velocity of the train, or the resistance, as 2
to 15; and consequently, the power which acts upon the piston
must be greater than the resistance of the train, which is moved
in the proportion of 15 to 2, omitting, as usual, the consideration
of friction, &c.   In like manner, in a watch or clock, the resistance
of the object moved is merely that which is opposed to the motion
of the hands on the dial-plate, while the moving power is the
energy of the main-spring, or of a descending weight.   In both
these cases it is obvious that the power is vastly greater than the
weight.

The machinery, therefore, may be stated generally as being the
means by which the force and motion of the power are modified,
so as to adapt them to the force or motion which is required
to be imparted to the object moved.

402. In all that precedes, it has been assumed that the point of
application of the power moves in the direction in which the power
acts, and that the motion imparted to the working point is in a
direction immediately opposed to the action of the weight or resistance.   This, in fact, is what generally takes place in the prac-

tical construction and operation of machinery; for it is evident that if the point of application of the power were not free to move in the direction in which the power acts, a part of the power would necessarily be lost; and that if the working point did not move in a direction immediately opposed to the weight or resistance, a part of the force transmitted to the working point would be inefficient. But as, in certain cases, these conditions would not be fulfilled, it will be useful to state how the principles which have been established in the present chapter must be modified in such cases.

If by the construction of the machinery the point of application of the power moves in a line different from that in which the power acts, then the effective part of the power will be found by the parallelogram of forces.

Let A (*fig.* 120.) be the point of application of the power, and

Fig. 120.

let B be the working point. Let A P represent the direction of the power, and B w the direction of the resistance or weight. Let A *p* be the direction in which the point of application is free to move, and let the working point B be free to move in the direction opposite to B *w*. Let right-angled parallelograms be formed, having for their diagonals A P and B w, representing the power and resistance, and having their sides in the directions A *p* and B *w*, in which the point of application and the working point are respectively free to move.

The power will then be equivalent to two forces represented by A *m* and A *p*; the latter, being in the direction in which alone the point of application can move, is alone effective; that part of the power represented by A *m* will necessarily be expended in pressure and strain upon the fixed points of the machine. In like manner the weight or resistance represented by B w is equivalent to two forces, B *w* and B *n*; the force B *w*, being in the direction against which alone the working point can act, is that portion of the weight or resistance which the working point will act against: the remainder of the weight will produce strain or pressure on the fixed point.

In the application of the principles determining the relation of the power and weight in cases of equilibrium which have been established in the present chapter, the effective portion only of the power and weight must be to be taken into account. Thus, the power is to be considered as represented by ʌ *p*, and the weight by ʙ *w*. These principles will be rendered more clearly intelligible when they have been illustrated in their application ·to the simple machines.

CHAP. II.

SIMPLE MACHINES.

403. Machines simple and complex.— Machines which are composed of two or more parts acting one upon another are called complex machines. Machines which consist only of one part are called *simple machines*. The several parts composing a complex machine are themselves simple machines. In a *complex machine* the effect of the power is transmitted successively through each of the parts composing it until it reaches the working point. The effect of complex machines is determined by combining together the separate effects of the simple machines of which they are composed.

To estimate the effects of machinery, therefore, it will be necessary, in the first instance, to explain the principles of simple machines.

404. Classification of simple machines. — Simple machines have been differently enumerated by different writers. If the object be to group in the smallest possible number of distinct classes those machines whose efficacy depends on the same principle, the simple machines may be comprised under the following three denominations : —

    I. A solid body turning on an axis.
    II. A flexible cord.
    III. A hard and smooth inclined surface.

Notwithstanding the infinite variety of machinery, and of the parts composing it, it will be found that those parts may invariably be brought under one or other of the above classes.

405. In a machine composed of a solid body turning on an axis, all the parts are carried round such axis as a common centre, and describe circles round it in the same time. It is evident that the magnitude of these circles, and consequently the velocities of the different parts, will be proportional to their respective distances from the axis in which their common centre lies.

. But since it has been already shown that when the power and weight are in equilibrium they must be inversely as the velocities of the points to which they are applied, it follows that any power and weight applied to such a machine will be in the inverse proportion of their distances from the axis when they are in equilibrium.

It must be understood, in the application of this important principle, that the power and weight are supposed to act in the direction of the motion of the parts to which they are respectively applied. If they do not act in this direction, then they must be resolved by the principle of the composition of force into two forces, one acting in the direction of the motion of the point of application, and the other in a direction passing through a point upon the axis. This has been already explained (402.).

406. The second class of simple machines includes all those in which a force is transmitted by means of flexible threads, ropes, or chains. The principle by which the effects of these machines is estimated is, that the tension throughout the whole length of the same cord, provided it be flexible and free from the effects of friction, must be the same. Thus, if a force acting at one end be balanced by a force acting at the other end, however the cord may be bent, or whatever course it may be compelled to take, by any cause which may affect it between its ends, these forces must be equal, provided the cord be free to move over any obstacles which may deflect it. This class includes all the various forms of pulleys.

407. The third class includes all those cases in which the weight or resistance is supported or moved upon a hard surface inclined to the direction in which the weight or resistance itself acts. The effects of such machines may be estimated by the principles already explained. The force of the weight or resistance being resolved into two other forces by the principle of the composition of force, one of these two forces will be perpendicular to the surface, and thus supported by its reaction; the other will be parallel to it, and will act against the power.

408. **Mechanic powers.** — The first class of simple machines above mentioned, consisting of a solid body revolving on an axis, is usually subdivided into two.

1st. The lever, which consists of a solid bar, straight or bent, resting upon a prop, pivot, or axis.

2nd. A cylinder connected with a wheel of much greater diameter moving round a centre or axis. This combination is called the wheel and axle.

The second class includes the pulley. The third class includes the simple machines, commonly known as the inclined plane, the wedge, and the screw; the last being, as will appear hereafter, nothing more than an inclined plane rolled round a cylinder.

The classes, therefore, of the simple machines, as they are generally received, and which are known as the mechanical powers, are the six following: —

    I. The lever.
    II. The wheel and axle.
    III. The pulley.
    IV. The inclined plane.
    V. The wedge.
    VI. The screw.

We shall accordingly explain these, and show the most important varieties and combinations of which they are susceptible.

THE LEVER.

409. A straight and solid bar turning on an axis is called a *lever*. The *arms* of the lever are those parts of the bar extending on each side of the axis. The axis is called the *fulcrum* or *prop*.

Levers are commonly divided into three kinds, according to the position which the fulcrum has in relation to the power and weight.

If the fulcrum be between the power and weight, as in *fig.* 121., the lever is of the first kind.

Fig. 121.

If the weight be between the fulcrum and power, as in *fig.* 122., the lever is of the second kind.

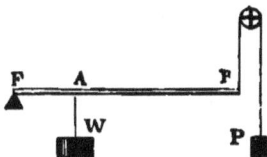 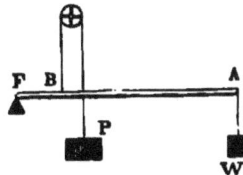

    Fig. 122.   Fig. 123.

If the power be between the fulcrum and weight, as in *fig.* 123., the lever is of the third kind.

410. Of whatever kind the lever may be, the conditions of equilibrium of the power and weight will be such that they are

inversely as their distances from the fulcrum, this being the general condition of equilibrium for all machines which turn round a fixed axis (405.). It follows, therefore, that in *figs.* 121, 122, and 123., we shall have

$$P : W :: FA : FB;$$

or, if *p* express the distance of the power from the fulcrum, and *w* the distance of the weight from the fulcrum, we shall have

$$P : W :: w : p;$$

or, what is the same,

$$P \times p = W \times w.$$

This statement, as will be perceived, is nothing more than a repetition of the general principle affecting machines which turn on an axis, in virtue of which forces upon them are in equilibrium when their moments round the axis are equal. The moment of the power is $P \times p$, and the moment of the weight is $W \times w$. The tendency of the power to turn the lever round its fulcrum in the direction of the power is expressed by the moment $P \times p$, and the tendency of the weight to turn the lever in the contrary direction is expressed by $W \times w$.

411. It follows, therefore, that the tendency of the power to turn the lever would be augmented either by increasing the amount of the power P, or by increasing its distance *p* from the fulcrum. In either case the effect will be increased in a corresponding proportion. Thus, if we remove the power to double its distance from the fulcrum, we shall double its effect; and if we remove it to half its distance, we shall diminish its effect one half. The distance of a force, whether power or weight, from the fulcrum, is called its *leverage*; and it is evident from what has been stated, that the effect of any force applied to a lever will be proportional to its leverage.

412. If the forces applied to a lever do not act perpendicular to it, their effect will be found by drawing from the fulcrum a perpendicular on their directions. This perpendicular will be their leverage. Thus in *fig.* 124, if the power act in the direction B P, draw F N perpendicular to the direction P B N; the power will have the same effect in turning the lever, as if it acted at N upon the lever N F. The moment of the power, therefore, in this case, will be found by multiplying it by F N, the perpendicular distance of its direction from the fulcrum. In general, therefore, the lever-

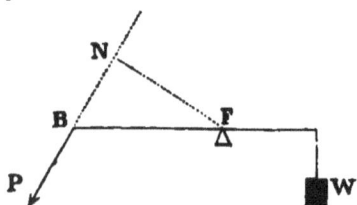

Fig. 124.

N

age of any force applied to such a machine is estimated by the perpendicular distance of the direction of such force from the fulcrum.

413. In a lever of the first kind, the power and weight may be equal, and will be so when their leverages are equal. The weight may be less than the power, and it will be so when it is at a greater distance from the fulcrum than the power.

In a lever of the second kind, the weight, being between the fulcrum and the power, must be at a less distance from the fulcrum than the power, and must consequently be always greater than the power.

In a lever of the third kind, the power being between the fulcrum and the weight, will be at a less distance from the fulcrum than the weight, and consequently in this case the power must always be greater than the weight.

414. **The balance.** — Numerous examples of levers of the first kind may be given. A balance is a lever of this kind with equal arms, in which the power and weight are necessarily equal. The dishes are suspended by chains or cords from points precisely at equal distances from the fulcrum, and being themselves adjusted so as to have precisely equal weights, the balance will rest in equilibrium when the dishes are empty. To maintain this equilibrium, it is evident that equal weights must be put into the two dishes; the slightest inequality would give a preponderance to one or the other dish.

To obtain all the necessary precision, the centre of the beam in well-constructed balances (*fig.* 125.) has angular-shaped steel prisms attached to each side of it, with their edges presented downwards. These, which are called *knife edges*, rest upon a hard surface, such as steel or agate. The hooks to which the chains supporting the dishes are attached are similarly supported. The instrument is so constructed that the centre of gravity of the beam and its appendages is a little below the knife edge on which it is suspended when the beam is

Fig. 125.

horizontal, the index attached to the centre of the beam then pointing vertically upwards. If the beam be inclined downwards on either side, the centre of gravity will deviate to the other side, and, by its tendency to take the lowest position, it will bring the beam back to the horizontal position when the force which disturbed it is removed.

By the conditions of equilibrium of the lever, as explained above, the beam can only rest in the horizontal position, with the index vertical, when the weights in the dishes are equal. The article to be weighed being placed in one dish, known weights are placed in the other, until the index points to the zero on the graduated scale behind it, which is so arranged that the index is then vertical. The weight of the article is then equal to the known weights in the other dish.

To verify the exactitude of a balance, let it be first observed whether the index is vertical, and the beam horizontal when the dishes are empty. The article to be weighed being then placed in one dish, and balanced by weights in the other, let the article and the weights be transposed, the former being placed in the dish containing the latter, and *vice versâ*. If the index is still vertical, and the beam horizontal, the balance is exact; if not, the arms are unequal, and the indications erroneous. Such a balance should be rejected.

415. **Sensibility of a balance.** — This term is used to express the facility with which the index is turned from its position by a preponderating weight placed in either dish, and it is greater or less, according as the distance of the centre of gravity below the point of support is less or greater. Let A B (*fig.* 126.) be the beam, c the point of support, and G the centre of gravity. If the beam be slightly turned from the horizontal position to the position A′ B′, the centre of gravity will take the position G′ and would be balanced by a weight upon A′, which would

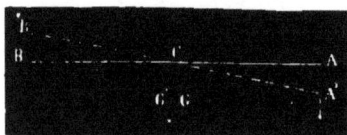

Fig. 126.

be less than that of the beam in the proportion of the distance of G′ from G c to that of A′ from G c. The nearer G is to c, the less will be this proportion, and therefore the less will be the weight which would suffice to turn the beam in a given degree from the horizontal position, and it is evident that the less this weight is, the more sensitive will be the balance.

In very sensitive balances the beam does not immediately come to rest when weights, whether equal or unequal, are put in the dishes, but oscillates for a long time slowly on the one side and the other. If the zero of the graduated arc on which the index

N 2

plays be exactly in the middle of the arc of oscillation, the weights in this case will be equal; but if the middle of the arc of oscillation be on either side of the zero, the preponderating weight will be on the same side.

However inaccurately a balance may be constructed, the exact weight of a body may be ascertained by it. For this purpose let the body to be weighed be placed in one dish, and let it be equi-poised by sand placed in the other. Leaving the sand undis-turbed, let the body be removed, and replaced by weights of known value which will equipoise the sand. The amount of these weights will then be the weight of the body.

416. A steelyard is a lever with unequal arms; the power being represented by a sliding weight, is adjusted so that its lever-age may be changed at pleasure. These and similar instruments are used for the purpose of weighing in commerce.

This instrument, sometimes called the Roman balance, is con-structed in various forms, one of the most common of which is represented in *fig.* 127. The weight q slides upon a graduated

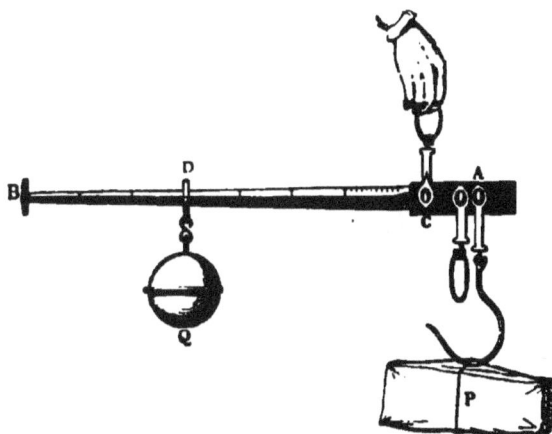

Fig. 127.

arm, and in some ascertainable position balances the article p which is to be weighed. The weight of p is just as many times that of q as d c is greater than a c.

It may happen that even when q is moved to the last division of the arm b c, the article p will still preponderate. In that case the steelyard is held by the ring nearer to a, which hangs down in the figure. By this means the distance of p from the fulcrum being less, the weight q, at a given distance from the fulcrum, will balance a proportionally greater weight.

417. **Letter balance.**—One of the most simple forms of these useful instruments, which gives exact indications, is shown in

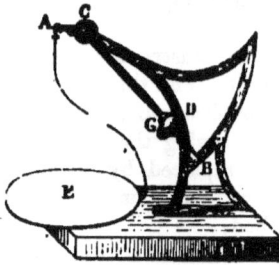

Fig 128.

fig. 128. It is a bent lever of the first kind. The dish E is suspended from a short arm A C, and is balanced by the arm C B, having a heavy knot at G, and a point at B, which moves upon a graduated arc. When A is depressed, C B rises, and its centre of gravity G moves to a greater distance from the vertical line through the point of support C, and takes a position in which it balances the weight placed in the dish E. The indications expressed on the arc show the weights in E, which correspond to the various positions of the index.

418. **Spring balances.** — Besides the instruments constructed upon the principle of the lever, various forms of weighing instruments are made, which depend on the elastic tension of springs.

Fig. 129.

Fig. 130.

One of these is shown in *fig.* 129. and another in 130., which will be easily understood upon inspection, without further explanation.

419. A crowbar is a lever of the first kind. In this instrument, when used, for example, to raise a block of stone, the fulcrum, *fig.* 131., is another stone F, placed near that which is to be raised, and the power of the hand H is placed at the other end of the bar. A poker applied to raise fuel is a lever of the first kind, the fulcrum being the bar of the grate.

The force exerted by the hand at B (*fig.* 132.) will in this case overcome a resistance greater than itself in the proportion of B C to

Fig. 131.

A C.. The instrument may be rendered still more efficacious by removing the point of support o to the line r q, or on the other hand a greater range of motion with less power may be obtained by removing it to the line m n.

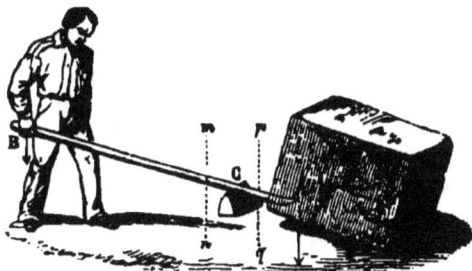

Fig. 132.

Scissors, shears, nippers, pincers, and other similar instruments are composed of two levers of the first kind, the fulcrum being the joint or pivot, and the weight the resistance of the substance to be cut or seized, the power being the fingers applied at the other end of the levers.

The brake of a pump is a lever of the first kind, the pump-rods and piston being the weight to be raised.

420. Examples of levers of the second kind, though not so frequent, are not uncommon.

An oar is a lever of the second kind. The reaction of the water against the blade is the fulcrum. The boat is the weight, and the hand of the boatman the power.

The rudder of a ship or boat is an example of this kind of lever, and explained in a similar. way.

The chipping-knife, (*fig.* 133.), is a lever of the second kind.

Fig. 133.

The end F attached to the bench is the fulcrum, and the weight the resistance of the substance R to be cut.

A door moved upon its hinges is another example.

Nutcrackers are two levers of the second

kind, the hinge which unites them being the fulcrum, the resistance of the shell placed between them being the weight, and the hand applied to the extremity being the power.

A wheelbarrow is a lever of the second kind, the fulcrum being the point at which the wheel presses on the ground, and the weight being that of the barrow and its load collected at their centre of gravity M (*fig.* 134.).

The same observation may be applied to all two-wheeled carriages which are partly sustained by the animal which draws them.

Fig. 134.

421. Levers of the third kind, acting, as has been explained, to mechanical disadvantage, the power being less than the weight, are of less frequent use. They are adopted only where rapidity and dispatch are required more than power.

The most striking examples of levers of the third kind are found in the animal economy. The limbs of animals are generally levers of this description. The socket of the bone is the fulcrum, a strong muscle attached to the bone near the socket is the power, and the weight of the limb, together with whatever resistance is opposed to its motion, is the weight. A slight contraction of the muscle in this case gives a considerable motion to the limb: this effect is particularly conspicuous in the motion of the arms and legs in the human body; a very inconsiderable contraction of the muscles at the shoulders and hips gives the sweep to the limbs, from which the body derives so much activity.

The treadle of the turning-lathe is a lever of the third kind. The hinge which attaches it to the floor is the fulcrum; the foot applied to it near the hinge is the power; and the crank upon the axis of the fly-wheel, with which its extremity is connected, is the weight.

- Tongs are levers of this kind, as also the shears used in shearing sheep. In these cases the power is the hand, placed immediately below the fulcrum or point where the two levers are connected.

422. The pressure on the fulcrum of a lever, when the power and weight are in equilibrium, is determined by the principle of the composition of forces. In a lever of the first kind, the resultant of the power and weight is a single force passing through

N 4

the fulcrum equal to their sum; consequently, the pressure on such point will be equal to the sum of the power and weight.

In a lever of the second and third kind, the power and weight, acting in contrary directions, will have a resultant equal to their difference passing through the fulcrum. This resultant will therefore express the pressure on the fulcrum.

423. In the rectangular lever, the arms are perpendicular to each other, and the fulcrum, **r** (*fig.* 135.), is at the right angle. The moment of the power in this case is **p** multiplied by **a f**, and that of the weight **w** multiplied by **b f**. When the instrument is in equilibrium, these moments must be equal.

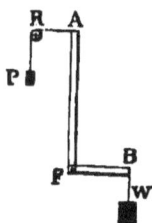

Fig. 135.

When the hammer is used for drawing a nail, it is a lever of this kind; the claw of the hammer is the shorter arm, the resistance of the nail is the weight, and the hand applied to the handle is the power.

424. When a beam rests on two props, **a b** (*fig.* 136.), and supports at some intermediate place, c, a weight w, this weight is distributed between the props in a manner which may be determined by the principles already explained.

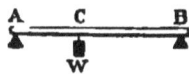

Fig. 136.

If the pressure on the prop **b** be considered as a power sustaining the weight w by means of the lever of the second kind **b a**, then this power multiplied by **b a** must be equal to the weight multiplied by c **a**. Hence the pressure on **b** will be the same fraction of the weight as the part **a** c is of **a b**. In the same manner it may be proved that the pressure on **a** is the same fraction of the weight as **b** c is of **b a**. Thus, if **a** c be one third, and therefore **b** c two thirds of **b a**, the pressure on **b** will be one third of the weight, and the pressure on **a** two thirds of the weight.

It follows from this reasoning, that if the weight be in the middle, equally distant from **b** and **a**, each prop will sustain half the weight. The effect of the weight of the beam itself may be determined by considering it to be collected at its centre of gravity. If this point, therefore, be equally distant from the props, the weight of the beam will be equally distributed between them.

According to these principles, the manner in which a load borne on poles is distributed between the bearers may be ascertained. As the efforts of the bearers and the direction of the weight are always parallel, the position of the poles relatively to the horizon makes no difference in the distribution of the weights between

them.   Whether they ascend or descend, or move on a level
plane, the weight will be similarly shared between them.

If the beam extend beyond the prop, as in *fig.* 137., and the
weight be suspended at a point not placed
between them, the props must be applied at
different sides of the beam.   The pressure
which they sustain may be calculated in the
same manner as in the former case.

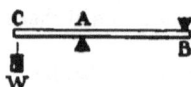

Fig. 137.

The pressure of the prop B may be con-
sidered as a power sustaining the weight w by means of the lever
B C.   Hence, the pressure of B multiplied by B A must be equal to
the weight w multiplied by A C.   Therefore, the pressure on B
bears the same proportion to the weight as A C does to A B.   In
the same manner, considering B as a fulcrum, and the pressure of
the prop A as the power, it may be proved that the pressure of A
bears the same proportion to the weight as the line B C does to A B.
It therefore appears that the pressure on the prop A is greater thar
the weight.

425. **Compound lever.** — A combination consisting of several
levers acting one upon
another, as represented
in *fig.* 138., is called a
compound lever,

Fig. 138.

The manner in which
the effect of the power
is transmitted to the weight may be investigated by considering
the effect of each lever successively.   The power at P produces an
upward force at P′, which bears to P the same proportion as P F to
P′ F.   Therefore, the effect at P′ is as many times the power as the
line P F is of P′ F.   Thus, if P F be ten times P′ F, the upward force
at P′ is ten times the power.   The arm P′ F′ of the second lever is
pressed upwards by a force equal to ten times the power at P.   In
the same manner this may be shown to produce an effect at P″ as
many times greater than that at P′ as P′ F′ is greater than P″ F′.

Thus, if P′ F′ be twelve times P″ F′, the effect at P″ will be
twelve times that at P′.   But this last was ten times the power,
and therefore the effect at P″ will be one hundred and twenty
times the power.   In the same manner it may be shown that the
weight w is as many times greater than the effect at P″, as P″ F″ is
greater than w F″.   If P″ F″ be five times w F″, the weight will
be five times the effect at P″.   But this effect is one hundred and
twenty times the power, and therefore the weight would be six
hundred times the power.

In the same manner, the effect of any compound system of levers
may be ascertained by taking the proportion of the weight to the

power in each lever separately, and multiplying these numbers together.

In the example given, these proportions are 10, 12, and 5, which, multiplied together, give 600. In *fig.* 138. the levers composing the system are of the first kind; but the principles of the calculation will not be altered, if they be of the second or third kind, or some of one kind and some of another.

426. **Weighing-machines.** — These are usually compound levers, and are constructed in very various forms, but all depend-

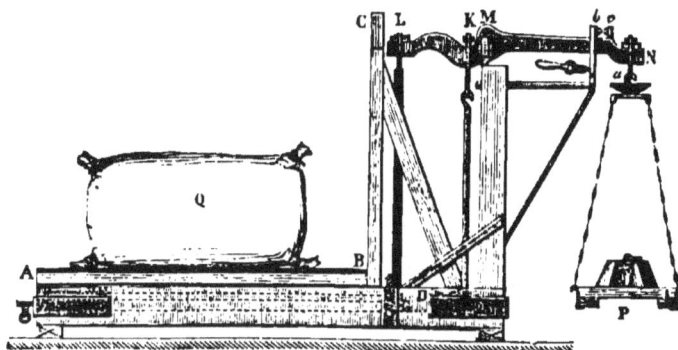

Fig. 139.

ing nearly on the same mechanical principles; one of these varieties is shown in *figs.* 139. and 140.

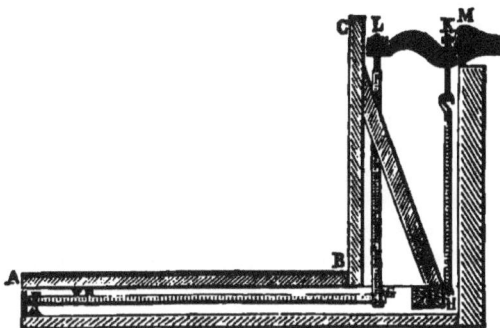

Fig. 140.

The platform A B, upon which the object to be weighed is placed, is supported by the short arm M L, of a lever L N, through the intervention of a compound lever, K H D C B E F G I, the longer arm of which, M N, supports a dish in which the counterpoising weight is placed.

The platform A B, is supported at two points, 1° by the lever E G L at E, and 2° by the arm L M at K, through the intervention of the rod K H, the diagonal piece D C, and the upright piece C B. The proportions of the levers are so regulated that the pressure at L is less than that at K, in the same proportion as K M is less than L M, and these two pressures together are balanced by the weight P.

Before commencing the operation the lever L M N is rendered horizontal by small weights or sand put into the dish a, and its horizontal position is ascertained by the coincidence of the two points b and c, the former of which is fixed, and the latter attached to the arm M N of the lever.

It is easy to understand that the proportion of the arms of the lever may be so arranged that each hundredweight of the article Q, may be balanced by a pound or any lesser denomination of weight placed in P.

427. **The knee lever.** — A form of compound lever, known as the knee lever, is much used in the arts. This combination con-

sists of a metal rod A B, *fig.* 141., having a fixed point of support A, on which it works. Another bar G C is jointed to it at C, a point intermediate between A and B. This bar C G is jointed at G to a plate, such as R, or any other object to which it is desired to trans-mit an intense force acting through a very limited space, as, for example, in the case of the printing press, where the paper is pressed upon the type by a plate which is driven upon it by a sudden and severe force. The handle B

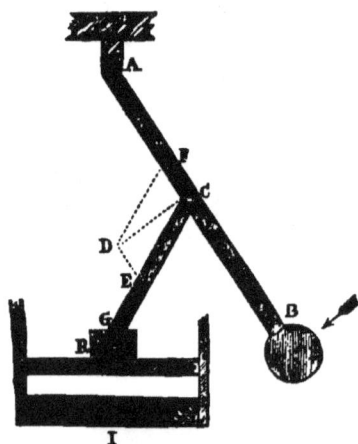

Fig. 141.

of the lever being pressed in the direction of the arrow, exerts a corresponding pressure on the point C, which is driven in the direction C D, perpendicular to A B. This motion C D is resolved into two by the parallelogram of forces, one in the direction C E, and the other in the direction C F; the latter exerts pressure on the fixed point A, and the other acts upon the plate R, by means of the joint G forcing it downwards. As the joint C advances, the angle A C G becomes more and more obtuse, and the component C E of the force acting at B bears a rapidly increasing proportion to the force itself, so that when the levers A C and C G come nearly into a right line, the pressure exerted at B is augmented at G in an almost infinite proportion.

428. **Key of Erard's pianoforte.** — In this instrument, the object is to convey from the point where the finger acts upon the key, to that at which the hammer acts upon the string, all the

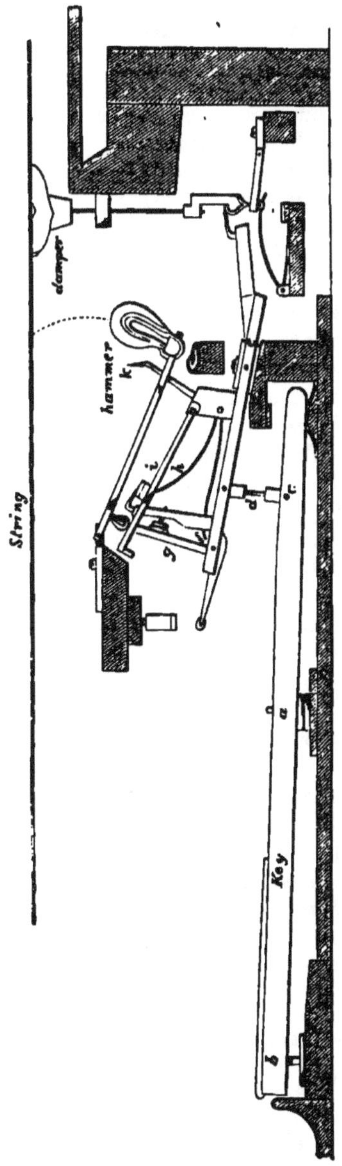

Fig. 142.

delicacy of action of the fin
ger, so that the piano may
participate, to a certain extent,
in that sensibility of touch
which is observable in the harp,
and which is the consequence
of the finger acting immediately
on the string in that instrument, without the intervention
of any other mechanism.

The combination of levers,
by which the action of the
finger is transmitted to the
string, in Erard's pianoforte, is
represented in *fig.* 142.

The key is represented at *b a c,*
the centre or pivot on which it
plays being *a,* and the ivory table
upon which the finger acts being
at *b.* The point to which its motion is communicated is at *c.*

The motion is transmitted by a
double-jointed piece *d* to an intermediate lever *e f,* the pivot of
which is at *e.* At the joint *f* is a
rod *g* called the *sticker,* which carries up the hammer to the string.
The hammer is supported by the
head of the sticker *g,* and at the
same time rests upon the oblique
lever *i,* which latter is acted upon
by the spring *h.*

When the key is pressed down
by the finger at *b,* the piece *d* is
raised, and by it the lever *e f* and
the sticker *g.* This lever raises the
lever *i,* and acts upon the rod of
the hammer at a point near the
joint, making the hammer rise
along the dotted curve so as to
strike the string.

The proportions of this combination of levers are such,
that, after the blow of the hammer on the string, the check *k*
comes forward and receives the
hammer in its fall, at about
one third of its original dis-

tance from the string; so that while the finger continues to keep down the key, the hammer remains at a distance from the string, equal to one third of its distance when the key is not depressed.

In the meantime, the spring h has given way under the weight of the hammer, and, under these circumstances, the key b being allowed to rise by the finger through one third of its play, and then again being depressed, another stroke of the hammer on the string will be produced; for in this case the hammer will be brought back to the level of the head of the sticker g, by which means it will be driven upwards upon the depression of the key.

In the combination of levers used in other pianofortes, the note cannot be repeated without allowing the key to rise to the position it has before it is depressed; consequently in this case, a repetition of the note is produced with one third of the motion of the finger which is necessary in other pianofortes.

429. **Power of a machine.**—That number which expresses the proportion of the weight to the equilibrating power in any machine, we shall call the *power of the machine*. Thus if, in a lever, a power of 1 lb. support a weight of 10 lbs., the power of the machine is 10. If a power of 2 lbs. support a weight of 11 lbs., the power of the machine is $5\frac{1}{2}$.

430. **Equivalent lever.**—As the distances of the power and weight from the fulcrum of a lever may be varied at pleasure, and any assigned proportion given to them, a lever may always be conceived having a power equal to that of any given machine. Such a lever may be called, in relation to that machine, the *equivalent lever*.

431. As every complex machine consists of a number of simple machines acting one upon another, and as each simple machine may be represented by an equivalent lever, the complex machine will be represented by a compound system of equivalent levers. From what has been proved, it therefore follows that the power of a complex machine may be calculated by multiplying together the power of the several simple machines of which it is composed.

WHEEL-WORK.

432. **The wheel and axle.**— The form of simple machine denominated the wheel and axle, consists of a cylinder which rests in pivots at its extremities, or is supported in gudgeons, and is capable of revolving between those pivots, or in those gudgeons. Attached to this cylinder, and supported on the same pivots or gudgeons, a wheel is fixed, so that the two revolve together with a common motion. The wheel is often supplied with a rope

or chain, which winds round the axle, and the power with another rope or chain which winds round the wheel.

Such an arrangement is represented in *fig.* 143., where w is

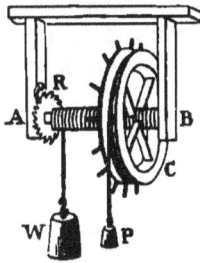

Fig. 143.

the weight, A and B the pivots or gudgeons, c the wheel, and P the power.

433. The condition of equilibrium is, according to what has been already proved, the inverse proportion of the power and weight to the diameters of the wheel and axle, that is to say, the power is to the weight as the diameter of the axle is to the diameter of the wheel. The weight is generally applied, as represented in the figure, by means of a rope coiled upon the axle.

434. The manner of applying the power is very various. Sometimes the circumference of the wheel is furnished with projecting points, as represented in *fig.* 143., to which the hand is applied when human force is the power. Examples of this are numerous · a familiar one is presented in the steering-wheel of a ship.

435. In the common *windlass* the power is applied by means of a winch D C, as represented in *fig.* 144. The arm B C of the winch represents the radius of the wheel, and the power is applied to D C at right angles to B C. In some cases no wheel is attached to the axle, but it is pierced with holes, directed towards its centre, in which long levers are incessantly inserted, and a continuous action produced by several men working at the same time, so that whilst some are transferring the levers from hole to hole, others are working the other levers, *fig.* 145.

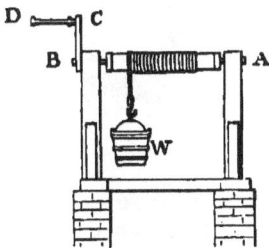

Fig. 144.

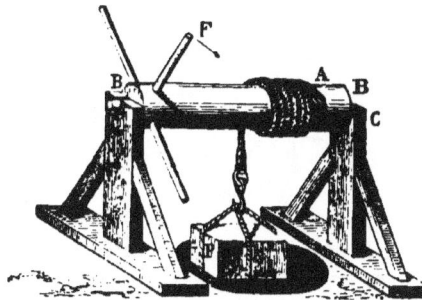

Fig. 145.

436. The axle is sometimes placed in a vertical position, the wheel or levers being moved horizontally. The *capstan* is an example of this. A vertical axis is fixed in the deck of the ship, the circumference being pierced with holes presented towards its centre.

These holes receive long levers, as represented in *fig.* 146. The men who work the capstan walk continually round the axle, pressing forward the levers near their extremities.

More generally the head of the capstan is circular, and pierced with many holes, in each of which a lever can be inserted, so that many men can work together when great power is required, as in weighing anchor, *fig.* 147.

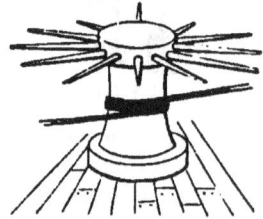

Fig. 147.

Fig. 146.

437. **The tread-mill, &c.** — In some cases the wheel is turned by the weight of animals placed at its circumference, who move forward as fast as the wheel descends, so as to maintain their position continually at the extremity of the horizontal diameter. The tread-mill, *fig.* 148., and certain cranes, such as *fig.* 149., are examples of this.

Fig. 148.

Fig. 149.

438. **French quarry wheels.** — In the neighbourhood of Paris, and other parts of France, stone for building is obtained in subterranean quarries, from which it is raised through deep vertical shafts, by means of large wheels, worked by the weight of men, on the same principle as that of the tread-mill. One of these quarry wheels, and the manner of working it, is shown in *fig.* 150.

From what has been already explained, it is evident that the effect of the power upon the weight would be augmented by diminishing the thickness of the axle, and diminished by increasing that thickness.

Fig. 150.

439. It sometimes happens that an invariable power has to act against a variable resistance, or a variable power against a constant resistance. In such a case, the effect of the wheel and axle as just described would vary: an augmentation of the power or diminution of the resistance would throw the power and weight out of equilibrium.

If, however, the axle were made to increase in thickness in the same proportion as the ratio of the power to the weight is augmented, then the change of such ratio would be compensated by a corresponding change in the leverage, and an equilibrium would

be maintained between the power and weight, notwithstanding
their variation. Numerous instances of this are presented in the
arts, some of which will be noticed hereafter.

440. When a weight or resistance of comparatively great
amount is to be raised by a very small power by means of the
simple wheel and axle, either of two inconveniences would ensue;
either the diameter of the axle would become too small to support
the weight, or the diameter of the wheel would become so great
as to be unwieldy in its operation. This has been remedied, without having recourse to a complex machine, by a simple expedient
represented in *fig.* 151. The axle of the windlass here consists of
two parts, one thicker than the other, and the rope by which the

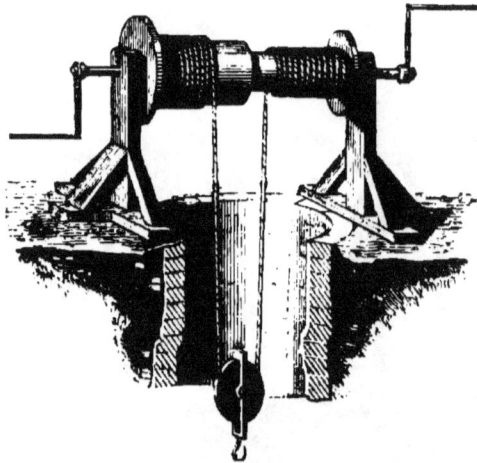

Fig. 151.

weight is raised rolls on the thicker while it rolls off the thinner.
In each revolution, therefore, the part which is rolled on exceeds
that which is rolled off by the difference between the circumference of the two parts of the axle. The effect, accordingly, is the
same as if an axle had been used, whose diameter is equal to the
difference between the diameters of the thicker and thinner part.
Since, then, without diminishing the thickness of the axle, we may
diminish without limit the difference between the thicker and
thinner parts, the ratio of the weight to the power may be augmented indefinitely without diminishing the strength of the axle.

441. When great power is required, wheels and axles may be
combined in a manner analogous to the compound lever already
explained. The power being supposed to act on the circumference

o

of the first wheel, its effect is transmitted to the circumference of the first axle ; this circumference acts on the circumference of the second wheel, and transmits motion thereby to the circumference of the second axle, which, in its turn, acts on the circumference of the second wheel, transmitting motion to the circumference of the third axle, and so on.

There is nothing different in the mechanical effect of such a combination from that of a system of compound levers, except that it admits more conveniently of a continuous action, and produces continued and regular motion. The relation between the power and the weight of resistance, when in equilibrium, is determined in exactly the same manner as in the case of the compound lever.

If the diameters of all the wheels be multiplied together, and the diameters of all the axles be also multiplied together, then the power will be to the weight as the product of the diameters of all the axles to the product of the diameters of all the wheels. Thus, if the diameters of all the axles be expressed by the numbers 2, 3, 4, and the diameters of all the wheels be expressed by the numbers 20, 25, and 30, then the ratio of the power to the weight will be as $2 \times 3 \times 4 = 24$ to $20 \times 25 \times 30 = 15000$.

442. The manner in which the wheels and axles act one upon another is very various. Sometimes a strap or cord is placed in a groove in the circumference of the axle, and carried round a similar groove in the circumference of the wheel. This, which is called an endless band, is represented in *figs.* 152. and 153.

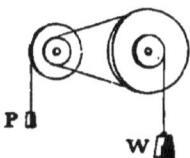 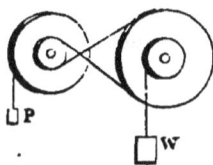

Fig. 152.   Fig. 153.

443. In the case represented in *fig.* 152. the wheels are driven in the same direction ; in that represented in *fig.* 153., they are driven in opposite directions. Examples of this method of transmitting the motion from wheel to wheel are presented in every department of the arts and manufactures. In the turning-lathe and the grinding-wheel a cat-gut cord carried round the treadle wheel imparts motion to the maundrell or the grindstone. In the great factories shafts, as shown in *fig.* 154., are carried along the ceilings of the room, round which, at certain points, endless straps are carried which are conducted round the wheels, thus giving motion to the lathes or other machines. One of the chief advan-

tages of this method of transmitting motion by wheels and axles is, that the bands by which the motion is conveyed may be placed at

Fig. 154.

any distance from each other, and even in any position with respect to each other, and may, by a slight adjustment, receive motion in either one direction or the other.

444. When the circumference of the axle acts immediately on that of the wheel, which it moves without the intervention of a strap or cord, means must be adopted to prevent them from moving in contact, (*fig.* 155.) without transmitting motion, which they would do if both surfaces were perfectly smooth and free from friction.

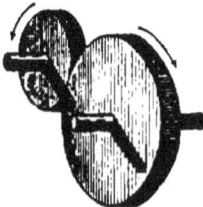

Fig. 155.

This is accomplished by different expedients. In cases where great power is not required, motion is communicated through a series of wheels and axles by rendering their surfaces rough, either by facing them with rough leather, or making them of wood cut across the grain. This method is used in spinning machinery, where a large buffed wheel, placed in a horizontal position, is surrounded by a series of small buffed rollers pressed close against it, each roller communicating motion to a spindle. As the wheel revolves, revolution is imparted to the rollers, the velocity of which exceeds that of the wheel in the same proportion as the diameter of the wheel exceeds that of the roller. This method is very convenient in cases where the motion of the rollers requires to be occasionally suspended, each roller being provided with a means by which it can be thrown out of contact with the wheel, and thus stopped.

445. The most frequent method of transmitting motion through a train of wheel-work is by the construction of teeth upon their circumference, so that the teeth of each falling between those of the other, the one wheel necessarily pushes forward the other. When teeth are used, the axles are usually called pinions, and the teeth raised upon them are called leaves.

446. In the formation of the teeth of wheels and pinions, expedients are adopted to prevent them from rubbing one upon another, when they move in contact with each other. A particular form is adopted for the teeth, in virtue of which the surfaces are applied one to the other with a rolling motion like that of a carriage wheel upon the road. By this expedient the rapid wear of the teeth, which would be produced by constant friction accompanied by pressure, is prevented.

447. In computing the mechanical effects of toothed wheels and pinions, the number of teeth may be substituted for their circumferences and diameters.

The condition of equilibrium will therefore be obtained by multiplying together the number of teeth in all the wheels, and the number of teeth in all the pinions, the power being to the weight, when in equilibrium, as the latter product to the former.

448. **Spur wheels.**—Teeth are formed upon the edges of wheels and pinions in different positions. When they are formed in the plane of the wheel so as to diverge from the axis as a centre, the wheels are called *spur wheels;* and a train of such wheels and pinions, such as *fig.* 156., is called spur gearing. By such gearing the several axes round which motion is produced are parallel one to another, as appears by the figure.

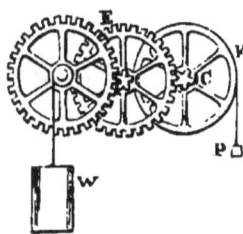

Fig. 156.

It is evident in this case that the wheel and pinion, which are upon the same axis, need not be in immediate juxtaposition, but may be separated one from the other by any length of the shaft. In this way, if the shaft be carried along the factory, a pinion at one end of a room may give motion to or receive it from a wheel upon the same shaft at the other end.

449. Crown wheels. — When the teeth are formed in the surface of a hoop or cylinder, so as to be directed parallel to the axis, as shown in *fig.* 157., the wheel is called, from its form, a *crown wheel.* If the teeth of such a wheel be engaged in those of a spur wheel, as in the figure, the axis of the latter, will be at right angles to that of the former.

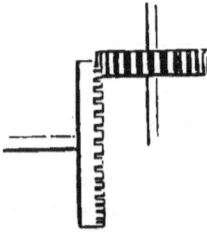

Fig. 157.

Sometimes the crown wheel works in a sort of cylindrical basket on a shaft at right angles to its axis, as shown in *fig.* 158.

450. Bevelled wheels are such as have their teeth inclined to the shaft, as shown in *fig.* 159. By this expedient motion may be imparted from one shaft to another inclined to it at any angle.

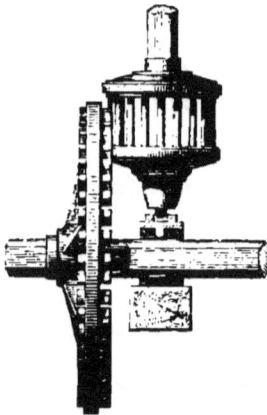

Fig. 158.

Fig. 159.

451. Rack and pinion. — When it is desired to impart a rectilinear motion of limited range by means of the rotation of an axle, the object is attained by the combination of a spur wheel or pinion with a straight bar having teeth of corresponding form and magnitude formed upon it. Such a toothed bar is called a *rack.* The mode of action of such a combination is shown in *fig.* 160.

Fig. 160.

The efficacy of the power is generally increased by the combination of two or more wheels and pinions, as shown in *fig.* 161., where the handle drives the pinion D which works in the wheel B,

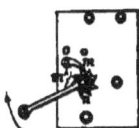

Fig. 162.

Fig. 161.

on the axle of which is the pinion which drives the rack A.

452. **Ratchet wheel.** — When it is desired to allow an axle or shaft to revolve in one direction, but not in the other, the object is accomplished by fixing upon it a wheel having teeth, n (*fig.* 162.), inclined in the direction contrary to that in which it is intended the shaft should be free to move. A catch, o m, falls between these teeth, and stops all motion of the wheel directed against it; but when the wheel turns in the other direction, the catches fall from tooth to tooth, producing a clicking noise.

PULLEYS.

453. Although the term pulley implies the combination of a rope and a wheel on which it runs, the practical effects of the simple machine so denominated depend altogether on the rope; the wheel being introduced for the mere purpose of diminishing the effects of friction and imperfect flexibility, the consideration of both of which are omitted in theory. If a rope were perfectly flexible, and were capable of being bent over a sharp edge, and of moving upon it without friction, we should be enabled by its means to make a force in any one direction overcome a resistance or communicate a motion in any other direction.

Thus, if a perfectly flexible rope, F s (*fig.* 163.), pass over a

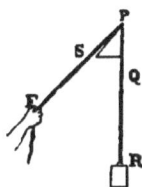

Fig. 163.

sharp edge P, and be connected with a weight R vertically. a force acting obliquely in the direction P F will raise the weight vertically in the direction R Q; but as no materials of which ropes can be made can render them perfectly flexible, and as in proportion to the strength by which they are enabled to transmit force their rigidity increases, it is necessary in practice to adopt means to remedy or mitigate those effects which attend the absence of perfect flexibility, and which would otherwise render cords practically inapplicable as machines.

But, beside the want of perfect flexibility, the surface of the rope is always rough, and often considerably so. This surface, in passing over an edge, would produce a degree of friction which would altogether stop its movement. If a rope were used in the manner represented in the figure, to transmit a force in one direction to a resistance in another, some force would be necessary to bend it over the angle P, which the two directions form one with the other; and, if the angle were sharp, the effect of such a force might be the rupture of the rope.

454. But if, instead of bending the rope at one point over a single acute angle, the change of direction were produced by successively deflecting it over several angles, each of which would be less sharp, the force necessary for the deflection and the liability of breaking the cord would be diminished. But such object will be still more effectually attained if the cord be deflected over the surface of a curve.

If the rope were applied merely to sustain a weight without moving it, a curved surface would therefore be sufficient to remove the inconvenience arising from imperfect flexibility; but when motion is required, the rope, in passing over such a surface, would be subject to great friction and rapid wear. This inconvenience is removed by causing the surface on which the rope runs to move with it, so that no more friction is produced than would arise from the curved surface itself rolling upon the rope. These objects are attained by the common pulley, which consists of a wheel called a sheave, fixed in a block turning on a pivot. A groove is formed in the edge of the wheel, in which the rope runs, the wheel revolving with it.

This apparatus is represented in *fig.* 164. Notwithstanding, however, that this expedient removes the effects of friction and rigidity to so great a degree as to render the use of the cord practically available, it must not be supposed that these effects are altogether overcome; they still produce some impediment to the full efficiency of the power, as will be explained more fully hereafter. For the present, however, we shall consider the rope as rendered by this expedient flexible and free from friction.

Fig. 164.

455. By means of a single sheave, a power acting in any one direction may be made to transmit its effects to a resistance in any other direction, provided the two lines of direction be in the same plane, and not parallel. Thus, let A B (*fig.* 165.) be the direction in which the power acts, and let C D be the direction in which the weight or resistance acts.

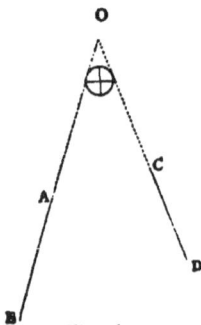

Fig. 165.

To transmit in this case the power to the weight, let the two directions B A and D C be prolonged until they meet, which they will do at o. In the angle o, formed by the two directions, let a sheave be placed, and let the power be connected with a rope in the direction B A. This rope, being carried from A over the sheave at o, must be brought down in the direction C D, and connected with the resistance.

*456. But if the direction of the power and resistance be parallel, as in *fig:* 166., then the effect of the power might be transmitted to the weight by means of a cord and two fixed pulleys, one placed over the direction of the force, and the other over the direction of the weight, as represented in the figure.

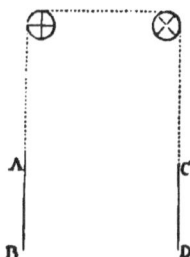

In fine, if the direction of the power and the weight be not placed in the same plane, then the effect of the power may still be transmitted to the weight by means of two pulleys. Let us suppose, for example, that a power acting in a given horizontal line is required to be transmitted to a weight acting at some distance from it, in a certain vertical line, which is not in the same plane with the direction of the power.

Fig. 166.

Let two fixed pulleys be placed at two points in any convenient positions on the lines of direction of the power and weight, and let a line be supposed to join these points. Let the axis of one of the pulleys be placed at right angles to the plane formed by the line of direction of the power and the line joining the two pulleys, and let the other be placed with the axis at right angles to the plane passing through the line of direction of the weight and the line joining the two pulleys. By this arrangement, the cord being passed successively over both pulleys, the effect of the power will be transmitted to the weight.

457. In all these cases, the same cord by which the weight is suspended being directly connected with the power, and its tension throughout its entire length being the same, the weight and the power must be equal when they are in equilibrium. This condition is also rendered manifest by the fact, that from the motion of the mechanism connecting it, the weight and power will move with the same velocity. It appears, therefore, that no mechanical advantage is gained by a single rope acting over one or more fixed pulleys; nevertheless, there is scarcely any engine, simple or complex, which is attended with more convenience.

458. In the applications of power, whether of man or animals, or arising from other natural forces, there are always some direc-

tions in which it may be exerted to greater convenience and advantage than others, and in many cases the power is capable of acting only in one particular direction. Any expedient, therefore, which can give the most advantageous direction to the moving power, whatever be the direction of the resistance opposed to it, contributes as much practical convenience as one which enables a small power to balance or overcome a great weight. •

459. By means of the fixed pulley a man may raise himself to a considerable height, or descend to any proposed depth. If he be placed in a chair or a basket attached to one end of a rope which is carried over a fixed pulley, by laying hold of this rope on the other side, he may, at will, descend to a depth equal to half of the entire length of the rope, by continually yielding rope on the one side, and depressing the basket or chair by his weight on the other. Fire escapes have been constructed on this principle, the fixed pulley being attached to some part of the building.

460. A single movable pulley is represented in *fig.* 167.; a cord is carried from a fixed point F, and passing through a block B attached to a weight w, passes over a fixed pulley c, the power being applied at P. We shall first suppose the parts of the cord on each side the wheel B to be parallel: in this case the whole weight w being sustained by the parts of the cords B c and B F, and these parts being equally stretched, each must sustain half the weight, which is therefore the tension of the cord. This tension is resisted by the power at P, which must therefore be equal to half the weight. In this machine, therefore, the weight is twice the power.

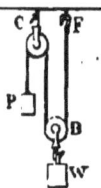

Fig. 167.

461. If the parts of the cord B c and B F be not parallel, as in *fig.* 168., a greater power than half the weight is therefore necessary to sustain it. To determine the power necessary to support a given weight in this case, take the line B A in the vertical direction, consisting of as many inches as the weight consists of ounces; from A draw A D parallel to B c, and A E parallel to B F: the force of the weight represented by A B will be equivalent to two forces represented by B D and B E. The number of inches in these lines respectively will represent the number of ounces which are equivalent to the tensions of the parts B F and B c of the cord. But as these tensions are equal, B D and B E must be equal, and each will express the amount of the power P, which stretches the cord at P c.

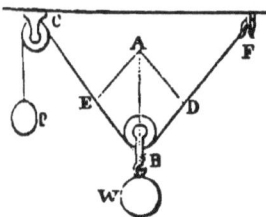

Fig. 168.

It is evident that the four lines A E, E B, B D, and D A are equal, and as each of them represents the power, the weight which is represented by A B must be less than twice the power which is represented by A E and E B taken together. It fol-lows, therefore, that as the parts of the rope which support the weight depart from parallelism, the ma-chine becomes less and less efficacious, and there are certain obliquities at which the equilibrating power would be much greater than the weight.

462. If several sheaves be constructed in the same movable block, the mechanical advantage may be proportionally augmented. In *fig.* 169. a system is represented in which three sheaves are inserted in the movable block bearing the weight, the same number being inserted in the fixed block: The cord is carried from the power first over the fixed sheave A, then over the movable sheave B, then over the fixed sheave C, the movable sheave D, the fixed sheave E, and the movable sheave F, and finally attached to the block at G.

Fig. 169.

Now, since the cord throughout its whole length is stretched with the same force, and since it is evident that at the part where the power is applied, this force of tension must be equal to the power, it follows that the six parts of the cord which support the weight will each be stretched by a force equal to the power, and that consequently the weight, when in equilibrium, must be equal to six times the power.

This condition may also be inferred from the fact, that if the power move the cord, its velocity will be six times that of the weight; for if six feet of the rope be drawn over the fixed pulley, these six feet must be equally distributed between the six parts of the rope which sustain the weight, and consequently each part must be raised through one foot, which is therefore the height through which the weight would be raised for every six feet through which the power passes.

463. In general it may therefore be inferred, that, in a pulley which consists of a single movable block containing one or more sheaves, the weight, when in equilibrium, will be just as many times the power as is represented by the number of cords, or the number of parts of the cord, which sustain the weight.

464. In the form of movable block represented in *fig.* 170., the cord, after passing successively over the sheaves, is finally attached to the lower block. This, by increasing the parts of the cords supporting the lower block by one, augments the efficiency of the instrument without increasing the number of sheaves.

465. Two of the most powerful forms of pulley, consisting of a single movable block, are represented in *figs.* 171. and 172. The combination represented in *fig.* 171. is called Smeaton's pulley,

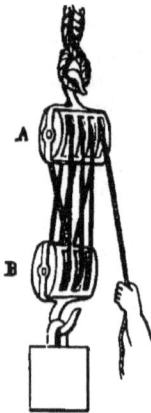 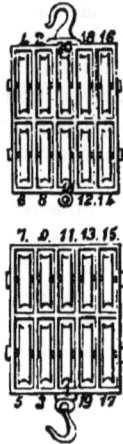 

Fig. 170.   Fig. 171.   Fig. 172.

having been invented by that celebrated engineer. The fixed and movable block contain each ten sheaves, and the order in which the rope is carried over them is represented in the figure by the numbers 1, 2, 3, 4, &c. The total number of parts of the cord supporting the lower block is in this case twenty, and consequently the power is to the weight as 20 to 1.

The form of pulley represented in *fig.* 172. is called White's pulley.

466. If instead of one movable block and a single rope, two or more movable blocks with independent ropes be used, the power of the pulley may be augmented on the same principle as in the case of compound levers or compound wheel-work.

Different combinations of this kind are represented in *figs.* 173, 174, 175, 176.

The figures which are annexed to the ropes in each case represent the power they respectively exert.

The first rope in *fig.* 173. is stretched by the force of the power only. The first movable block being supported by two parts of this first rope will exert a force equal to double the power on the second rope; and the second movable block being supported by two parts of this rope will exert a force upon the third rope equal to four times the power; and in the same way it follows that the third block will exert a force in supporting the weight equal to eight times the power. In such a system, the addition of each movable block doubles the mechanical effect.

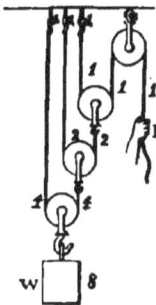
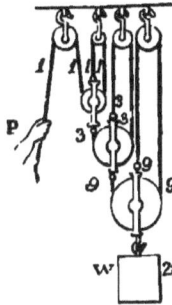
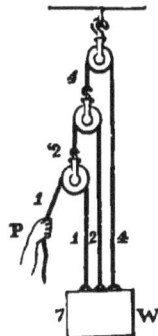
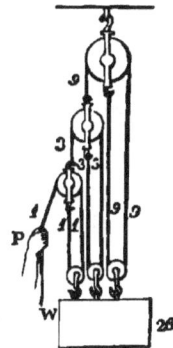

Fig. 173.　　　　Fig. 174.　　Fig. 175.　　　Fig. 176.

But without augmenting the number of movable blocks, but only adding fixed blocks, the effects may be augmented in a three-fold instead of a two-fold proportion, as represented in *fig.* 174., where each successivemovable block is supported by three parts of the same cord. In *fig.* 175. the ends of the cords, instead of being attached to fixed points, are attached to the weight. In this case, the weight is supported by each of the several cords, these cords being stretched by different forces. The first is stretched with a force equal to the power, the second with a force equal to double the power, and the third with a force equal to four times the power, and so on. In such a system, the mechanical effect for the same number of blocks is greater than in that represented in *fig.* 173., where three blocks only support a weight four times the power; whereas in the system represented in *fig.* 175., three blocks support seven times the power. The effect of this system may be still further increased by attaching blocks to the weight, as represented in *fig.* 176., and carrying the ropes to the pulleys above. The effect produced is indicated by the numbers fixed to the cords.

467. From its portable form, cheapness of construction, and the facility with which it may be applied in almost every situation, the pulley is one of the most useful of the simple machines. The mechanical advantage, however, which it appears in theory to possess, is considerably diminished in practice, owing to the stiffness of the cordage and the friction of the wheels and blocks. By these means it is computed that in most cases so great a proportion as two thirds of the power is lost. The pulley is much used in building when weights are to be elevated to great heights; but its most

extensive application is found in the rigging of ships, where almost every motion is accomplished by its means.

In all these examples of pulleys, we have supposed the parts of the rope sustaining the weight, and each of the movable pulleys, to be parallel to each other. If they be subject to considerable obliquity, the relative tensions of the different ropes must be estimated according to the principle applied in 461.

INCLINED PLANE. — WEDGE AND SCREW.

468. **Effect of an inclined surface.** — A hard surface pressed against a weight or resistance in a direction at right angles to it, would support it, and the whole amount of such weight or resistance, would in this case press upon the surface. But if, instead of being at right angles to it, it were placed in an oblique direction, then the weight or resistance would be resolved by the parallelogram of forces into two, one of which would act perpendicularly to the plane and produce pressure upon it, and the other would be parallel to the plane, and be free to produce motion.

Let A B, *fig.* 177, be such a surface, and let w be a body producing some resistance, or having a tendency to move in the direc-

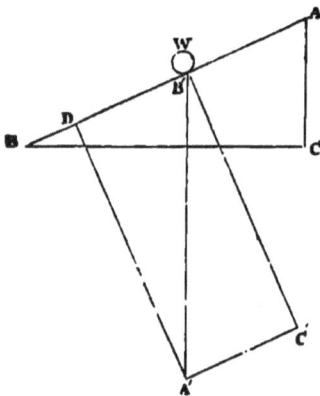

Fig. 177.

tion B′ A′ oblique to B A. Let the whole force with which w would move if not supported by the plane be expressed by B′ A′. Then, taking B′ A′ as the diagonal of a parallelogram one of whose sides is B′ C′ at right angles to B A, and the other B D in the direction of the plane, the force B′ A′ will be equivalent to two forces, one represented by B′ C′, and the other by B′ D. The former, being perpendicular to the plane, will be resisted by its reaction, and the other only will take effect. To support the weight, therefore, in this case, would require a force acting parallel to the plane, and opposite to the force represented by B′ D. If we take on the plane the length B A equal to B′ A′, and draw A C parallel to B′ A′, and B C perpendicular to it, then the triangle A B C will be in all respects equal and similar to A′ B′ C′; in fact, it may be considered as the same triangle, but in a different position. Since, therefore, A′ B′, B′ C′, and A′ C′ represent respectively the whole force of the body w, its pressure on the plane, and its tendency to move in the direction of the plane, these three forces will be exactly represented in their effects by the lines A B, B C, and A C.

469. We have here taken the general case, and supposed the body w to exercise a force in any direction whatever; but if we apply the principle to the case of a heavy body resting upon a plane inclined to the vertical direction, then the machine becomes what is commonly called the inclined plane. A B is called the *length* of the plane, A C its *height*, and B C its *base*.

From what has been just proved, then, it follows, that if a weight be placed upon an inclined plane, the weight consisting of as many pounds as there are inches in the length of the plane, the pressure on the plane will consist of as many pounds as there are inches in the base, and the tendency to move down the plane would be balanced by as many pounds as there are inches in the height.

470. The apparatus represented in *fig.* 178. is intended to represent this experimentally. The weight placed upon the plane is a roller so formed as to move freely upon it. A string is attached to it, which being carried parallel to the plane is conducted over a fixed pulley, and supports a dish bearing a weight. On comparing this weight, including the weight of the dish, with the weight of the roller upon the plane, and by varying the angle of elevation of the plane, we find that in every case the weight necessary to produce equilibrium will be expressed by the height of the plane, the entire weight of the roller being expressed by its length.

Fig. 178.

It is evident from what has been just explained that the less the elevation of the plane is, the less will be the power requisite to sustain a given weight upon it, and the greater will be the pressure upon it; for the less the elevation of the plane is, the less will be its height and the greater will be its base.

471. **Inclined roads.** — Roads which are not level may be considered as inclined planes, and loads drawn upon them in carriages, regarded in reference to the powers which impel them, are subject to all the conditions which have been established for inclined planes.

The inclination of the road is estimated by the height corresponding to some proposed length: thus we say, a road rises one foot in twenty-five, or one foot in thirty; meaning that if twenty-five or thirty feet of the road be taken as the length of an inclined plane, the corresponding height of such plane would be one foot; and if twenty-five or thirty feet be measured upon the road, the difference of the level of the two extremities will be one foot. According to this method of estimating the inclination of roads, the power required to sustain a load upon them, friction apart, is always proportional to this rate of elevation. If a road rise one foot in twenty, then a power of one ton will be sufficient to sustain twenty tons; and so on.

472. **Inclined planes on railways.**—When a power is employed in moving a load upon a road thus inclined, the action of the power may also be regarded in another point of view. Let us suppose a railway train weighing 200 tons moving up an inclined plane, which rises at the rate of one in two hundred; what mechanical effect does the moving power produce in moving up 200 feet of such a plane ? First, it acts against the friction, the atmospheric and other resistances to which it would be exposed if the plane had been level. Secondly, it is employed in raising the entire weight of the train through the elevation which corresponds to 200 feet in length, that is, through one perpendicular foot.

The mechanical effect, therefore, is precisely the same as if the load of 200 tons had been first moved along a level plane 200 feet long, and then elevated up a step one foot high; but instead of being called upon to make this great exertion of raising 200 tons directly through one perpendicular foot, the moving power is enabled gradually to accomplish the same object by a longer continuance of a more feeble exertion of force, such exertion being spread over 200 feet instead of being condensed into a single foot.

473. In all that precedes, we have assumed that the power acts parallel to the plane; in some cases, however, it acts obliquely to it.

Let w p, *fig.* 179., be the direction of the power. Taking that

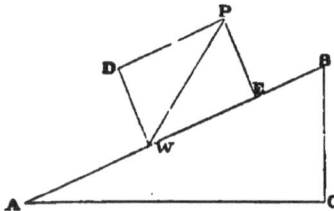

Fig. 179.

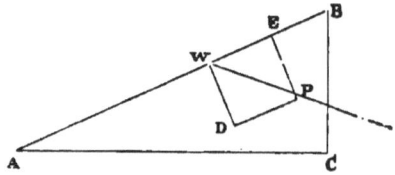

Fig. 180.

as the diagonal of a parallelogram, it will be equivalent to w D and w E, the former perpendicular, and the latter parallel to the plane. w D will have the effect of diminishing the pressure on the plane, and w E will be efficient in drawing the weight up the plane. In

some cases the direction of the power is below the plane, as in
fig. 180. In this case, as before, the power w p is resolved into
two forces, w ʙ parallel to the plane, and w ᴅ perpendicular to it.
The latter augments the pressure of the weight on the plane, and
the former is efficient in drawing it up the plane.

474. The method of drawing up or letting down barrels from or
into a cellar, represented in *fig.* 181., and that of drawing them

Fig. 181.

upon a dray, shown in *fig.* 182., are cases of the inclined plane
which will be easily understood.

Fig. 182.

475. **Double inclined plane.** — It sometimes happens that a
weight upon one inclined plane is raised or supported by another
weight upon another inclined plane. Thus, if ᴀ ʙ and ᴀ ʙ′,
fig. 183., be two inclined planes, forming an angle at ᴀ, and w w

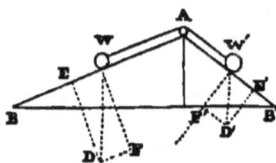

Fig. 183.

be two weights placed upon these planes, and connected by a cord passing over a pulley at A, the one weight will either sustain the other, or one will descend, drawing the other up. To determine the circumstances under which these effects will ensue, draw the lines W D and W′ D′ in the vertical direction, and take upon them as many inches as there are ounces in the weights respectively. W D and W′ D′ being the lengths thus taken, and therefore representing the weights, the lines W B and W′ B′ will represent the effects of these weights respectively down the planes. If W B and W′ B′ be equal, the weights will sustain each other without motion; but if W B be greater than W′ B′, the weight W will descend, drawing the weight W′ up; and if W′ B′ be greater than W B, the weight W′ will descend, drawing the weight W up. In every case W F and W′ F′ will represent the pressures upon the planes respectively.

476. **Self-acting planes.**—It is not necessary, for the effect just described, that the inclined planes should, as represented in the figure, form an angle with each other. They may be parallel, or in any other position, the rope being carried over a sufficient number of wheels, placed so as to give it the necessary deflection. This method of moving loads is frequently applied in great public works where railroads are used. Loaded waggons descend one inclined plane, while other waggons, either empty or loaded, so as to permit the descent of those with which they are connected, are drawn up the other.

477. **The wedge.**—When the weight is not moved upon the plane, but is stationary, the plane being itself moved under the weight, the machine is called a wedge.

Let D E, *fig.* 184, be a heavy beam, secured in a vertical posi-

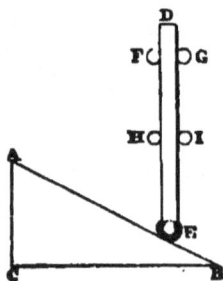

Fig. 184.

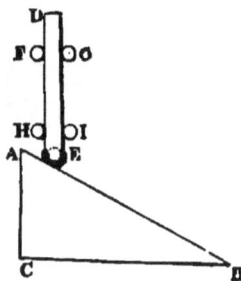

Fig. 185.

tion between guides, F G and H I, so that it is free to move upwards

P

and downwards, but not laterally.   Let A B C be an inclined plane, the extremity of which is placed beneath the end of the beam.   A force applied to the back of this plane A C, in the direction C B, will urge the plane under the beam so as to raise the beam to the position represented in *fig.* 185.   Thus, while the inclined plane is moved through the distance C B, the beam is raised through the height C A.

It follows, therefore, that in the case of the wedge the velocity of the resistance is to the velocity of the weight as the base of the inclined plane, which forms the wedge, is to its height.

478. Wedges, however, are more generally formed of two inclined planes, connected base to base, as represented in *fig.* 186.   In this case, the back of the wedge is the sum of the heights of the two inclined planes, and the length of the wedge is their common base.   The force, therefore, which drives the wedge is to the resistance with which it equilibrates, as half the back of the wedge is to its length.

Fig. 186.

479. This theory of the wedge is not applicable in practice with any degree of accuracy.   This is owing chiefly to the enormous disproportion which friction in these machines bears to the power ; but independently of this there is another difficulty in the theory of this machine.

480. The power commonly used in the case of a wedge is not pressure, but percussion.   The force of a blow is of a nature so different from continued force, such as the pressure of weights, that it admits of no numerical comparison with the resistance offered by cohesion, to overcome which it is generally applied. We cannot properly state the proportion which a blow bears to a weight.   The wedge is almost invariably urged by percussion, while the resistances which it has to overcome are as constantly forces of another kind.   Although, however, no exact numerical computation can be made, yet it may be stated in general that the wedge is more and more powerful as the angle is more acute.

481. The cases in which wedges are most generally used in the arts and manufactures, are those in which an intense force is required to be exerted through a very small space.   This instrument is therefore used for splitting masses of timber or stone, for raising vessels in docks, when they are about to be launched, by being driven under their keels, in presses where the juice of seeds, fruits, or other substances are required to be extracted, as, for example, in the oil mill, in which the seeds from which the oil is

extracted are introduced into hair bags, which being placed between planes of hard wood are pressed by wedges. The pressure exerted by the wedges is so intense that the dry seeds are converted into solid masses as hard and compact as the most dense woods. Wedges have been used occasionally to restore to the perpendicular, edifices which have inclined owing to the sinking of their foundations.

482. **Practical examples.**— All cutting and piercing instruments, such as knives, razors, shears, scissors, chisels, nails, pins, needles, &c., are wedges. The angle of the wedge in all these cases is more or less acute, according to the purpose to which it is applied. Chisels intended to cut wood have their edge at an angle of about 30°; for cutting iron from 50° to 60°, and for brass about 80° to 90°. In general, tools which are urged by pressure admit of being sharper than those which are driven by percussion. The softer or more yielding the substance to be divided is, the more acute the wedge may be constructed.

483. **Utility of friction.**— In many cases the efficiency of the wedge depends on that which is entirely omitted in its theory, viz., the friction which arises between its surface and the substance which it divides. This is the case when pins, bolts, or nails are used for binding the parts of structures together, in which case, were it not for the friction, they would recoil from their places and fail to produce the desired effect. Even when the wedge is used as a mechanical engine the presence of friction is absolutely indispensable to its practical utility.

The power, as has already been stated, generally acts by successive blows, and is therefore subject to constant intermission, and but for the friction the wedge would recoil between the intervals of the blows with as much force as it had been driven forward. Thus the object of the labour would be continually frustrated. The friction in this case is of the same use as a ratchet-wheel, but is much more necessary, as the power applied to the wedge is much more liable to intermission than in the cases where ratchet-wheels are generally used.

484. **The screw.**—In ascending a steep hill it has been the practice of road engineers, instead of making an inclined plane directly from the base to the summit, to carry the road round the hill, gradually rising as it proceeds. If we desire to ascend with ease to the top of a high column, we could do so if a path or ledge were formed on the outer surface, gradually winding round and round the column from the bottom to the top. Such a path would be, in fact, an inclined plane carried round the column. But it will be evident that such an arrangement would constitute *a screw*.

This will be rendered still more apparent by the following contrivance.

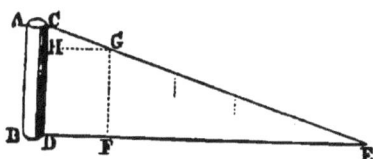

Fig. 187.   Fig. 188.

Let A B, *fig.* 187., be a cylindrical roller, and let C D E be an inclined plane cut in paper, the height of which C D is equal to the length of the roller. Let the edge C D be pasted on the roller, and then let the roller be turned so that the paper shall be wrapped round it. When it makes one revolution of the roller, the portion of the edge C G will have made one spiral coil; the next revolution will make an equal spiral coil, and so on until all the paper has been rolled upon the roller, when the edge of the paper so coiled will show a regular spiral line round the roller, as represented in *fig.* 188. Taking C H G, *fig.* 187., as the inclined plane thus rolled round the roller, it is evident that C H is its height, and H G its base. But C H is the distance between two successive coils of the spiral, and H G is the circumference of the roller. The coils of the spiral are called the *threads* of the screw, and the distance C H between the successive coils is called the *distance between the threads.*

485. In the application of the screw, the weight of resistance is not, as in the inclined plane and wedge, placed upon the surface of the plane or thread. The power is usually transmitted by causing the screw to move in a concave cylinder, on the interior surface of which a spiral cavity is cut, corresponding exactly to the thread of the screw, and in which the thread will move by turning round the screw continually in the same direction. This hollow cylinder is usually called the nut or concave screw. The screw surrounded by its spiral thread is represented in *fig.* 189.

Fig. 189.

486. There are several ways in which the power is transmitted to the resistance by means of a screw; but by whatever means it may be so transmitted, it is evident that the screw will move the resistance in a single revolution through a space equal to the distance between two contiguous threads. The comparative velocities, therefore, of the power and weight will always be found in this class of simple machines by comparing the space described by the power, in imparting one revo-

lution to the screw with the distance between two contiguous threads.

487. The most common manner of urging the screw is by a lever attached to its head, as represented in *fig.* 190. at B F. Sup-

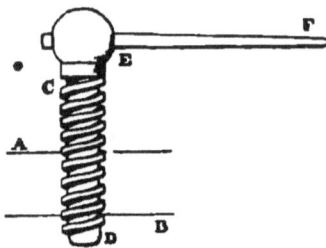

posing the power to be applied at F, it will in producing one revolution of the screw, and therefore in moving the resistance through a space equal to the distance between two contiguous threads, make one complete revolution in a circle whose radius is the length of the lever on which it acts. The velocity, therefore, of the power will be to the velocity of the weight as the circumference of the circle described by the power is to the distance between two contiguous threads; and consequently, the condition of equilibrium between the power and weight will be this, that the power is to the weight as the distance between the contiguous threads is to the circumference described by the power.

Fig. 190.

488. The great mechanical force exerted by the screw will hence be easily understood. There is no limit to the smallness of the distance between the threads, except the strength which is necessary to be given to them, and there is no limit to the magnitude of the circumference to be described by the power, except the necessary facility of moving in it. We can conceive the power acting by a lever, and therefore moving through a great circumference while the screw moves through a comparatively minute space; and consequently, in such case, the power will be so much less than the resistance, as the distance between the threads is less than the circumference described by the power.

489. The manner of acting upon the resistance by means of the screw is very various. Sometimes the nut is fixed and the screw movable; sometimes the screw is fixed and the nut movable; sometimes the nut, though incapable of revolving, can be moved progressively; and sometimes the screw is incapable of revolving, but is moved progressively. These conditions admit of various combinations which are severally adopted in practice. In *fig.* 190., the nut A B being supposed to be fixed, if the lever F be turned, the end D of the screw will descend or ascend, according to the direction in which B F is turned, and will act upon the resistance accordingly. If the screw be fixed, the nut may be moved upon it, either by turning the nut or the screw. In either case the nut will ascend or descend, according to the direction of the motion.

P 3

In each revolution it will move through a space equal to the distance between two contiguous threads.

If we suppose the nut A B, *fig.* 190., to be incapable of ascending or descending, but to be capable of revolving, then, by turning it round, the screw which plays in it will ascend or descend through a space equal to the distance between two contiguous threads for every revolution made by the nut. •

On the other hand, the apparatus may be so arranged that the screw, though capable of revolving, is incapable of a progressive motion, and the nut, though capable of a progressive motion, is incapable of revolving. In this case, when the screw is made to revolve, the nut in which it plays will be moved upwards or downwards, through a space equal to the distance between two threads, by the revolution of the screw.

The screw is generally used in cases where severe pressure is to be exercised through small·spaces; it is, therefore, the agent in most presses.

In *fig.* 191. the nut is fixed, and by turning the lever which passes through the head of the screw a pressure is exercised upon any substance placed upon the plate immediately under the end of the screw. In *fig.* 192. the screw is incapable of revolving, but

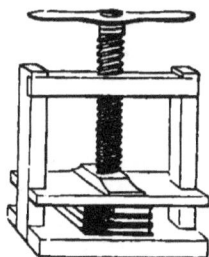

Fig. 191.

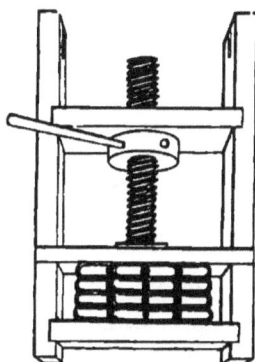

Fig. 192.

is capable of advancing in the direction of its length. On the other hand, the nut is capable of revolving, but does not advance in the direction of the screw. When the nut is turned by means of the lever inserted in it, the screw advances in the direction of its length, and urges the board which is attached to it downwards, so as to press any substance placed between it and the fixed board below.

490. **Examples.**—In cases where liquids or juices are to be

expressed from solid bodies, the screw is the agent generally employed. It is also used in coining, where the impression of a die is to be made upon a piece of metal, and in the same way in producing the impression of a seal upon wax or other substance adapted to receive it. When soft and light materials, such as cotton, are to be reduced to a convenient bulk for transportation, the screw is used to compress them, and they are thus reduced into hard dense masses. In printing, the paper is sometimes urged by a severe and sudden pressure upon the types by means of a screw.

491. A screw may be cut upon a cylinder by placing the cylinder in a turning-lathe, and giving it a rotatory motion upon its axis. The cutting point is then presented to the cylinder and moved in the direction of its length at such a rate as to be carried through the distance between the intended threads while the cylinder revolves once. The relative motions of the cutting point and the cylinder being preserved with perfect uniformity, the thread will be cut from one end to the other. The shape of the threads may be either square, as in *fig.* 189., or triangular, as in *fig.* 191.

492. If the lever by which the power acts on the screw were capable of indefinite increase, or the thread of indefinite fineness, there would be no limit to the mechanical effect of the instrument; but to both of them there are practical limits. The lever cannot be increased so as to render the operation of the power unwieldy and impracticable, and the thread cannot be diminished beyond that limit which will give it sufficient strength; and the cases in which the greatest mechanical efficacy is needed, are precisely those in which the thread requires to be strongest.

493. **Hunter's screw.** — To obtain an indefinite augmentation of the power of the screw, without diminishing the strength of the thread, Mr. Hunter proposed an arrangement which is known by his name, as the Hunterian screw. This (represented in *fig.* 193.) consists in the use of two screws, the threads of which may have any strength and magnitude, but which have a very small difference of breadth.

While the working point is urged forward by that which has the greater thread, it is drawn back by that which has the less; so that during each revolution, the screw, instead of being advanced through a space equal to the magnitude of either of the

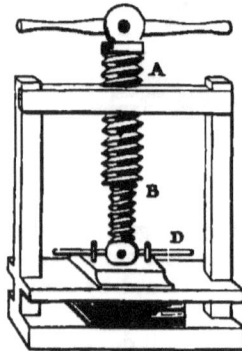

Fig. 193.

P 4

threads, moves through a space equal to their difference. The mechanical power of such a machine will be the same as that of a single screw having a thread whose magnitude is equal to the difference of the magnitudes of the two threads just mentioned.

Thus, without inconveniently increasing the sweep of the power on the one hand, or on the other diminishing the thread until the necessary strength is lost, the machine will acquire an efficacy limited by nothing but the smallness of the difference between the two threads.

494. **Micrometer screw.** — The slow motion which may be imparted to the end of a fine screw by a considerable motion of the power, renders it an instrument peculiarly well adapted to the measurement of very minute motions and spaces, the magnitude of which could scarcely be ascertained by any other means. To explain the manner in which it is applied: suppose a screw to be so cut as to have fifty threads in an inch, each revolution of the screw will advance its point through the fiftieth part of an inch. Now, suppose the head of the screw to be a circle whose diameter is an inch, the circumference of the head will be something more than three inches: this may be easily divided into a hundred equal parts, distinctly visible. If a fixed index be presented to this graduated circumference, the hundredth part of a revolution of the screw may be observed by noting the passage of one division of the head under the index. Since one entire revolution of the head moves the point through the fiftieth of an inch, one division will correspond to the five-thousandth of an inch. In order to observe the motion of the point of the screw in this case, a fine wire is attached to it, which is carried across the field of view of a powerful microscope, by which the motion is so magnified as to be distinctly perceptible.

495. **Endless screw.** — When the thread of a screw acts in the teeth of a wheel, the screw being fixed so that it can revolve without advancing, it is called an endless screw. It has the effect of giving motion to the wheel in the same manner as does a pinion, one tooth of the wheel being driven forward by each revolution of the screw. Such an arrangement is applied, for example, to tighten the strings of double bases, *fig.* 194.

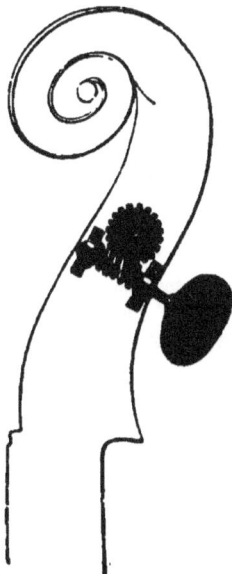

Fig. 194.

CHAP. III.

THE PENDULUM, THE BALANCE WHEEL, AND THE FLY WHEEL.

496. Of the class of mechanical contrivances whose purpose is to render motion uniform, the principal are the pendulum, the balance wheel, and the fly.

497. **Simple pendulum.**—The pendulum consists of a heavy mass attached to a rod, the upper extremity of which rests upon a point of support in such a manner as to have as little friction as possible. Such an instrument will remain at rest when its centre of gravity is in the vertical line immediately under the point of suspension or support. But if the centre of gravity be drawn from this position on either side, and then disengaged, the instrument will swing horizontally from the one side to the other of the position in which it would remain at rest, the centre of gravity describing alternately a circular arc on the one side or the other of its position of rest. If there were neither friction nor atmospheric resistance, this motion of vibration or oscillation on either side of the position of equilibrium would continue for ever; but, in consequence of the combined effects of these resistances, the distances to which the pendulum swings on the one side and on the other are continually diminished, until, after the lapse of an interval more or less protracted, it comes to rest.

Suppose a small ball of lead suspended by a fine silken string, the length of which is incomparably greater than the diameter of the leaden ball. Such an arrangement is called the *simple pendulum.*

Let s, *fig.* 195., be the point of suspension; let s b be the fine silken thread by which the ball b is suspended, and the weight of which, in the present case, is neglected. Let b be the position of the ball when in the vertical under the point of suspension s. In that position the ball would remain at rest; but if we suppose the ball drawn aside to the position A, it will, if disengaged, fall down the arc A b, of which the centre is s, and the radius the length of the string. Arriving at b, it will have acquired a certain velocity, which, in virtue of its inertia, it will have a tendency to retain, and with this velocity it will commence to move through the arc b A'. Supposing neither

Fig. 195.

the resistance of the atmosphere nor friction to act, the ball will rise through an arc B A' equal to B A; but it will lose the velocity which it had acquired at B, for it is evident that it will take the same space and the same time to destroy the velocity which has been acquired, as to produce it. Thus, the velocity at B, being acquired in falling through the arc A B, will be destroyed in rising through the equal arc B A'.

Having arrived at A', the ball, being brought to rest, will again fall from A' to B, and at B will have again acquired the same velocity which it had obtained in falling from A to B, but in the contrary direction; and in the same manner it may be explained that this velocity will carry it from B to A. Having arrived at A, the ball, being again brought to rest, will fall once more from A to B, and so the motion will be continued alternately between A and A'.

The motion of the pendulum from A to A', or from A' to A, is called an *oscillation*, and its motion between either of those points and B is called a semi-oscillation, the motion from B to A or from B to A' being called the ascending semi-oscillation, and the motion from A or A' to B, the descending semi-oscillation.

The time which elapses during the motion of the ball between A and A' is called the *time of one oscillation*.

It is evident, from what has been stated, that the time of moving from either of the extremities A A' of the arc of oscillation to the point B, is half the time of an oscillation. If, instead of falling from the point A, the ball had fallen from the point c, intermediate between A and B, it would have then oscillated between c and c'; two points equally distant from B, and the arc of oscillation would have been c c', more limited than A A'. But in commencing its motion from c, the declivity of the arc down which it falls towards B would be evidently less than the declivity at A; consequently, the force which would accelerate it, commencing its motion at c, would be less than that which would accelerate it, commencing its motion at A. The ball, therefore, commencing its motion at A, would be more rapidly accelerated than when it commences its motion at c.

The result of this is, that, although the arc A B may be twice as long as the arc c B, the *time* which the ball takes to fall from A to B will not be sensibly different from the time it takes to fall from c to B, provided that the arc of oscillation A B A' is not considerable.

It was at first supposed, as we have just stated, that, whether the oscillations were longer or shorter, the times would be absolutely the same. Accurately speaking, however, this is not the case; but if the total extent of the oscillation A A' do not exceed

5° or 6°, the time of oscillation in it may be considered, practically, the same as in the lesser arcs.

498. This important principle may be easily experimentally verified. Let two small leaden balls be suspended from the same point of support, but one being in advance of the other, so that in oscillating the two balls shall not strike each other. This being done, let one of the balls be drawn from its point of rest through an angle less than 3°, and let it be disengaged. It will oscillate as described above. Let the other ball be now drawn from its point of rest through a much less angle, and let it be so disengaged that it shall commence its oscillation at the same moment with the commencement of one of the oscillations of the other ball.

Let it, in short, be so managed, that when the one ball is at A, the other shall be at c; and that both shall commence their descending motion towards B at the same moment. It will be then found that their oscillations will be synchronous for a considerable length of time, that is to say, the balls will arrive at A' and c', respectively, at the same instant; and returning, will simultaneously arrive at A and c respectively.

If, in this case, the oscillation of the ball A were made through an arc, even as great as 10°, that is to say, 5° on each side of the vertical, the oscillation of the ball c being made through an arc of 2°, it would be found that 10001 oscillations of the latter would be equal to 10000 oscillations of the former, so that the actual difference between their times of oscillation would not exceed the ten thousandth part of such time.

499. It might be expected that the time of oscillation of different pendulums would depend, more or less, upon the weight of the matter composing them, and that a heavy body would oscillate more rapidly than a lighter one. Both theory and experience, however, prove the result to be otherwise. The force of gravity which causes the pendulum to oscillate acts separately on all the particles composing its mass; and if the mass be doubled, the effect of this force upon it is also doubled; and, in short, in whatever proportion the mass of the pendulum be increased or diminished, the action of the force of gravity upon it will be increased or diminished in exactly the same proportion, and consequently the velocity imparted by gravity to the pendulous mass at each instant will be the same.

It is easy to verify this by experiment. Let different balls of small magnitude, of metal, ivory, and other materials, be suspended by light silken strings of the same length, and made to oscillate; their oscillations will be found to be equal.

500. If pendulums of different lengths have similar arcs of oscilla-

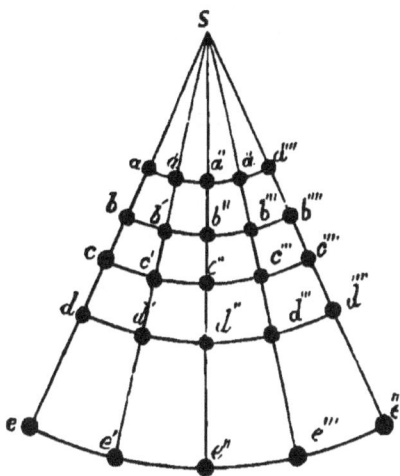

Fig. 196.

tion, the times of oscillation of those which are shorter will be less than the times of oscillation of those which are longer. Let $a, b, c, d,$ and $e,$ *fig.* 196., be five small leaden balls, suspended by light silken strings to the point of suspension s, and let all of those be supposed to form pendulums, having the same angle of oscillation. The arc of oscillation of the ball a will be $a\,a''''$, that of b will be $b\,b''''$, that of $c,$ $c\,c''''$, and so on. In commencing to fall from the points $a, b, c, d, e,$ towards the vertical line, these five balls are accelerated equally, inasmuch as the circular arcs down which they fall are all equally inclined at this point to the vertical line. The same will be true if we take them at any corresponding points, such as $a', b', c', e'.$ It may therefore be concluded that throughout the entire range of oscillation of each of these five pendulums, they will be impelled by equal accelerating forces.

Now it has been shown that when bodies are impelled by the same or equal accelerating forces, the spaces through which they move are proportional to the squares of the times of their motion; therefore it follows that the lengths of these arcs of oscillation are proportional to the squares of the times. But the lengths of these arcs are evidently in the same proportion as the lengths of the pendulums; that is to say, the arc $a\,a''''$ is to $b\,b''''$ as s a is to s b, and the arc $b\,b'''$ is to $c\,c''''$ as s b is to s c, and so on.

It follows, therefore, that the squares of the times of oscillation of pendulums are as their lengths, or, what is the same, the times of oscillation are as the square roots of their lengths. This principle is easily verified experimentally.

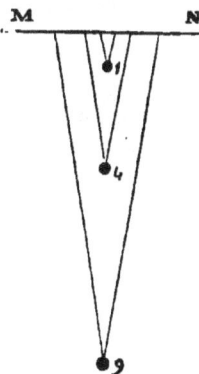

Fig. 197.

501. **Experimental illustration.** — Let three small leaden balls be suspended vertically under each other by means of loops of silken threads, as represented in *fig.* 197.,

and in such a manner that they can all oscillate in the same plane at right angles to the plane of the diagram, the suspending loops not interfering with each other.

Let the loops be so adjusted that the distance of the ball 1 below the line м n shall be 1 foot, the distance of the ball 4, 4 feet, and the distance of the ball 9, 9 feet.

Let the ball 9 be put in a state of oscillation through small arcs, and let the ball 4 be then drawn from its vertical position, and disengaged so as to commence one of its oscillations with an oscillation of the ball 9; and in the same manner let the ball 1 be started simultaneously with one of the oscillations of the ball 9.

It will be found that two oscillations of the one-foot pendulum are made in exactly the same time as a single oscillation of the four-foot pendulum; consequently, the time of each oscillation of the latter will be double that of the former, while its length is fourfold that of the former.

In the same manner, while the one-foot pendulum makes three oscillations, the nine-foot pendulum will make one, and, consequently, the time of oscillation of the latter will be three times that of the former, while its length is nine times that of the former.

502. By this principle, the length of a pendulum which would oscillate in any proposed time, or the time of oscillation of a pendulum of any proposed length, can be ascertained, provided we know the length of a pendulum which oscillates in any given time. Thus, suppose L to be the length of a pendulum which oscillates in the time т. Let it be required to determine the length of a pendulum L', which would oscillate in any other time т'. We shall have the following proportion:

$$L : L' :: T^2 : T'^2.$$

From this proportion, if L and т be both given, we can find the time т' of oscillation of the other pendulum if L' be given; or we can find the length L' if т' be given.

In the first case we have

$$T'^2 = T^2 \times \frac{L'}{L};$$

in the second we have

$$L' = L \times \frac{T'^2}{T^2}.$$

503. Since the force which produces the oscillation of a pendulum is the accelerating force of gravity urging the pendulous body alternately from the extremities of the arc of oscillation to

the middle point of that arc, it is evident that if this force were increased in its intensity, the velocity with which the pendulous body would be precipitated to its lowest position would be increased, and consequently the time of oscillation diminished; and if, on the other hand, the impelling force of gravity were diminished, the force urging the pendulous body being enfeebled, it would be moved with a diminished velocity, and consequently the time of oscillation would be increased. It follows, therefore, that the same pendulum will oscillate more slowly or more rapidly, according as the force of gravity which acts upon it is diminished or increased.

504. But it is not enough to state that a variation in the force of gravity will change the time of oscillation of the pendulum. It is required to ascertain in what proportion it will produce this change; that is to say, if the force of gravity acting on the pendulum be augmented in any given ratio, in what corresponding ratio will the time of oscillation of such pendulum be diminished. It is proved in the theory of accelerating forces that under such circumstances, the squares of the times of oscillation will vary in the inverse proportion of the force; that is to say, in whatever ratio the force of gravity be augmented, the squares of the times of oscillation of the pendulums will be diminished.

505. **Pendulum indicates variation of gravity in different latitudes.** — But as the squares of the times of oscillation are proportional to the lengths of the pendulums, it follows from this that the lengths of the pendulums which oscillate in the same time under the influence of different accelerating forces will be proportional to these forces; and that consequently, if in any two places it be found that the pendulums which oscillate in the same time have different lengths, it must be inferred that the forces of gravity in these two places are in the exact proportion of these lengths.

It is in virtue of this principle that the pendulum supplies means of determining the variation of the forces of gravity upon different parts of the earth's surface.

506. **Compound pendulum.** — We have hitherto supposed that the pendulous body is a heavy mass of indefinitely small magnitude, suspended by a wire or string having no weight. These are conditions which cannot be fulfilled in practice. Every real pendulous body has a definite magnitude, its component parts being at different distances from the point of suspension; the rod which sustains it has considerable weight, and all the points of this rod, as well as those of the pendulous mass itself, are at different distances from the point of suspension. In estimating, therefore, the effect of pendulums, it is necessary to take into account this circumstance.

Let us suppose a, b, c, d, e, f, g (*fig.* 198.) to be as many small heavy balls connected by independent strings, the weight of which may be neglected, with a point of suspension s, and let these seven balls be supposed to vibrate between the positions s M and s M'. Now, if these balls were totally independent of each other, and

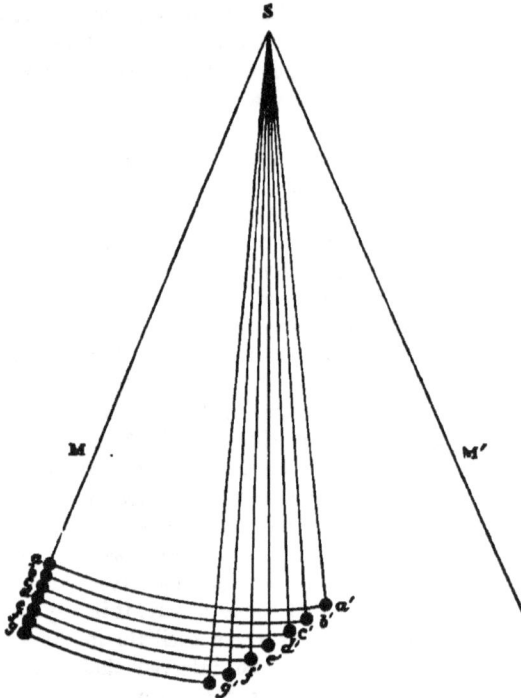

Fig. 198.

connected with the point of suspension by independent strings, they would all vibrate in different times, those which are nearer the point s vibrating more rapidly than those which are more distant from it. If, therefore, they be all disengaged at the same moment from the line s M, those which are nearest to s will get the start of those which are more distant, and at any intermediate position between the extremes of their vibration they will assume the positions a', b', c', d', e', f', g'. That which is nearest to the point s, and which is the shortest pendulum, will be foremost, since it has the most rapid vibration. The next in length, b, will follow it, and so on; the most remote from s, being the longest pendulum, will be the last in order.

Now if, instead of supposing these seven balls to be suspended by independent strings, we imagine them to be fixed upon the same wire, so as to be rendered incapable of having any independent motion, and compelled to keep in the same straight line, then it is evident that while the whole series vibrates with a common motion, those which are nearest to the point of suspension will have a tendency to accelerate the motion of those which are more distant, while those which are more distant will have a tendency to retard the motion of those which are nearer

These effects will produce a mutual compensation; b and c will vibrate slower than they would if they were moving freely, while e and f will evidently move more rapidly than if they were moving freely.   Among the series, there will be found a certain point which will separate those which are moving slower than their natural rate from those which are moving faster than their natural rate; and a ball placed at this point would vibrate exactly as it would do if no other balls were placed either above or below it. Such a ball would, as it were, be the centre which would divide those which are accelerated from those which are retarded.

507. **Centre of oscillation.** — Such a point has been denominated the *centre of oscillation*.   It is evident, that a pendulous mass, of magnitude more or less considerable, will vibrate in the same time as it would do if the entire mass were concentrated at its centre of oscillation, and formed there a material point of insensible magnitude.   By the length of a pendulum, no matter what be its form, is always to be understood the distance of its centre of oscillation from its point of suspension.

508. The centre of oscillation has the following remarkable and important quality, which is established by the higher mathematics, and verified by experiment : — If a pendulum be inverted and suspended by its centre of oscillation, its former point of suspension will become its new centre of oscillation, and the time of vibration will remain the same as before.   This property is usually expressed by stating that the " centres of suspension and oscillation are interchangeable."

This property can be verified by experiment.   If the centre of oscillation of any pendulous body, ascertained by mathematical calculation, be taken as a point of suspension, it will be found that the time of oscillation of the pendulum will be the same as it was with the first point of suspension.   Since the length of a pendulum, and, therefore, the time of its oscillation in a given place and subject to a given intensity of the force of gravity, depends upon the distance between its point of suspension and its centre of oscillation, it is evident that any variation which may take place in this distance will cause a corresponding change in the time of oscillation ; and if the pendulum be applied to a chronometer, it will cause a variation in the rate of that instrument, and a corresponding error in its indications.

509. Now, it is found in practice that all pendulums are subject to a change, more or less, in their form and magnitude, in consequence of the change of temperature of the atmosphere to which they are exposed. With this change they expand and contract, and with every expansion and contraction the distance between their centres of oscillation and suspension will be varied, unless expedients be adopted to counteract such an effect. Although the variations produced by these causes are not sufficiently great to render it necessary to provide a correction for them in common time-pieces, yet, in cases where extreme accuracy is required, expedients have been adopted to prevent the consequent error.

510. These expedients are called compensation pendulums. The principle upon which all these depend is the combination of two substances in the structure of the pendulum, which expand in unequal degrees for the same change of temperature; and they are so arranged, that while the expansion of the one increases the distance of the centre of oscillation from the point of suspension, the expansion of the other has the contrary effect, and the dimensions of these two substances are so adjusted that the increase of distance produced by the one shall be exactly equal to the diminution of distance produced by the other; so that the result is that the centre of oscillation remains at the same distance from the point of suspension, and therefore the time of oscillation of the pendulum remains unaltered.

511. The first, the most important, and the most universal use of the pendulum, is as a measure of time. The uniformity of the rate of its vibration is the property which renders it so eminently qualified for this purpose. A pendulum vibrating alone, independently of any mechanism, would measure the time which elapses during its motion. It would be only necessary for an observer to sit by it and count the number of its oscillations. If the time of one oscillation were previously known, then the number of oscillations performed in any interval would at once give the length of such interval. But, in order to supersede the attention and vigilance of such an observer, a train of wheel-work is placed in connection with the pendulum, the movement of which it regulates; and in connection with this train of wheel-work are fixed the dial-plate and the hands of the clock, by which the number of oscillations of the pendulum which take place in a day, or in any part of a day, are indicated and registered.

512. When the same pendulum is transported to different parts of the earth's surface, it is found that the rate of its vibration varies, and this variation is proved to take place even after pre-

cautions have been taken to keep the centres of oscillation and suspension at the same distance from each other. Now this change in the rate of vibration under such circumstances can only be explained by a change in the intensity of the force of gravity by which the pendulum is moved. It is found that when the pendulum is carried towards the terrestrial equator, the time of its vibrations is longer; and that when it is carried towards the pole, the time of its vibrations is shorter: the inference deduced from which is, that the force of gravity diminishes as we approach the equator, and increases as we approach the pole.

If the earth had the form of an exact sphere, did not revolve on its axis, and was of uniform density, the force of gravity of all parts of its surface would be the same, and no such variation in the rate of a pendulum could take place when transported from one point of the surface of the earth to another. But if the earth be an exact sphere, revolving upon its polar axis in 23 h. 56 min., then the effect of such motion of rotation would be to produce a certain small diminution of the intensity of the force of gravity in approaching the equator, and an increase in such intensity in approaching the pole. The amount of such diminution or increase produced by such rotation is capable of calculation, and, being computed and compared with such change of intensity of the force of gravity indicated by the variations of a pendulum, is found not to correspond exactly with it. This absence of complete correspondence indicates another cause affecting the force of gravity besides the rotation of the earth.

If the earth be not an exact sphere, but have a form of which a turnip and an orange are exaggerated representations, called in geometry an oblate spheroid, such a form, combined with the rotation of the earth, would produce a further effect in varying the force of gravity in proceeding towards the equator or towards the pole. Now it is found, by calculation, that a certain degree of this form, combined with the diurnal rotation of the earth, would produce exactly that variation in the force of gravity going towards the equator and going towards the pole which is indicated by the variation in the time of vibration of the same pendulum.

513. Hence it appears that the pendulum becomes an instrument by which not only the doctrine of the diurnal rotation of the earth is verified and corroborated, but by which the departure of the earth from an exact globular form is also established. From what has been stated, it will appear that the length of a pendulum which vibrates seconds in different parts of the earth will be different; the force of gravity in lower latitudes being less than in higher, the length of the pendulum which vibrates seconds will be proportionally less.

514. Pendulum measures the velocity of falling bodies. — As the pendulum thus supplies a measure of the intensity of the force of gravity, it necessarily also affords the means of calculating the height from which a body falling freely would descend in a second if it moved in vacuo. The method of determining this by the pendulum is susceptible of much greater accuracy than that which has been already indicated by Atwood's machine. To find the space through which a body will fall at any place in a second of time, let the length of a pendulum which vibrates seconds in that place, expressed in inches, be multiplied by 4·9348, and the product will express in inches the height through which a body would fall in a second in that place, independently of the resistance of the air.

515. Curious property of a swing. — It is well known to all persons who have amused themselves with exercising in this species of pendulum, that by humouring the motion by certain changes of attitude, the oscillation to and fro, which, if they remain stationary, would be gradually diminished and would ultimately cease, by reason of the resistance of the air and friction, can be not only sustained, but increased in its range. This is explicable by the well established principles which govern the motion of pendulous bodies.

Let A (*fig.* 199.) be a small weight suspended upon a fine thread

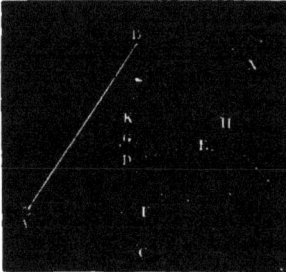

Fig. 199.

from the centre B, and let it be supposed to fall from the position A to the vertical line B C ; when it arrives at C, it will have exactly the same velocity in the horizontal direction as a body would have after having fallen vertically from F to C. Now let us suppose that at the moment of arriving at C, the body A is suddenly raised to some point such as D, nearer to B, retaining nevertheless the horizontal velocity it had acquired. With this velocity it will begin to move up the arc D N, and it will continue to rise in that arc, until it arrives at a point H whose height D K, above D, is equal to C F.

Now it is evident that, in this case, the angle H B D, through which it will swing from the vertical B D, will be much greater than the angle A B C. This will be still more evident if it be considered that if B E be drawn, making the angle E B D equal to A B C, the height D G will be less than C E, and therefore less than D K in the same proportion as D B is less than C B ; and since D G is less than D K, the angle D B E, and therefore the angle A B C, is less than the angle D B H.

Q 2

It follows, therefore, that if by any expedient the centre of gravity of a pendulous body be lowered in each descending semi-oscillation, and raised in each ascending one, the range of the oscillation will be constantly augmented; but this, as will be appa-

Fig. 200.

rent, is precisely what the swinger does when, by humouring his attitude with the changes of oscillation, he increases the range of

Fig. 201.

the swing. If he stand upon the swing board he lowers his body, as shown in *fig.* 200., in each descending motion of the swing.

and raises it, as in *fig.* 201., in each ascending motion; in the one case he lowers the centre of gravity, and in the other he raises it.

If he sit upon the swing board, he lies back with his body at right angles to the cord and his legs downwards in descending, and with his body upwards and his legs at right angle to the cord in ascending. By these changes of attitude, the centre of gravity, as before, is lower in descending than in ascending.

In both cases the same effect is produced as we have described above, in supposing the body A (*fig.* 199.) to be lower upon the string while it descends than while it ascends.

516. **The actual length of a pendulum which vibrates seconds** is found by accurate experiments to be 39 139 inches at London, 39·021 inches at the Line, and 39·215 at Spitzbergen, lat. 79° 49′ 58″ N.

517. **The balance wheel** applied to regulate watchwork is a light steel wheel A B C, (*fig.* 202.) nicely balanced upon its axis. Under this wheel a highly elastic spiral spring, made of very fine steel wire, is placed, the inner end of which is attached to the centre of the wheel, and the outer end *e* is attached to the fixed plate of the watch under the wheel. The wheel naturally assumes a certain position of equilibrium, from which if it be disturbed in either direction, by turning it so as to coil or uncoil the spring, it will, when disengaged, return to its position of equilibrium; but at the moment of arriving at that position it will have acquired such a velocity as will cause it to go on in virtue of the inertia of the wheel, and it will thus swing to the other side of the position of equilibrium. In this manner the wheel will swing alternately on the one side and the other of its position of equilibrium, with a motion altogether analogous to the motion of a pendulum; and this analogy is rendered still more close by the fact that the spiral spring has a property which renders the oscillations of the wheel, whatever be their range, great or small, isochronous; that is to say, they will be performed in exactly equal times. It is therefore by this property that the balance wheel enjoys the same regulating power as the pendulum.

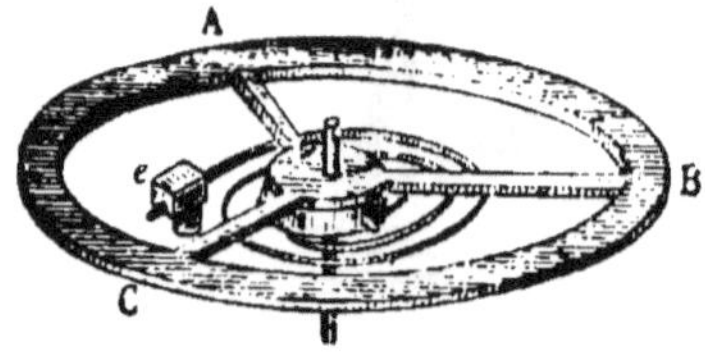

Fig. 202.

518. **The fly** is a regulator, consisting of a wheel having four arms, upon each of which a thin oblong vane of metal is so fixed that it can be turned with its plane at any desired inclination to the axis; the wheel is accurately balanced so as to be in equili-

Q 3

brium upon its axis, whatever be the position of the vane. When a motion of revolution is imparted to such a wheel the four vanes acting against the atmosphere will be resisted by it with a force which will be greater as the velocity of revolution is augmented, and it is demonstrable that the rate of increase of this resistance is proportional to the square of the velocity of the wheel :—thus, with a double velocity, there will be a fourfold resistance ; with a triple velocity, a ninefold resistance, and so on. An apparatus for illustrating experimentally the effect of such a wheel is represented in *fig.* 203., where two such wheels, A and B, are mounted on

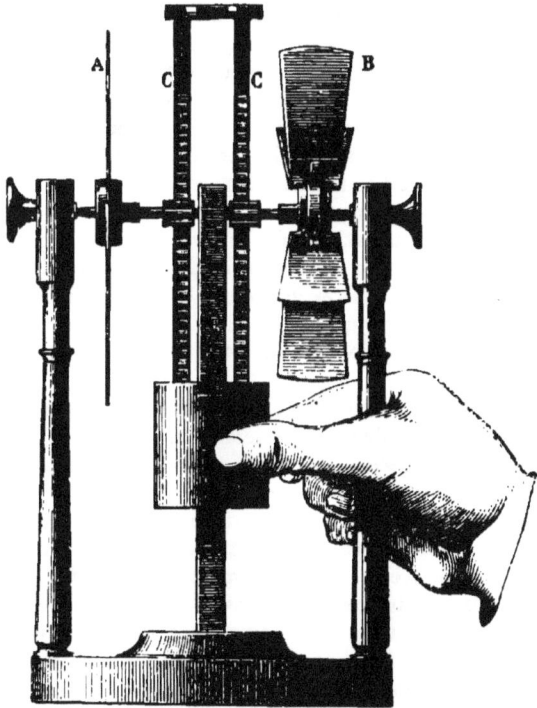

Fig. 203.

horizontal axles so as to be capable of turning with very little friction. Upon each of these axles is fixed a small pinion, engaged in the teeth of a rack c ; to the lower ends of these racks is attached a heavy weight. When this weight is pushed up to the pinions, the lowest teeth of the racks will be engaged in those of the latter, and if the weight be then let go, it will fall, drawing the racks with it, and imparting revolution to the two pinions, so that when the weight shall have dropped upon the bottom of the stand, the two pinions, and consequently the two flies, will have received precisely the same velocity of rotation; but the racks having terminated just before the arrival of the weight upon the stage, the pinions are no longer engaged in their teeth, and will therefore be free to revolve, as well as the flies, with the velocity acquired. As represented in the figure, the vanes of the fly B are presented with their surfaces at right angles to the direction of the motion, while the vanes of A are presented with their edges to the motion; it is evident, therefore, that one must encounter more resistance than the other from the air, and this is proved by the fact that while B is very soon brought to rest, A will continue to move for a considerable time. The experiment may be varied by inclining the vanes at various angles, when it will be found that the more near their position is to that represented at B, the sooner the wheel will be brought to rest by the resistance; and the nearer it is to the position shown in A, the longer will be the continuance of the motion. The velocity imparted to the wheels may also be varied by varying the weight attached to the racks.

The practical application of the fly as a regulator depends upon the two properties here explained: first, that with a given position of the vanes, the resistance of the air will increase indefinitely with the velocity; and, secondly, that this resistance will vary with the inclination of the vanes to the direction of the motion.

Let us suppose, for example, that a fly be applied to equalise the accelerating force produced by a descending weight; the fly is then mounted on an axis which receives motion, in common with the other parts of the mechanism to which it is applied, from the weight. At the commencement, the motion will be accelerated because the weight will be greater than the resistance, and consequently a continually increasing velocity will be imparted to the fly; according to what has been explained, the fly will therefore encounter a resistance which, increasing as the square of the velocity, will soon become equal to the force of the descending weight, and then all acceleration must cease, and the motion must become absolutely uniform, since any, even the least, increase of velocity would produce a resistance greater than the moving power; the uniform velocity thus attained will continue so long as the moving power continues to act with the same force upon the mechanism.

If it be desired to increase the uniform velocity which a given moving force will thus impart, it is only necessary to turn the vanes of the fly so as to present them more obliquely to the direction of the motion.

The fly possesses a property as a regulator which renders it available in many cases in which neither the pendulum nor the balance wheel would be applicable: these two regulators, as we shall more fully explain hereafter, produce, not an uniform, but an intermitting motion, as will be readily comprehended by observing the motion of the seconds-hand of a time-piece, whether it be regulated by a pendulum or a balance wheel; such a hand, as every one knows, moves not continuously, but by starts; and the same is true of the minute and hour hands, only that in these the intermissions are made through spaces so minute that they are not observable. Now, in many cases, such an intermitting motion is altogether inadmissible, as, for example, in the apparatus for illustrating the motion of falling bodies represented in *fig.* 48., where the fly placed at the top regulates the rotation of the cylinder, which must be continuous and uniform.

Musical boxes, self-acting pianofortes, barrel organs, played by mechanism, are all examples where the regulator must be a fly, and where neither pendulum nor balance wheel would be admissible.

CHAP. IV.

REGULATION AND MODIFICATION OF FORCE AND MOTION.

519. REGULATION and uniformity are two of the conditions most universally indispensable in the operation of machinery; sudden changes of speed are always a source of loss of power, often injurious and sometimes destructive to the apparatus, and never fail to produce imperfection in the articles fabricated.

Much attention, therefore, has been directed to, and much mechanical ingenuity expended on, contrivances for insuring these conditions of regularity and uniformity in the movement of machinery, by removing those causes of inequality which can be avoided, and by compensating those which cannot.

520. Irregularity in the motion of machinery will result in any one of the following causes : —

1°. When a varying power is opposed to a uniform resistance.

2°. When a uniform power is opposed to a varying resistance.

3°. When the power and resistance both vary, but not proportionally to each other.

4°. When the power is not transmitted with uniform effect to the working point in the successive positions assumed by the machine.

521. Varying power opposed to uniform resistance. —The force of the prime mover is seldom regular. The force of water varies with the copiousness of the stream; the force of wind is proverbially capricious ; the power of steam varies with the intensity of combustion in the furnace; and the force of animal power, depending on the temper and health, is difficult of control, human labour being the most unmanageable of all. No machine works so irregularly as one that is manipulated.

In some cases the prime mover is subject, by the very conditions of its existence, to constant variation ; as, for example, where it is a main spring, which gradually loses its energy as it is relaxed. In some cases, the prime mover is liable to intermission, and is totally suspended during certain intervals. An example of this is presented in the single-acting steam-engine, where the force of the steam acts only during the descent of the piston, but is suspended during its ascent.

522. Uniform power opposed to varying resistance. — In almost all the applications of machinery, the load or resistance is subject to continual fluctuation. In mills, a multiplicity of parts are liable to be occasionally and irregularly disengaged, and to have their operations suspended. In large factories for spinning, printing, dyeing, &c., a great number of separate spinning-machines, looms, presses, and other engines, are usually driven by a common power, such as a water-wheel or a steam-engine. In such cases, the number of machines worked at the same time must necessarily vary according to the employment supplied to the factory, and to the fluctuating demand for the articles produced. Under such circumstances, the velocity with which the machinery is moved would suffer corresponding changes, increasing with each diminution, and being retarded with each increase of the resistance.

523. In many cases, the variation of the power and the variation of the resistance are both from their conditions inevitable, and yet a uniform effect is indispensable. It is evident that this can only be insured by a class of contrivances which have for their object to proportion the power to the resistance, by either causing a diminution or increase of resistance to diminish or augment the supply of power ; or, on the other hand, by causing the variation of the power to act in a corresponding manner upon the resistance or load. In a word, uniformity of action in machinery can only be insured by providing means by which the power and the resistance, no matter what be their respective variations, shall always be proportional to each other.

Whenever the power is less than that which is in equilibrium with the resistance, the motion will be retarded, and if this condition continue, it will ultimately stop; and whenever the power is greater than that determined by the condition of equilibrium, the motion will be accelerated; and if this condition should continue, the acceleration would continue until the machine would be destroyed by its own momentum.

524. There is scarcely any machine in which the energy of the power is transmitted uniformly to the resistance in all the phases of the mechanism. In all machines the moving parts assume in succession a variety of positions, in each of which their effect to transmit the power to the resistance is different; and thus the effective energy of the machine in acting against the resistance is subject to continual fluctuation. It is not easy to convey, without numerous examples, a general idea of those causes of inequality to those who are not familiar with machinery. It will, however, be more clearly understood when we come to explain the methods of equalising the action of the power and the resistance.

525. **Regulators.**—The class of contrivances which have for their object to render the power and resistance proportionate to each other are called *regulators*. They generally act upon that point of the machine which commands the supply of the power by means of some mechanical contrivances, which check the quantity of the moving principle conveyed to the machine whenever the motion becomes accelerated, and increase the supply whenever it becomes retarded. In a water-wheel, for example, this is accomplished by acting upon the shuttle, in a wind-mill by the adjustment of the sails, and in a steam-engine by acting on a valve called the throttle valve placed in the main pipe, through which steam flows from the boiler to the cylinder.

526. **The governor.**— One of the most interesting and instructive examples of this class

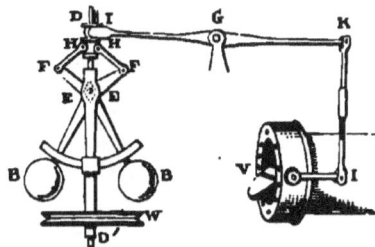

Fig. 204.

of contrivances is called the *governor*. This expedient, which was long used to regulate mill work and other machinery, owes its beautiful adaption to the steam-engine to the ingenuity of Watt. It consists of two heavy balls, B B, *fig.* 204., attached to the extremities of rods B F jointed at E, and passing through a mortice in the vertical stem D D'.

When the balls B are driven from the axis by the centrifugal force arising from their rotation, their upper arms E F are caused to increase their divergence in the same manner as the blades of scissors are opened by separating the handles. These acting upon the ring H, by means of the short rods F H, draw it down the vertical axis from D towards E. A contrary effect is produced when the balls B are brought closer to the axis, and the divergence of the rods B E diminished. A horizontal wheel W is attached to the vertical axis D D', having a groove to receive a rope or strap upon its rim. This strap passes round the wheel or axis, by which motion is transmitted to the machinery to be regulated so that the spindle or shaft D D' will be always made to revolve with a speed proportionate to that of the machinery.

As the shaft D D' revolves, the balls B are carried round it with a circular motion, and consequently acquire a centrifugal force which causes them to recede from the axle, and therefore to depress the ring H. On the edge or rim of this ring is formed a groove, which is embraced by the prongs of a fork I at the extremity of one arm of a lever whose fulcrum is at G. The extremity K of the other arm is connected by some means with the part of the machine which supplies the power. In the present instance we shall suppose it a steam-engine, in which case the rod K I communicates with a flat circular valve v placed in the principal steam pipe, and so arranged that when K is elevated as far as by their divergence the balls B have power over it, the passage of the pipe will be closed by the valve v, and the passage of steam entirely stopped; and, on the other hand, when the balls subside to their lowest position, the valve will be presented with its edge in the direction of the tube, so as to interrupt no part of the steam.

The property which renders this instrument so well adapted to its purpose is, that there is but one velocity at which the balls can remain in equilibrium.

527. **Fusee.** — The effect of a power of variable energy may be rendered uniform by transmitting it to the working point, through the agency of a leverage of corresponding variation so regulated, that, in the same proportion as the power diminishes in energy, the leverage shall increase, and *vice versâ*. A well-known example of this occurs in the construction of certain watches, where the moving power, being a main-spring inclosed in a barrel, has a gradually diminished energy as the spring is relaxed. The chain as it is discharged from the barrel is coiled upon a conical spiral, called a fusee, represented in *fig.* 205. The leverage by which the force of the spring is transmitted being the semi-diameter of the fusee; and the motion commencing from the top, or narrowest end, it follows that when the energy of the spring is greatest the leverage is least; and as the chain coils upon the barrel containing the spring, and is discharged from the fusee, the radius

of each part of the fusee which discharges the chain gradually increases. The form of the fusee is such that this increase of

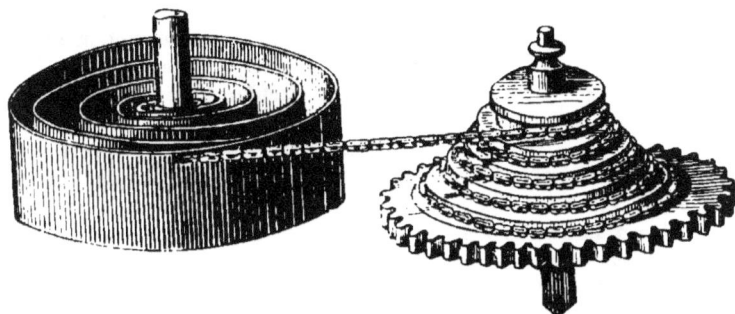

Fig. 205.

leverage is in the exact proportion of the diminished force of the spring.

528. The several parts of a machine have certain periods of motion, in which they pass through a variety of positions, return-ing constantly to similar positions after equal intervals.

In the different positions assumed by the moving parts during these periods, the effect of the power transmitted to the working point is different; and cases even occur in which for a moment this effect is altogether interrupted, and the machinery is then in a pre-dicament in which the power loses all effect upon the resistance. The consequence of this would be, that, supposing the power and resistance to be both uniform, the action of the former upon the latter would be subject to periodical variation, being at one time more and at another time less than what would be necessary to keep the whole in equilibrium.

Under these circumstances, it is possible to suppose that the movement of the machine may continue, and even that its average rate may be uniform; but its motion would be subject to periodical variations, being alternately accelerated and retarded. This would be attended not only with an injurious effect upon the work pro-duced by the machine, but would be also detrimental to the machine itself, whose moving parts would be subject to continual starts and strains arising from the alternate reception and destruction of momentum.

529. **Crank.** — To render these general observations more clearly intelligible, we shall take, as an example, the action of a common crank, used in steam-engines and many other ma-chines.

A crank is nothing more than a double winch. It is represented

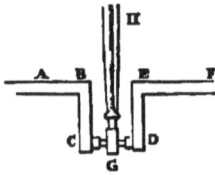

Fig. 206.

complete with both its arms in *fig.* 206. Attached to the middle of c d, by a joint, is a rod, which is the means of imparting the effect of the power to the crank. This rod is driven by an alternate motion like the brake of a pump. The bar c d is carried with a circular motion round the axis a f. ·

Let the machine viewed in the direction a b e f of the axis he conceived to be represented in *fig.* 207., where a represents the centre round which the motion is to be produced, and o the point where the connecting rod o h is attached to the arm of the crank. The circle through which o is to be urged by the rod is represented by the dotted line. In the position represented in *fig.* 207., the rod acting in the direction h o has its full power to turn the crank o a round the centre a. As the crank comes into the position represented in *fig.* 208., this power is diminished; and when the point o comes immediately below a, as in *fig.* 209., the force in the direction h o has no effect in

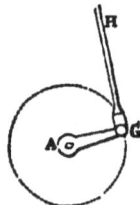

Fig. 207.

Fig. 208.

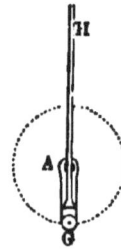

Fig. 209.

turning the crank round a, but, on the contrary, is entirely expended in pulling the crank in the direction o a, and therefore only acts on the pivots or gudgeons which support the axle.

At this crisis of the motion, therefore, the whole effective energy of the power is annihilated.

After the crank has passed to the position represented in *fig.* 210., the direction of the force which acts upon the connecting rod is changed, and now the crank is drawn upward in the direction o h. In this position the moving force has some efficacy to produce rotation round a, which efficacy continually increases until the crank attains the position shown in *fig.* 211., when its power is greatest. Passing from this position, its efficacy is continually diminished until the point o comes immediately above the axis a (*fig.* 212.). Here again the power loses all its efficacy to turn the axle. The force in the direction G H or H G can obviously produce no other effect than a strain upon the pivots or gudgeons.

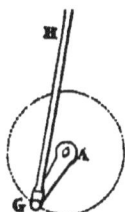 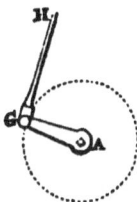 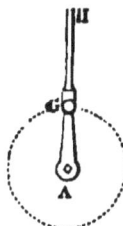

Fig. 210.   Fig. 211.   Fig. 212.

It will be evident from this that the action of the power transmitted to the working point G is very variable. At the dead points represented in *figs.* 209. and 212., the machine, if depending solely upon the moving power, must come to rest, for at both points the whole effect of the power would be exerted in producing pressure on the axle and gudgeons of the crank. Through a small space at either side of those dead points, the effect transmitted to G, though not absolutely nothing, is almost evanescent, so that it may be considered that through a small arc at either side of each of the dead points the machine is still inert.

It must, however, be considered that, in virtue of its inertia, the motion which the machinery had previously to its arrival at its dead points has a tendency to continue ; and if the resistance of the load and the effects of friction be not too great, this disposition to preserve its state of motion will extricate the machinery from the mechanical dilemma in which it is involved in these cases by the particular disposition of its parts. Although, however, the motion will not therefore be actually suspended, on the arrival of the crank at the dead points, it will be greatly retarded ; and, on the other hand, when the power acquires its greatest activity, as it does in the position represented in *figs.* 207. and 211., it will be unduly accelerated.

530. **Fly wheel.** — These irregularities are equalised by fixing upon the axis of the crank, or at any other convenient part of the machine, a fly wheel, which is a massive ring of metal, connected with a central box or nave by comparatively light spokes, and turning on an axis with but little friction. If any force be applied to it, with that force, making some slight deduction for friction, it will move and will continue to move until some obstacle retard it, which obstacle will receive from it as much force as the fly wheel loses.

The effect of such a wheel applied to the parts moved by the crank will equalise the inequality which has just been described. When the crank assumes the position represented in *figs.* 207. and

Fig. 213.

211., where the power has full play upon it, the effect of the power is partly transmitted to the machine, and partly received by the movable rim of the fly wheel, to which it imparts increased momentum. There is here, it is true, an acceleration of the motion, but one which is comparatively small, inasmuch as the great mass of the fly wheel receives the momentum without sensible increase of speed. When the crank gets into the predicament represented at the dead points (*figs.* 209. and 212.), the momentum of th. fly wheel, received when the crank acted with the most advantage, immediately conveys its force to the working-point G, extricates the machine, and carrying the crank out of the neighbourhood of the dead point, brings the power again to bear upon it.

531. It happens frequently in the practical application of machinery, that the moving power is much too intense, or much too feeble, for the resistance; and in the one case contrivances are required by which it may be greatly attenuated, and in the other by which it may be greatly augmented.

Main-spring. — In the case of watchwork, the resistance to be overcome is nothing more than that presented by the hands which move upon the dial-plate. In this case the moving power is the force of the fingers, by which once in twenty-four hours the main-spring is wound up. The main-spring itself must be regarded, in this case, as a mere depository for the power exerted in winding

up the watch, and not as a prime mover. The force which is thus
deposited once in twenty-four hours in the main-spring is delivered
gradually and regularly, by such spring, to the fusee, and transmitted, through the system of wheels, to the hands.

In the case of clocks moved by main-springs instead of a weight,
this attenuation of the moving power is still more extensive.
These, being wound up, will frequently go for fifteen days. In
this case, therefore, the mechanical force exerted by the hand in
winding them up, and which is developed in less than a minute, is
spread over fifteen times twenty-four hours by the mechanism of
the clock.

532. **Clockwork moved by a weight.** — In the case of clocks
which move by a descending weight, the original moving force is
also the application of the human hand in winding up the weight.
The weight, being lifted a height of three or four feet, descends
slowly through the same height, imparting its descending force
gradually and regularly to the clockwork. In this case, therefore,
the descending force of a weight through a small height is so
attenuated as to impart a motion to the hands which will continue
sometimes for a month or longer.

533. It is frequently required, on the contrary, to impart to the
resistance a force vastly greater in intensity than the moving power.
Numerous examples of this have been already given in illustrating
the simple machines. In all cases where the leverage of the power
is greater than the leverage of the resistance, there will be an
augmented intensity of mechanical action in the same proportion ;
and this intensity, by combining levers or other simple machines,
may be augmented without any practical limit.

534. **Hammers, sledges, &c.** — But in some cases a force is
required more intense than can be obtained even by these means.
In such cases, it becomes necessary to convert the continued agency
of the moving power into one which acts instantly and by intermission. If, for example, it be required to cause a nail to penetrate
a beam of wood, we should attempt in vain to accomplish this by
producing any pressure, however great, on the head of the nail.
A few blows of a hammer, nevertheless, easily effect this. In this
case, the moving power is the hand, or other force which raises the
hammer. The mass of the hammer, in falling on the head of the
nail, imparts instantly to the nail the entire force which was exerted
in lifting it, but with this difference, that such force, in raising the
hammer, was developed in a certain definite time, whereas it is
discharged upon the head of the nail in an instant.

The same observations apply to all cases in which percussion is
used. In all these cases the force is developed in a definite time,
but is discharged upon the resistance in an instant.

535. Inertia supplies means of accumulating force.—In some cases where a severe instantaneous action is required, the moving power is accumulated by means of the inertia of matter. A mass of matter retains, by virtue of its inertia, the whole amount of any force which may be given to it, except that part of which friction and the atmospheric resistance deprives it. To render this method of accumulating force intelligible, let us first imagine a polished level plane, on which a heavy globe of metal, also polished, is placed. It is evident that the globe will remain at rest on any part of the plane without a tendency to move. Suppose, then, a slight impulse be given to it, which will cause it to move with any given velocity, for example, three feet per second. It will then continue to move with this velocity, for any length of time, except so far as it may be impeded by the resistance already mentioned.

Let us then imagine a second impulse given to it equal in force to the former : this will increase its velocity to six feet per second ; a third impulse will augment it to nine feet per second, and so on. Now there is no limit to the number of impulses which may be successively given to the moving body, provided only space were given for its motion. Thus, ten thousand repetitions of the impulse would make the body move at the rate of thirty thousand feet a second. If the body to which these impulses were transmitted were a cannon-ball, it might, by the constant application of a feebly impelling force, be made to move at length with as much force as if it were impelled from a powerful piece of ordnance. The force with which such a ball would strike a buttress might be sufficient to reduce it to ruins ; and yet such force may be nothing more than the accumulation of a number of feeble forces, not beyond the power of a child to exert, which are stored up and preserved in the moving mass, and then brought to bear at the same moment on the resistance against which the force is directed. It is the same for any number of actions exerted successively and during a long interval, brought into operation at one and the same moment.

But the case here supposed cannot actually occur, because we have not in general practical means of moving a body for a considerable time in the same direction, without much friction, and without encountering other obstacles which would impede its progress. If, however, a leaden ball be attached to the end of a string and whirled rapidly round, a great force would be given to it, and it will strike a board with such intensity as to penetrate it.

536. A weapon called a life-preserver consists of a piece of lead sometimes attached to the end of a piece of cane or whalebone, with which a blow may be given with great force. · Innumerable

R

examples of the application of this principle will present themselves to every mind. Flails used in threshing, clubs, canes, whips, and all instruments used for striking, axes, hatchets, cleavers, and all instruments which act by a blow, present examples of this principle.

537. Where very intense force is required, as, for example, in certain presses, two heavy balls are attached to the ends of a horizontal lever A B, with equal arms, *fig.* 214. This lever works a screw, at the lower end of which is the working point. A rapid motion is imparted to the balls by the hand, and the working point is driven against the resistance by the accumulated momentum acquired by the balls, augmented by the leverage of the arms to which they are attached, and the mechanical force of the screw.

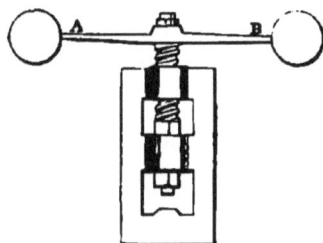

Fig. 214.

538. The surprising effects produced by the accumulation of force are apt to lead to erroneous suppositions, that instruments thus acting by inertia have the effect of actually augmenting the amount of moving power. When the quality of inertia, however, is rightly understood, such an error cannot occur. The instruments by which force is thus accumulated, so far from augmenting the effect of the moving power, must to some extent diminish it; inasmuch as they are liable to friction and atmospheric resistance, by which more or less force is intercepted. An accumulator of force, whatever be its form, can never have more force than has been applied to put it in motion. Whether it be a falling weight, a revolving mass, a string which is coiled up, or air which is condensed, it cannot develop a greater amount of force than that which is imparted in raising it if it be a weight, in putting it in motion if it be a moving mass, in winding it up if it be a spring, or in compressing it if it be air. The only difference between the power which is imparted to these agents, and the effects which they produce respectively upon the resistance, is in the time during which the effects are developed. The power is in general imparted slowly, while the effects are produced instantaneously.

539. Mills for rolling metals, or for punching boiler-plates, supply striking examples of this. The water-wheel, or steam-engine, or whatever other power be used, is allowed for some time to act upon the fly wheel alone, no load being placed upon the machine. When a sufficient momentum has been imparted to the mass of metal forming the fly wheel, the metal to be rolled or

FLY WHEEL. 243

pierced is submitted to the machine, and is immediately flattened
or perforated by it, depriving at the same time the fly wheel of a
corresponding quantity of its momentum.

In the same manner, a force may be obtained by the arms of
men acting on a fly for a few seconds, sufficient to impress an
image on a piece of metal by an instantaneous stroke. The fly is
therefore the principal agent in coining-presses.

Some presses used in coining have flies with arms four feet long,
bearing a hundredweight at each of their extremities. If such a
velocity be imparted to such an arm that it shall make one revolution per second, the die will be driven against the metal with the
same force as that with which 3¼ tons would fall from the height
of 16 feet, which is an enormous power if the simplicity and com-
pactness of the machine be considered.

540. **Position of the fly wheel.** — The place to be assigned to
a fly wheel relatively to the other parts of the machinery is determined by the purpose for which it. is used. If it be intended to
equalise the action, it should be near the working point. Thus,
in a steam-engine, it is placed near the crank which turns the axle,
by which the power of the engine is transmitted to the object it is
finally designed to affect. On the contrary, in hand-mills, such as
those commonly used for grinding coffee, &c., it is placed upon the
axis of the winch by which the machine is worked.

541. The open work of fenders, fire grates, and similar ornamental articles constructed in metal, is produced by the action of
a fly in the manner already described.

The cutting tool, shaped according to the pattern to be executed,
is attached to the end of the screw, and the metal being held in a
proper position beneath it, the fly is made to urge the tool downwards with such force as to stamp out pieces of the required
figure. When the pattern is complicated, and it is necessary to
preserve with exactness the relative situation of its different parts,
a number of punches are impelled together, so as to strike the entire piece of metal at the same instant, and in this manner the most
elaborate open work is executed by a single stroke of the hand.

542. **Modification of motion.** — Although the simple machines, classed under the general denomination of *mechanic powers*,
include in principle all the solid parts which can enter into the
composition of any piece of mechanism, there are still many expedients which, on account of their simplicity of construction and
extensive utility in practice, merit more particular notice.

In the arts and manufactures, the kind of motion produced by
a given force is often of much greater importance than either the
velocity imparted to the working-point, or the amount of the
useful effect obtained from the moving power: the latter may

affect the quantity of work done in a given time; but the former, in particular cases, is essential to the performance of the work, in any quantity whatever.

In the practical application of machinery, the object aimed at is often to communicate to the working-point some peculiar sort of motion adapted to the uses to which the machine is applied, and it rarely, indeed almost never, happens that the moving power is capable of this sort of motion. Expedients must, therefore, be discovered by means of which the motions which the moving power is capable of directly producing can be converted into those which are necessary for the purposes to which the machine is applied.

The varieties of motion which most commonly present themselves in the practical application of mechanics may be divided into *rectilinear* and *rotatory*. In rectilinear motion the several parts of the moving body proceed in parallel straight lines with the same speed. In rotatory motion the several points revolve round an axis, each performing a complete circle, or similar parts of a circle, in the same time.

Each of these may again be resolved into continued and reciprocating. In a continued motion, whether rectilinear or rotatory, the parts move constantly in the same direction, whether that be in parallel straight lines, or in rotation on an axis. In reciprocating motion the several parts move alternately in opposite directions, tracing the same spaces from end to end continually. Thus, there are four principal species of motion which more frequently than any others act upon, or are required to be transmitted by, machines: —

　　1. *Continued rectilinear motion.*
　　2. *Reciprocating rectilinear motion.*
　　3. *Continued circular motion.*
　　4. *Reciprocating circular motion.*

These will be more clearly understood by examples of each kind.

Continued rectilinear motion is observed in the flowing of a river, in a fall of water, in the blowing of the wind, in the motion of an animal upon a straight road, in the perpendicular fall of a heavy body, in the motion of a body down an inclined plane.

Reciprocating rectilinear motion is seen in the piston of a common syringe, in the rod of a common pump, in the hammer of a pavier, the piston of a steam-engine, the stampers of a fulling-mill.

Continued circular motion is exhibited in all kinds of wheel-work, and is so common, that to particularise it is needless.

Reciprocating circular motion is seen in the pendulum of a clock, and in the balance wheel of a watch.

We shall now explain some of the contrivances by which a power having one of these motions may be made to communicate either the same species of motion changed in its velocity or direction, or any of the other three kinds of motion.

543. **Continued rectilinear motion.** — The most obvious expedient for producing this motion is by means of one or more pulleys. It has been already shown that, by a single fixed pulley, a moving power acting in the direction of any line may impart continued rectilinear motion to a resistance in any other line, provided only that the two directions can be connected by a fixed pulley. This, however, can only be done when lines drawn in the two directions intersect each other. In that case a fixed pulley, being placed in the angle formed by the two lines, with its axis perpendicular to them, will serve to guide a rope, one part of which will be directed to the resistance, and the other to the moving power.

If, however, the direction of the power do not intersect that of the resistance, they must either be parallel or in different planes.

If they be parallel, as, for example, when the power acts vertically downwards, while the weight is moved vertically upwards, the object will be attained by two fixed pulleys having their axes horizontal, one placed directly above the power, and the other directly above the resistance.

If the direction of the power and that of the weight be in different planes, the object may also be attained by two fixed pulleys, placed at any two points arbitrarily taken. The axis of the pulley o (*fig.* 215.), which guides that part of the rope directed to the power P, must in that case be at right angles to the direction of the power and that of the line joining the pulleys; and the axis of the pulley o', which guides that part of the rope directed to the resistance w, must be at right angles to the direction of the line joining the pulleys and that of the resistance.

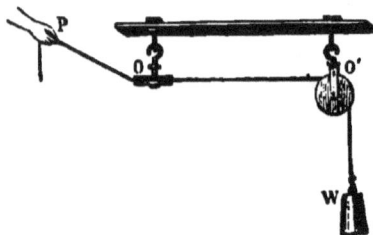

Fig. 215.

If it be necessary to change the velocity, any of the systems of pulleys described in 453., *et seq.*, may be used in addition to the fixed pulleys.

By the wheel and axle any one continued rectilinear motion may be made to produce another in any other direction, and with any other velocity. It has been already explained that the proportion of the velocity of the power to that of the weight is as the

R 3

diameter of the wheel to the diameter of the axle. The thickness of the axle being therefore regulated in relation to the size of the wheel, so that their diameters shall have that proportion which subsists between the proposed velocities, one condition of the problem will be fulfilled. The rope coiled upon the axle may be carried, by means of one or more fixed pulleys, into the direction of one of the proposed motions, while that which surrounds the wheel is carried into the direction of the other by similar means.

By the wheel and axle a continued rectilinear motion may be made to produce a continued rotatory motion, or *vice versâ*. If the power be applied by a rope coiled upon the wheel, the continued motion of the power in a straight line will cause the machine to have a rotatory motion. Again, if the weight be applied by a rope coiled upon the axle, a power having a rotatory motion applied to the wheel will cause the continued ascent of the weight in a straight line.

Continued rectilinear and rotatory motions may be made to produce each other, by causing a toothed wheel to work in a *rack* Such an apparatus is represented in *fig.* 216. In some cases the teeth of the wheel work in the links of a chain. The wheel is then called a *rag-wheel* (*fig.* 217.) Straps, bands, or ropes, may communicate rotation to a wheel, by their friction in a groove upon its edge.

A continued rectilinear motion is produced by a continued circular motion in the case of a screw.

Fig. 216.   Fig. 217.

The lever which turns the screw has a continued circular motion, while the screw itself advances with a continued rectilinear motion.

The continued rectilinear motion of a stream of water acting upon a wheel produces continued circular motion. In like manner the continued rectilinear motion of the wind produces a continued circular motion in the arms of a windmill.

Cranes for raising and lowering heavy weights convert a circular motion of the power into a continued rectilinear motion of the weight.

544. **Reciprocating rectilinear motion.** — Continued circular motion may produce reciprocating rectilinear motion, by a great variety of ingenious contrivances.

Reciprocating rectilinear motion is used when heavy stampers are to be raised to a certain height, and allowed to fall upon some object placed beneath them. This may be accomplished by a wheel bearing on its edge curved teeth, called *wipers*. The stamper is furnished with a projecting arm or peg, beneath which the wipers are successively brought by the revolution of the wheel. As the wheel revolves the wiper raises the stamper, until its extremity passes the extremity of the projecting arm of the stamper, when the latter immediately falls by its own weight.

It is then taken up by the next wiper, and so the process is continued. A similar effect is produced if the wheel be partially furnished with teeth, and the stamper carry a rack in which these teeth work. Such an apparatus is represented in *fig.* 218.

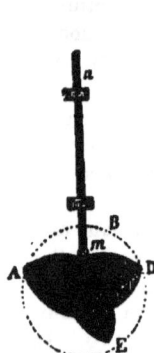

It is sometimes necessary that the reciprocating rectilinear motion shall be performed at a certain varying rate in both directions. This may be accomplished by the arrangement represented in *fig.* 219.

Fig. 218.　　　　Fig. 219.

A wheel turns uniformly in the direction A B D E. A rod *m n* moves in guides, which only permit it to ascend and descend perpendicularly. Its extremity *m* rests upon a path or groove raised from the face of the wheel, and shaped into such a curve that as the wheel revolves the rod *m n* shall be moved alternately in opposite directions through the guides, with the required velocity. The manner in which the velocity varies will depend on the form given to the groove or channel raised upon the face of the wheel, and this may be shaped so as to give any variation to the motion of the rod *m n* which may be required for the purpose to which it is to be applied.

The *rose-engine* in the turning-lathe is constructed on this principle. It is also used in spinning machinery.

It is often necessary that the rod to which the reciprocating motion is communicated shall be urged by the same force in both directions. A wheel partially furnished with teeth, acting on two racks placed on different sides of it, and both connected with the bar or rod to which the reciprocating motion is to be communicated, will accomplish this. Such an apparatus is represented in *fig.* 220., and needs no further explanation.

Another contrivance for the same purpose is shown in *fig.* 221., where A is a wheel turned by a winch H, and connected with a rod or beam moving in *guides* by the joint *a b*. As the wheel A is turned by the winch H, the beam is moved between the guides alternately in opposite directions, the extent of its range being

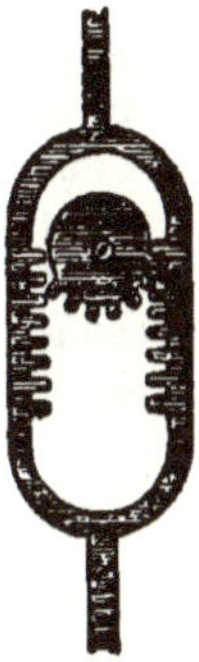

Fig. 220. 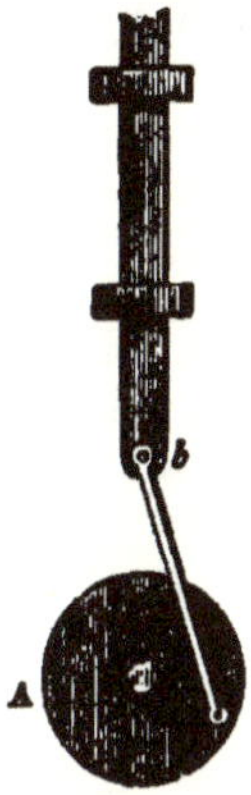   Fig. 221. 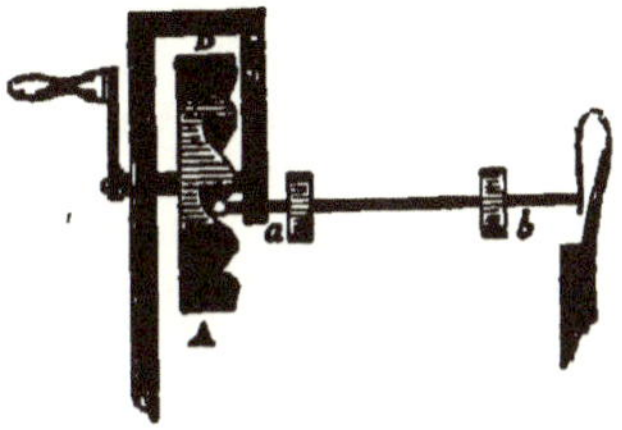   Fig. 222.

governed by the length of the diameter of the wheel. Such an apparatus is used for grinding and polishing plane surfaces, and also occurs in silk machinery.

An apparatus applied by M. Zureda in a machine for pricking holes in leather is represented in *fig.* 222. The wheel A B has its circumference formed into teeth, the shape of which may be varied according to the circumstances under which it is to be applied. One extremity of the rod *a b* rests upon the teeth of the wheel upon which it is pressed by a spring at the other extremity. When the wheel revolves, it communicates to this rod a reciprocating rectilinear motion.

Leupold applied this mechanism to move the pistons of pumps. Upon the vertical axis of a horizontal hydraulic wheel is fixed another horizontal wheel, which is furnished with seven teeth in the manner of a crown wheel. These teeth are shaped like inclined planes, the intervals between them being equal to the length of the planes. Projecting arms attached to the piston rods rest upon the crown of this wheel; and, as it revolves, the inclined surfaces of the teeth, being forced under the arm, raise the rod upon the principle of the wedge. To diminish the obstruction arising from friction, the projecting arms of the piston rods are provided with rollers, which run upon the teeth of the wheel. In one revolution of the wheel each piston makes as many ascents and descents as there are teeth.

545. **Continued circular motion.** — Wheelwork furnishes numerous examples of continued circular motion round one axis,

producing continued circular motion round another. If the axles be in parallel directions, and not too distant, rotation may be transmitted from one to the other by two spur wheels (448.) ; and the relative velocities may be determined by giving a corresponding proportion to the diameter of the wheels.

If a rotatory motion is to be communicated from one axis to another parallel to it, and at any considerable distance, it cannot in practice be accomplished by wheels alone, for their diameters would, in that case, be too large. In this case, the motion is communicated from wheel to wheel, by endless bands. If the two wheels are desired to turn in the same direction, the bands are placed as in *fig.* 223. If in contrary directions, they are crossed, as in *fig.* 224.

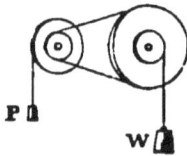

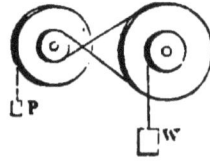

Fig. 223.

Fig. 224.

In this case, as in toothed wheels, the relative velocities are determined by the proportion of the diameters of the wheels.

In some cases it happens that the strain upon the wheels is too great to allow of the motion being transmitted by a band. In

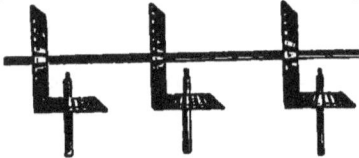

Fig. 225.

such cases, the motion may be transmitted from one shaft to the other, by means of bevelled wheels. In *fig.* 225. an arrangement is shown in which a revolving shaft imparts motion to a series of other shafts, perpendicular to it. Each of the latter might impart motion to other shafts, at any convenient distance, by like means.

In *figs.* 226. and 227. another arrangement is shown, by which

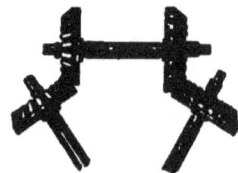

Fig. 226.

Fig. 227.

a shaft w imparts motion to another shaft w', by the intervention of a third shaft *w w'* on which bevelled wheels are fixed. The velocities may in this case be varied in any desired proportion by establishing a corresponding relation between the magnitudes of the wheels.

The methods of transmitting rotation from an axis or shaft to another at right angles to it, by spur and crown wheels, and by an endless screw and a spur wheel, have been already explained in 450.

546. **Universal joint.** — The axis to which rotation is to be given, or from which it is to be taken, is sometimes variable in its position. In such cases, an ingenious contrivance, called a *universal joint*, invented by the celebrated Dr. Hook, may be used. The two shafts or axles A B, *fig.* 228., between which the motion is to be communicated, terminate in semicircles, the diameters of which, C D and E F, are fixed in the form of a cross, their extremities moving freely in bushes placed in the extremities of the semicircles. Thus, while the central cross remains unmoved, the shaft A and its

Fig. 228.

semicircular end may revolve round C D as an axis; and the shaft B and its semicircular end may revolve round E F as an axis. If the shaft A be made to revolve without changing its direction, the points C D will move in a circle whose centre is at the middle of the cross. The motion thus given to the cross will cause the points E F to move in another circle round the same centre, and hence the shaft B will be made to revolve. •

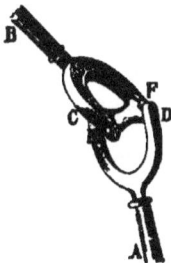

This instrument will not transmit the motion if the angle under the directions of the shafts be less than 140°. In this case a double joint as represented in *fig.* 229. will answer the purpose. This consists of four semicircles united by two crosses, and its principle and operation is the same as in the last case.

Universal joints are of great use in adjusting the position of large telescopes, where, while the observer continues to look through the tube, it

Fig. 229.

is necessary to turn endless screws or wheels whose axes are not in an accessible position.

The cross is not indispensably necessary in the universal joint. A hoop, with four pins projecting from it at four points equally distant from each other, or dividing the circle of the hoop into four equal arches, will answer the purpose. These pins play in the bushes of the semicircles in the same manner as those of the cross.

The universal joint is much used in cotton-mills, where shafts are carried to a considerable distance from the prime mover, and great advantage is gained by dividing them into convenient lengths, connected by a joint of this kind.

547. **Alternate circular motion.** — In the practical application of machinery, it is often necessary to connect a part having a continued circular motion with another which has a reciprocating or alternate motion, so that either may move the other. There are many contrivances by which this may be effected.

One of the most remarkable examples of it is presented in the escapements of watches and clocks.

A beam vibrating on an axis, and driven by the piston of a steam-engine, or any other power, may communicate rotatory motion to an axis by a connecter and a crank.

A wheel A (*fig.* 230.) armed with wipers, acting upon a sledge-hammer B, fixed upon a centre or axle C, will, by a continued rotatory motion, give the hammer the reciprocating motion necessary for the purposes to which it is applied. The manner in which this acts must be evident on inspecting the figure.

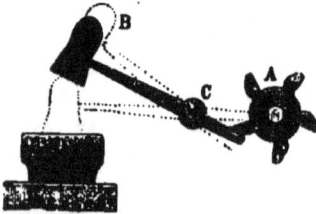

Fig. 230.

The large shears used in factories for cutting plates of metal are worked upon a similar principle. The lower edge of the shears (*fig.* 231.) is fixed, and the upper movable upon a joint.

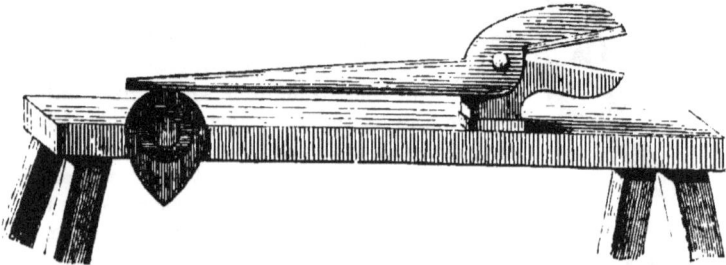

Fig. 231.

Under the extremity of the movable arm is placed a wheel, the contour of which has the form shown in the figure. In the

position there shown, the arm is in its lowest position, and therefore the blades of the shears diverge as much as possible. When
the wheel turns, so as to raise the movable arm, the shears will be
closed. In this way, by the continual revolution of the wheel, the
shears will be alternately closed and opened.

The treddle of the lathe furnishes an obvious example of a
vibrating circular motion producing a continued circular one.
The treddle acts upon a crank, which gives motion to the principal
wheel, in the same manner as already described in reference to the
working beam and crank in the steam-engine.

By the following ingenious mechanism an alternate or vibrating
force may be made to communicate a circular motion continually
in the same direction : — Let A B (*fig.* 232.) be an axis receiving
an alternate motion from some
force applied to it, such as a
swinging weight. Two ratchet
wheels, *m* and *n*, are fixed on this
axle, their teeth being inclined in
opposite directions. Two toothed
wheels c and D are likewise placed
upon it, but so arranged that they
turn upon the axle with a little
friction. These wheels carry two
catches *o p*, which fall into the teeth of the ratchet wheels *m n*,
but fall on opposite sides conformably to the inclination of the
teeth already mentioned. The effect of these catches is, that if
the axis be made to revolve in one direction, one of the two
toothed wheels is always compelled (by the catch *against* which
the motion is directed) to revolve with it, while the other is permitted to remain stationary in obedience to any force sufficiently great
to overcome its friction with the axle on which it is placed. The
wheels c and D are both engaged by bevelled teeth with the wheel E.

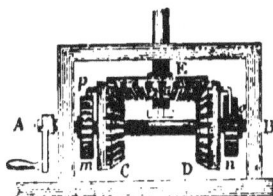

According to this arrangement, in whichever direction the axis
A B is made to revolve, the wheel E will continually turn in the
same direction; and therefore, if the axle A B be made to turn
alternately in the one direction and the other, the wheel E will not
change the direction of its motion. Let us suppose the axle A B is
turned against the catch *p*. The wheel c will then be made to
turn with the axle. This will drive the wheel E in the same
direction. The teeth on the opposite side of the wheel E being
engaged with those of the wheel D, the latter will be turned upon
the axle; the friction, which alone resists its motion in that direction, being overcome. Let the motion of the axle A B be now
reversed. Since the teeth of the ratchet wheel *n* are moved
against the catch *o*, the wheel D will be compelled to revolve with

Fig. 232.

the axle. The wheel E will be driven in the same direction as before, and the wheel C will be moved· on the axle A B, and in a contrary direction to the motion of the axle, the friction being overcome by the force of the wheel E. Thus, while the axle A B is turned alternately in the one direction and the other, the wheel E is constantly moved in the same direction.

It is evident that the direction in which the wheel E moves may be reversed by changing the position of the ratchet wheels and catches.

It is often necessary to communicate an alternate circular motion, like that of a pendulum, by means of an alternate motion in a straight line. A remarkable instance of this occurs in the steam-engine. The moving force in this machine is the pressure of steam, which impels a piston from end to end alternately in a cylinder. The force of this piston is communicated to the working beam by a strong rod, which passes through a collar in one end of the piston. Since it is necessary that the steam included in the cylinder should not escape between the piston rod and the collar through which it moves, and yet that it should move as freely and be subject to as little resistance as possible, the rod is turned so as to be truly cylindrical, and is well polished. It is evident that, under these circumstances, it must not be subject to any lateral or cross strain, which would bend it towards one side or the other of the cylinder. But the end of the beam to which it communicates motion, if connected immediately with the rod by a joint, would draw it alternately to the one side and the other, since it moves in the arc of a circle, the centre of which is at the centre of the beam. It is necessary, therefore, to contrive some method of connecting the rod and the end of the beam, so that while the one shall ascend and descend in a straight line, the other may move in the circular arc.

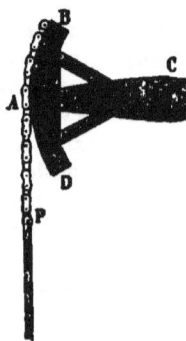

The method which first suggests itself to accomplish this is, to construct an arch-head upon the end of the beam, as in *fig.* 233. Let C be the centre on which the beam works, and let B D be an arch attached to the end of the beam, being a part of a circle having C for its centre. To the highest point B of the arch a chain is attached, which is carried upon the face of the arch B A, and the other end of which is attached to the piston rod. Under these circumstances it is evident that when the force of the steam impels the piston downwards, the chain P A B will draw the end of the beam down, and will, therefore, elevate the other end.

Fig. 233

When the steam-engine is used for certain purposes, such as pumping, this arrangement is sufficient. The piston in that case is not forced upwards by the pressure of steam. During its ascent it is not subject to the action of any force of steam, and the other end of the beam falls by the weight of the pump-rods drawing the piston, at the opposite end A, to the top of the cylinder. Thus the machine is, in fact, passive during the ascent of the piston, and exerts its power only during the descent.

If the machine, however, be applied to purposes in which a constant action of the moving force is necessary, as is always the case in manufactures, the force of the piston must drive the beam in its ascent as well as in its descent. The arrangement just described cannot effect this; for although a chain is capable of transmitting any force, by which its extremities are drawn in opposite directions, yet it is, from its flexibility, incapable of communicating a force which drives one extremity of it towards the other. In the one case the piston first *pulls* down the beam, and then the beam *pulls* up the piston. The chain, because it is inextensible, is perfectly capable of both these actions; and, being flexible, it applies itself to the arch-head of the beam, so as to maintain the direction of its force upon the piston continually in the same straight line. But when the piston acts upon the beam in both ways, in pulling it down and pushing it up, the chain becomes inefficient, being, from its flexibility, incapable of the latter action.

The problem might be solved by extending the length of the piston rod, so that its extremity shall be above the beam, and using two chains; one connecting the highest point of the rod with the lowest point of the arch-head, and the other connecting the highest point of the arch-head with a point on the rod below the point which meets the arch-head when the piston is at the top of the cylinder, *fig.* 234.

The connection required may also be made by arming the arch-

Fig. 234.　　　　　　　　　　Fig. 235

head with teeth, *fig.* 235., and causing the piston rod to terminate in a rack. In cases where, as in the steam-engine, smoothness of motion is essential, this method is objectionable; and under any circumstances such an apparatus is liable to rapid wear.

The method contrived by Watt, for connecting the motion of the piston with that of the beam, is one of the most ingenious and elegant solutions ever proposed for a mechanical problem. He conceived the notion of two straight rods, A B, C D, *fig.* 236., moving on centres or pivots A and C, so that the extremities B and D would move in the arcs of circles having their centres at A and C. The extremities B and D of these rods he conceived to be connected with a third rod B D united with them by pivots on which it could turn freely. To the system of rods thus connected let an alternate motion on the centres A and C be communicated: the points B and D will move upwards and downwards in the arcs expressed by the dotted lines, but the middle point P of the connecting rod B D will move upwards and downwards without any sensible deviation from a straight line.

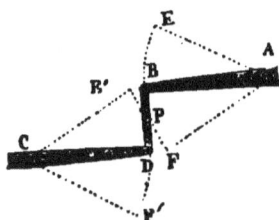

Fig. 236.

To prove this demonstratively would require some abstruse mathematical investigation. It may, however, be rendered in some degree apparent by reasoning of a looser and more popular nature. As the point B is raised to E, it is also drawn aside towards the right. At the same time the other extremity D of the rod B D is raised to E′, and is drawn aside towards the left. The ends of the rod B D being thus at the same time drawn equally towards opposite sides, its middle point P will suffer no lateral derangement, and will move directly upwards. On the other hand, if B be moved downwards to F, it will be drawn laterally to the right; while D being moved to F′ will be drawn to the left. Hence, as before, the middle point P sustains no lateral derangement, but merely descends. Thus, as the extremities B and D move upwards and downwards in circles, the middle point P moves upwards and downwards in a straight line.

An elegant contrivance, by which a continued motion of rotation can be made to produce a reciprocating rectilinear motion, is shown in *fig.* 237., where a wheel A B is surrounded on the inside with teeth, and a wheel C of half the diameter revolves within it; the wheel A B being fixed. While the centre of C moves in a circle round the centre of A B, any point taken on the circumference of C will move alternately across A B, in the direction of

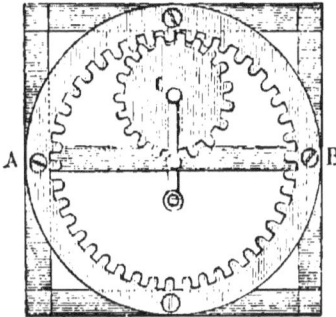

Fig. 237.

one of its diameters. Thus, that point of c which, in the figure, is at the highest point of A B, will move along the vertical diameter of A B, to the lowest point, while c rolls over half the circumference of A B; and the same point will again traverse the same vertical diameter of A B upwards, while the wheel c rolls over the other semicircle of A B.

Another method of producing a reciprocating rectilinear motion by a reciprocating circular motion is shown in *fig*. 238. The beam A B vibrates on the centre c, and a cord applies itself to a groove in the semicircular arch D E F. This cord is extended over the grooves of two fixed rollers M N; it is evident that as the beam A B vibrates, any point L upon the cord, between M N, will move alternately right and left in a straight line.

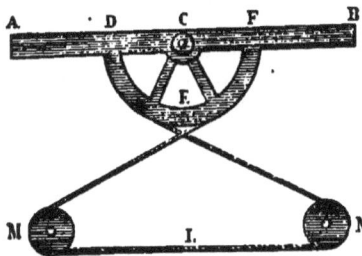

Fig. 238.

By the arrangement shown in *fig*. 239., the alternate circular motion of the unequal toothed sectors will impart reciprocating rectilinear motions to two racks. These motions will be unequal, and will be in the same ratio as the radii of the two circles.

By a mechanical contrivance represented in *fig*. 240., called the lever of La Garousse, a reciprocating circular motion is made to produce an intermitting rectilinear motion. The lever A B vibrates on the centre c, the arm which ascends drawing up one of the hooks E, while the arm which descends passes the other hook E' from one tooth to another of the vertical beam F G. By the continuance of this operation, the beam F G is elevated by an intermitting motion.

Fig. 239.

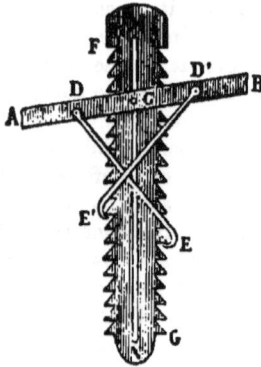

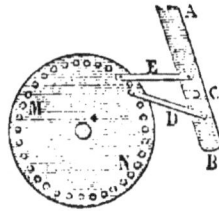

Fig. 240.   Fig. 241.

In *fig.* 241. an expedient is represented by which a similar
effect is produced upon a wheel M N. A lever A B vibrates on a
centre c, while two arms D and E extend from it, and act upon
pins projecting from the wheel M N. The arms D and E are thus
alternately pushed forwards and drawn backwards: that which is
pushed forwards drives before it one of the pins, and that which is
drawn backwards falls from one pin to the other.

548. **Joints.**—It frequently happens in mechanical combinations, that it is required to make temporary connections between
separate parts, or so to unite them together that they shall admit
of having their relative position changed at pleasure. In some
cases one, and in others both, of the parts thus connected are
susceptible of receiving motions more or less limited in their direction. In some cases the motion of one part necessarily produces
some corresponding motion in the other; and in others, the motion
of one is quite independent of the other, which may be fixed. The
class of contrivances which fulfil these purposes are called *joints.*

The universal joint already explained belongs to this class. One
of the arms thus connected may be inclined to the other in almost
any direction whatever, when the joint is constructed as in *fig.* 229.
But in thus changing its direction, no corresponding motion is imparted to the other arm. If, however, a twisting or rotatory motion round its own longitudinal axis be given to either arm, a
corresponding rotatory motion round its longitudinal axis will be
imparted to the other. In this manner, for example, a screw may
be turned by a twisting motion given to an arm having a direction
inclined at any angle to the longitudinal axis of the screw.

549. **Gimbals.** — This is a mechanical contrivance, which bears
a close relation to the universal joint. An example of it, familiar
to every one, is presented by the apparatus for suspending a ship's

8

compass (*fig.* 242.). The object is to keep the suspended body vertical, whatever be the derangements to which the points of suspension are liable.

A brass hoop is supported by two pins, projecting from points of its external surface which are diametrically opposed to each other. Another brass hoop is supported within the former, also by two pins projecting from points of its external surface, diametrically opposed one to the other, which play in two holes made in the former hoop at the extremities of that diameter, which is at right angles to the diameter at the extremities of which the first points of support are placed.

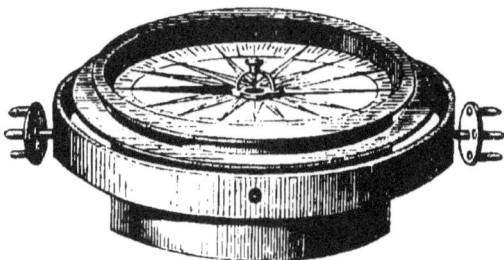

Fig. 242.

A similar mode of suspension is applied to the marine barometer.

550. **Ball and socket.** — The joint thus called has some, but not all, of the qualities of the universal joint. The extremity of one of the arms is attached to a solid ball of metal, and that of the other to a hollow ball, the internal diameter of which is equal to the diameter of the solid ball. If one of these balls were contained within the other, the arm of the solid ball, passing through a hole made to fit it in that of the hollow ball, the two arms would be incapable of any change whatever, in their relative position, and the joint so formed would form a rigid and invariable connection. But if the hole in the hollow ball, through which the arm of the solid ball passes, be much larger than the arm, as shown in *fig.* 243., then the arm will have a certain play, the solid ball moving within the hollow ball, and the extent of such play would obviously depend

Fig. 243.

on the magnitude of the opening. A joint thus constructed is that which is called the *ball and socket joint.*

This joint is represented in *fig.* 244.

It is evident that the inclination which can be given to the arms will be limited by half the circumference of a section of the socket.

The greatest play which such a joint can have will be produced when the socket exceeds an exact hemisphere by just enough to enable it to retain the ball within it; and it follows, therefore, that the angle under the arms thus connected, when they are most inclined, will be a little greater than 90°.

If either arm be fixed, the play of the other will be limited, by the directions of the sides of a cone, whose vertex is at the centre of the ball, and whose base is limited by the part of the ball left uncovered by the socket.

The arms connected by this joint are moved independently of each other. Either may be turned round its longitudinal axis, without imparting a corresponding motion to the other.

Fig. 244.

The socket is usually made of such a magnitude, that the ball moves in it with a certain degree of friction sufficient to retain the arms in whatever position may be given to them.

551. **Cradle joint.** — By this joint, either arm connected is allowed to move in a given plane round the other arm. The joint is usually formed by a pin passing through the centre of circular discs of metal formed at the end of the arms, having equal diameters, and placed concentrically face to face; sometimes, one such disc is attached to one arm passing between two, which are attached to the other : such a joint is represented in *fig.* 245.

It is by a joint of this kind that the legs of compasses are connected.

552. **Hinge.** — This is, in fact, a variety of the cradle joint. Its construction is rendered so familiar by the lids of

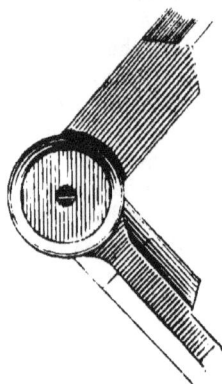

Fig. 245.

s 2

boxes, and the suspension of doors, that it will not require further explanation.

553. **Trunnions.** — The method of supporting heavy bodies, which require to be freely movable in a vertical plane, thus denominated, consists of two strong cylindrical pins, projecting from its sides, which rest in semi-cylindrical grooves formed at equal heights, in two pillars, so that a line, connecting the trunnions, shall be horizontal. A body thus supported will be free to assume every possible direction, in a vertical plane at right angles to the line joining the supports of the trunnions.

554. **Axles.** — These are, in general, the centres round which wheels revolve; but the wheels may either turn *upon* or *with* the axles.

The wheels of ordinary carriages turn upon their axles, the axles being strong iron bars, extending horizontally and transversely under the body of the vehicle. Their extremities, which pass through the centre of the wheels, have a cylindrical or slightly conical form. The centre part of the wheel in which the spokes are inserted, is called the *nave*, and the central part within it, into which the extremity of the axle passes, is called the *box*. The inner surface of the box has a form corresponding to that of the axle, but a little larger, so as to admit the play of some lubricating fluid between them, by the interposition of which the actual contact of the metallic surfaces of the axle and the box is prevented, no matter what be the pressure by which they are urged one against the other.

In railway carriages, the wheels and axle form one solid piece; the cylindrical or conical extremities of the axle project outside the wheels, and certain parts called *bearings* rest upon them. In this case, therefore, the extremities of the axle revolve with the wheels under the bearings. Reservoirs of lubricating matter, called the *grease boxes*, are placed immediately over the axles, by which the grease is let continually down upon the axles.

555. **Telescope joint.** — This mode of connection, which is of very extensive use in mechanics, is rendered familiar to every one, by the manner in which the tubes of telescopes and opera-glasses are adjusted. It is used, in general, where a rod or pillar requires to have its length varied occasionally, according to the circumstances in which it is used.

556. **Bayonet joint.** — The joint thus called takes its name from the simple and well known mechanical contrivance by which the handle of a bayonet is fastened on the end of a musket barrel. Its simplicity and efficiency have rendered it of extensive use in practical mechanics. As in the telescope joint, one tube passes within another; but in this case they are held in a fixed position

by means of a pin which projects from the inner tube, and passes through a rectangular opening in the outer tube.

Fig. 246. represents the position of the pin when the inner tube is pressed into the outer, until the pin comes against the edge of the opening and stops its further progress. In this position,

 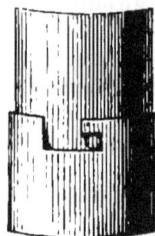

Fig. 246.   Fig. 247.

however, the inner tube might be detached from the outer by any force tending to draw it out. To prevent this, it is turned round its axis, until the pin enters the rectangular opening in the outer tube, as shown in *fig.* 247. It cannot be detached then without two successive motions being given to it; one round, and the other parallel to the axis of the tube.

557. **Clamps and adjusting screws.** — When parts of mecha-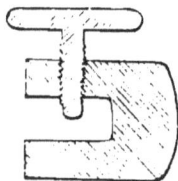nism usually separated require to be temporarily connected, the object is attained by clamps, which are made in an infinite variety of forms according to the circumstances in which they are applied. An example of this class of contrivances is shown in *fig.* 248. The ends of the parts to be united being placed one upon the other, are introduced into the rectangular opening shown in the figure, and the screw is then turned until it urge them by a pressure sufficiently strong to hold them together.

Fig. 248.

558. **Couplings.** — In the practical operation of machinery, it is frequently necessary to be enabled to suspend at pleasure, or to recommence, the motion of a wheel or wheels. This is accomplished by a class of contrivances called *couplings.*

When the motion of a wheel is imparted to another by an endless band of leather stretched tightly round them, the motion can be suspended or resumed, at pleasure, by making the band loose, and giving it the requisite tightness by pressing upon it a roller at any intermediate point between the wheels; so long as the pressure of the roller is maintained, the rotation of one wheel will be imparted to the other; but the moment the roller is removed, the communication of the motion will be suspended.

In some cases, the communication of the motion is discontinued or altered by removing the band from one or other of the wheels by lateral pressure. By this expedient, the velocity of the motion imparted may be modified. Thus, if several wheels be fixed side by side, on the same axis, having different diameters, the strap may be shifted from one to another, the velocity of rotation imparted being increased in the same proportion as the diameter of the wheel receiving the motion is diminished.

Couplings are sometimes constructed by providing two wheels upon the same shaft, one turning *upon* and the other *with* the shaft. Let us suppose that the moving power keeps the wheel which turns upon the shaft constantly in revolution, but does not act upon that which turns with the shaft. In that case, it is evident that no rotation would be imparted to the shaft. Now let us suppose that the wheel w w (*fig.* 249.) is that which turns *upon* the

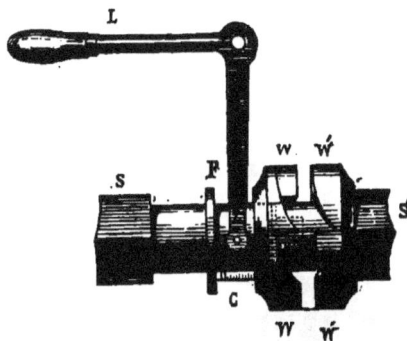

Fig. 249.

shaft, that attached to it is a collar c embraced by a fork F at the end of a lever L, that the wheel with the collar is capable of sliding longitudinally on the shafts, and that its surface towards s' is cut into a sort of angular teeth corresponding with similar ones formed in the face of the wheel w' w', which turns *with* the axle. When, by means of the lever L and the fork F, the wheel w w is moved towards the wheel w' w', so that the teeth are inserted the one within the other, the two wheels thus forming a single one, the wheel w w will impart its motion to w' w', and therefore to the shaft s s'.

On the contrary, when it is desired to suspend the motion of the shaft, the wheel w w is drawn by the lever L F towards s, and disengaged from w' w'.

A coupling by which the rotation of a shaft can be suspended or reversed at pleasure is shown in *fig.* 250., where w and w' are two bevelled wheels, both driven by a third bevelled wheel ʀ, and therefore, made to revolve in contrary directions. Let us suppose that these wheels w and w' turn *upon* the shaft s s', and carry at their centres two small wheels ɴ ɴ', the faces of which are cut into angular teeth; let ᴍ be two similar wheels similarly cut, presenting their opposite faces towards those of ɴ ɴ'; let these wheels ᴍ revolve *with* the shaft, but so as to be capable of sliding upon it; let the fork ꜰ, of the lever ʟ ꜰ, embrace between its prongs the collar ᴍ. By pressing this lever in the one direction or the other, the wheels ᴍ may be coupled, either with ɴ or ɴ', so as to revolve with them, and thus impart to the shaft s s' a motion in the one direction or the other.

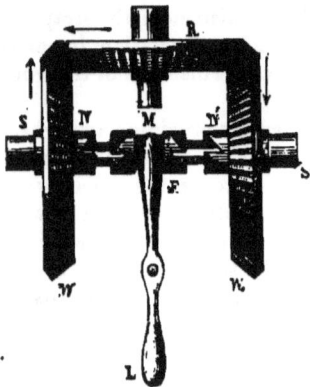

Fig. 250.

If the lever ʟ ꜰ be kept in the vertical position, the wheels ᴍ not being connected with either ɴ or ɴ', the motion will be suspended.

Another method of coupling is represented in *fig.* 251., where

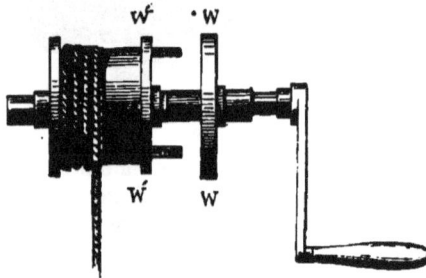

Fig. 251.

the wheel w w has two or more holes in it, corresponding with pins projecting from w' w'; the former revolves *with* and the latter *upon* the axle; the wheel w w slides upon the axle, so that when pressed towards w' w', the pins enter the holes and the wheel w' w' revolves with w w

CHAP. V.

RESISTING FORCES.

559. A physical agent capable of imparting motion to a quiescent body is called a force. It is evident that such an agent would also be capable of increasing the velocity of a body already in motion if it were applied in the direction of the motion, or diminishing the velocity, or even altogether destroying the motion and bringing the body to rest if it were applied in an opposite direction.

This principle is not, however, convertible : although it follows that an agent capable of imparting motion is also capable of diminishing or destroying it, it does not follow that an agent capable of diminishing or destroying motion is also capable of imparting or increasing it.

Forces of the class to which we now refer are capable of diminishing the velocity of a body in motion, and of bringing it to rest, but they are incapable of imparting motion to a body at rest, or of augmenting any motion it may have. The former class of forces, which are capable of producing or increasing motion, may be described for distinction *active forces*, and the latter *passive forces*.

The force of gravity, for example, comes under the former class. A body freely suspended, being disengaged and submitted to the action of gravity, is put in motion, and its motion is continually accelerated as it moves downwards, until it encounters some obstacle which brings it to rest.

If the same body be projected upwards with the velocity with which it strikes the ground, the force of gravity will then gradually diminish its motion until it rises to the height from which it fell, where its motion will altogether be destroyed.

Of the passive or resisting forces, the most important are friction and the resistance of fluid media, such as air or water. All bodies moving at or near the surface of the earth are subject to some, or all, of these forces, and, consequently, all terrestrial motions whatever are liable to constant retardation ; and, to be maintained, require the constant agency of some impelling force to repair the loss produced by the resisting forces to which they are exposed.

The smallest attention to the phenomena which form the subject of mechanical inquiries, will render manifest the great importance of investigating and comprehending the effects of resisting forces.

In the preceding part of this volume, the construction and properties of machinery have been explained, on the supposition that the moving force of the power is transmitted to the working point with undiminished effect. In order to disembarrass the questions of their complexity, and present them in the most simple and intelligible form, machines have been considered as absolutely free from the effects of all resisting forces ; surfaces moving in contact have been considered to be perfectly free from friction ; axles were regarded as mathematical lines ; pivots as mathematical points ; ropes as perfectly flexible ; and, in a word, the effect of the moving power has been considered as absolutely undiminished by any resistance whatever, in its transmission through the machinery to the working point.

It is scarcely needful to observe, that none of these suppositions are perfectly true. The surfaces of the machinery which move in contact are never perfectly smooth ; axles have always definite thickness, and move in sockets never perfectly polished; ropes have considerable rigidity, and this rigidity is necessarily greater in proportion to their strength. Much has been accomplished, it is true, by a variety of expedients, to diminish these resistances ; highly polished surfaces and effective lubricants have been applied to obtain additional smoothness ; but, still the surfaces in contact continue to be studded with small asperities, which, coming constantly in opposition to each other in their motion, produce considerable resistance and, robbing the moving power of a great part of its efficacy, transmit it with proportionally diminished intensity to the working point.

To estimate therefore, correctly, the practical effects of any machinery, it is essential that we should calculate the effect of this resistance, and subduct it from that effect of the power which has been computed on the theoretical principles established in the preceding chapters ; the overplus of effect after this deduction is all that part of the power which can be regarded as practically available.

560. **Friction.** — The effect of friction on a power supporting a weight or resistance at rest is different from its effect when the weight or resistance is moved. In the one case, friction assists ; in the other, it opposes the power.

Let us suppose, for example, a power P supporting a weight w, by means of a single movable pulley. From what has been already proved, it is evident that if the power P be half the weight of w, they would be in equilibrium, and the power would keep the weight at rest if the pulley were subject to no friction ; and in that case the slightest diminution of the power would cause the weight to descend, and draw the power upwards. But if the pulley be subject, as it always is in practice, to friction, then a small diminution of the power will be resisted by this friction, and the weight will not descend and overcome the power until the diminution of the power shall become so great as to enable the weight to overcome the friction.

561. It follows, therefore, that when a pulley or any other machine is subject to friction, a less power is sufficient to support a weight at rest than would be necessary if there were no friction; and the greater the friction is, the less will be the power necessary to support the weight. It is in this sense that friction is said to aid the power, when the weight or resistance is supported at rest. But if the power be required, not merely to support the weight, but to raise it, then we shall find that the friction, instead of aiding, opposes the power.

Let us suppose, for example, the power P acting on the weight through the intervention of a single movable pulley, the power being half the weight. In the absence of friction, the slightest addition to the power will cause it to descend, and to raise the weight; but when the machine is subject to friction, then the power will not descend until it shall receive such an addition as will be sufficient to overcome the friction.

562. This circumstance modifies materially the conditions of equilibrium. Representing by P that amount of the power applied to any machinery whatever which would keep the weight W in equilibrium in the absence of friction, let f express the addition which must be made to P, in order to enable it to overcome the friction and put the weight in motion; then f will also express the amount by which the power must be diminished, in order to enable the weight to prevail over it and to descend. It is evident, therefore, that any power which is less than P $+f$, and greater than P $-f$, would keep the weight in equilibrium and at rest.

563. It may therefore be inferred, generally, that when a machine of any kind is used simply to sustain a weight or to balance a resistance, the friction, acting in common with the power, becomes a mechanical advantage. In many instances this resisting force constitutes the entire efficiency of the instrument. Thus, when screws, nails, or pegs are used to bind together the parts of any structure, their friction with the surface with which they are in contact prevents their recoil, and gives them their entire binding power. In the ordinary use of the wedge itself we have another striking example of the mechanical advantage of friction. When the wedge is used for any purpose, as, for example, to split timber, it is urged forward by percussion, the action of the moving power being only instantaneous, and being totally suspended between each successive blow.

But for the resisting force of the friction which takes place between the surface of the wedge and the surface of the timber, the wedge would react after each blow, and render abortive the action of the moving power. The friction, therefore, in this case plays the part of a ratchet wheel, preventing the reaction of the wedge, and making good the action of the power.

564. Notwithstanding the disadvantages which attend the presence of friction in machines, it is an agent eminently useful in giving stability to structures, and in giving efficiency to the movements of almost all bodies, natural and artificial. Without friction, most structures, natural and artificial, would fall to pieces. The stones and bricks used in building owe to the mutual friction of their surfaces a great part of their solidity. Manual exertion would become impracticable if no friction existed between the limbs and the objects upon which they act. The friction between the foot and the ground gives a purchase to the muscular force, so as to enable it to produce a progressive motion. Without friction, every effort of the foot to propel the body forward would be attended with a backward action, so that no progressive motion would ensue. The difficulty of moving upon ice, or upon ground covered with greasy or unctuous matter, illustrates this. Without friction we could not hold any body in the hand. The difficulty of holding a lump of ice is an example of this. Without friction, a locomotive engine could not propel its load; for if the rail and the tire of the driving-wheels were both absolutely smooth, one would slip upon the other, without affording the necessary purchase to the steam power.

565. **Sliding and rolling.** — Friction is manifested in different ways, according to the kind of motion which one surface has upon the other. When one surface slides upon the other in the manner of a sledge, the friction is called sliding or rubbing friction. When a body rolls upon another, so that different points of such body come into successive contact with each other, it is called rolling friction.

566. The laws which regulate friction are derived exclusively from experiments, independently of theory. There are no simple or general principles from which they can be deduced by mathematical reasoning. It is a matter of regret that, even amongst the best conducted experiments that have been made, considerable discrepancies are observable, and that differences of opinion prevail between the most respectable authorities respecting many particulars connected with the properties and laws of these resisting forces.

Although these laws, so far as they are known, depend thus

wholly on experiment, yet the general principles of science, as applied to them, are far from being useless. They serve as a guide in the selection of the experiments which are best adapted to develop those laws which are the subject of inquiry, as well as to show the inconclusiveness of some experiments on which reliance might otherwise be placed, and thus enable us further to deduce from the results of experimental inquiries numerous useful prac tical results.

567. There are two methods by which the quantity of friction produced when two surfaces are moved one upon another can be ascertained.

1st. The surfaces being rendered perfectly flat, let one be fixed in a horizontal position on a table T T (*fig.* 252.), and let the other be attached to the bottom of a box B E, adapted to receive weights so as to vary the pressure.

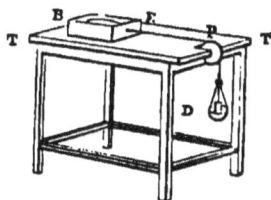

Fig. 252.

Let a flexible cord be attached to this box, and being carried parallel to the table, let it pass over a fixed pulley at P, and have a dish suspended to it at D.

If no friction existed between the surfaces, the smallest weight suspended from D would cause the box B E to move with a uniformly accelerated motion along the table towards the point; but the resistance of friction renders it necessary, before motion can take place or be maintained, that the weight D shall be equal to the amount of this friction. If the weight D and the friction be equal, then, the power and the resistance being in equilibrium, the box B E, if put in motion, will move towards the point with any velocity, continued uniform, which may be imparted to it. If the weight D be greater than the friction, then the motion of the box towards P will be accelerated; and if the weight be less than the friction, then any motion which may be given to the box will be retarded, and will soon cease altogether.

The determination, therefore, of the weight acting at D, which represents the exact amount of the friction, will depend upon the velocity given to the box in the direction B P being maintained uniform.

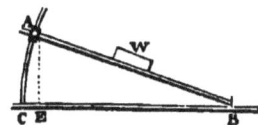

Fig. 253.

2ndly. Let one of the surfaces be attached, as before, to a flat plane A B (*fig.* 253.), but instead of being horizontal, let it be inclined, and so arranged that the inclination may be varied at pleasure. The box w being constructed as before, and placed upon the plane, let such an elevation be given to the plane that the box shall be capable of moving down it with an uniform velocity, without acceleration or retardation. If the elevation be greater than this, the motion of the box down the plane will be accelerated, and the gravity of the plane will be greater than the friction; if it be less, the motion of the box will be retarded, and the gravity will be less than the friction.

568. That particular inclination of the plane corresponding to the friction of any given surface, which renders the gravity of the plane equal to the friction, is called the *angle of repose*. According to the principles already explained, it follows that in these cases, if the length of the plane A B represent the total weight w, the gravity down the plane, which is equal to friction, will be represented by the height A E, and the pressure upon the plane will be represented by the base B E, and consequently the ratio of the friction to the pressure will be that of the height A E to the base B E. Experiments conducted according to both these methods have given nearly the same results, which may be summarily stated to be as follows.

569. The proportion of the friction to the pressure, when the *quality* of the surface is given, is always the same, no matter how the weight or the magnitude of the surface may be varied, except in extreme cases, when the proportion of the pressure to the surface is very great or very small.

570. If the surfaces in contact be placed with their grains in the same direction, the friction will be greater than if their grains cross each other. Smearing the surfaces with unctuous matter diminishes the friction, probably by filling the cavities between those minute projections which produce the friction.

571. The pivots of pendulums or balances are usually made of steel, and rest upon hard polished stones, different surfaces being used for the purpose of diminishing the amount of friction. Brass sockets are generally used for iron axles on the same principle.

572. **Lubricants.** — In the selection of lubricants, those of a viscous nature are selected, in the case of the rough surfaces of softer bodies, and those which are more fluid are applied to the smoother surfaces of harder bodies. Thus, when metal moves upon wood, tallow, tar, or some solid grease is generally used; but when metal moves upon metal, oil is preferred; and the harder and the smoother the metal, the finer the oil. Finely pulverised plumbago is found to be a very efficient agent in diminishing friction, especially as applied to the axles of carriages and the shafts of machinery.

573. **Rolling friction.** — The friction which attends a rolling motion is very much less than that which would attend a sliding or rubbing motion with the same surfaces and under the same pressure. Hence it is that rollers are used with so much success as an expedient for diminishing friction. A roughly chiselled block of stone, weighing 1080 lbs., was drawn from the quarry on the surface of the rock by a force of 758 lbs. It was then laid upon a wooden sledge, and drawn upon a wooden floor, the tractive force being 606 lbs. When the wooden surfaces moving upon one another were smeared with tallow, the tractive force was reduced to 182 lbs.; but when the load was, in fine, placed upon wooden rollers three feet in diameter, the tractive force was reduced to 28 lbs.

574. **Sledges.** — Although it is therefore obvious, on general principles, that the friction of sliding considerably exceeds that of rolling, vehicles supported on straight and parallel edges, sliding over the surface of the ground, are often found more convenient than wheel carriages. In climates where snow and frost prevail during the winter, sledges supersede all other carriages. In New

Fig. 254.

York and other cities of North America, public vehicles, such as omnibuses, hackney coaches, and all other conveyances, are taken off the wheels, and placed upon sledges, on which they are drawn along the streets and roads.

At all seasons and in all climates sledges are, for certain purposes, more convenient than wheel carriages; thus, draymen are provided with small ones to take barrels into narrow streets, where the approach of the dray would be difficult or impracticable.

Fig. 255.

575. **Use of rollers.** — When heavy weights, such as large blocks of stone, are required to be moved through short distances, the application of rollers is attended with great advantages; but when loads are to be transported to considerable distances, the

Fig. 256.

process is inconvenient and slow, owing to the necessity of continually replacing the rollers in front of the load, as they are left behind by each progressive advancement (*fig. 256.*).

Fig. 257

The wheels of carriages may be regarded as rollers which are continually carried forward with the load. In addition to the friction of the rolling motion on the road, they have, it is true, the friction of the axle in the nave; but, on the other hand, they are free from the friction of the rollers with the under surface of the load or the carriage in which the load is transported. The advantage of wheel carriages in diminishing the effects of friction is sometimes attributed to the slowness with which the axle A (*fig.* 257.) moves within the box, compared with the rate at which the wheel moves over the road; but this is erroneous. The quantity of friction does not in any case vary considerably with the velocity of the motion, but least of all does it in that particular kind of motion here considered.

Castors (*figs.* 258, 259.) placed on the feet of tables and other articles of furniture facilitate their movement from one place to another by substituting rolling for sliding friction.

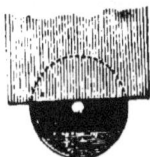

Fig. 258.

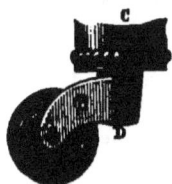

Fig. 259.

Fig. 260.

576. **Friction rollers** (s s, *fig.* 260.) are sometimes interposed between an axle and its bearings, to diminish the friction attending their motion one upon the other.

577. **Friction wheels.** — In certain cases where it is of great importance to remove the effects of friction, a contrivance called friction wheels or friction rollers is used. The axle of a friction wheel A (*fig.* 261.), instead of revolving within a hollow cylinder which is fixed, rests upon the edges of wheels B B, which revolve with it: the species of motion thus becomes that in which the friction is of least amount.

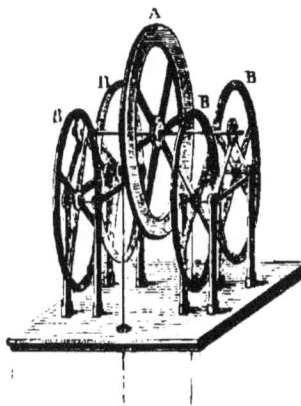

Fig. 261.

578. In carriages, the roughness of the road is more easily overcome by large wheels than by small ones ; hence we see wheels of very great magnitude used for carrying beams of timber of extraordinary weight. The animals drawing these, notwithstanding their weight, do not manifest any considerable exertion. The cause of this arises, partly from the carriage-wheels bridging over the cavities in the road, instead of sinking into them, and partly because, in surmounting obstacles, the load is elevated less abruptly.

579. **Line of draught.** — If a carriage were capable of moving on a road absolutely free from friction, the most advantageous direction in which the tractive force could be applied, would be parallel to the road; but when the motion is impeded by friction, as in practice it always is, it is better that the line of draught should be inclined to the road, so that the drawing force may be exerted partly in lessening the pressure on the road, by in some degree elevating the carriage, and partly in advancing the load.

It can be established by mathematical reasoning, that the best line of draught, in all cases, is determined by the angle of repose, that is to say, the traces should form an angle with the road equal to the elevation of a plane which would exactly overcome the friction. Hence it appears that the smoother the road, and the more perfect the carriage, and consequently the less the friction, the more nearly parallel to the road the line of draught should be.

In wheel carriages, there exist two sources of friction: one which prevails between the tires of the wheel and the road on which they run, the amount of which depends on the quality of the road; and the other which prevails between the axle and the nave of the wheel in which it turns. This latter is sliding friction; but the rubbing surface is small, being the line of contact of the axle with the nave or socket.

580. From the structure of the axle and the nave, this source of friction admits of being almost indefinitely diminished, by the application of lubricants and other expedients. The other resistance, depending on the action of the tires of the wheels, amounts, on well-paved roads, to about $\frac{1}{40}$th of the load. On gravelled roads it is but $\frac{1}{15}$th; and when a fresh layer of gravel has been laid, it is increased to the $\frac{1}{8}$th of the load.

It is found, however, on a well macadamized road, when in good order, that the resistance does not exceed the $\frac{1}{35}$th or $\frac{1}{40}$th of the load.

581. **Railway.** — The most perfect modern road is the iron railway, by which the resistance due to friction is reduced to an extremely small amount.

The rolling portion of the wheels is in this case diminished by substituting for the surface iron bars called rails, supported upon cross beams of timber (*fig.* 262.), at distances apart corresponding to that of the wheels, which are formed with ledges or flanges projecting from their tires. These falling within the rails (*fig.* 263.) confine the vehicle to the rails, the even surfaces of which in contact with the tires produce very little resistance.

Various experiments have been made, with a view to determine

T

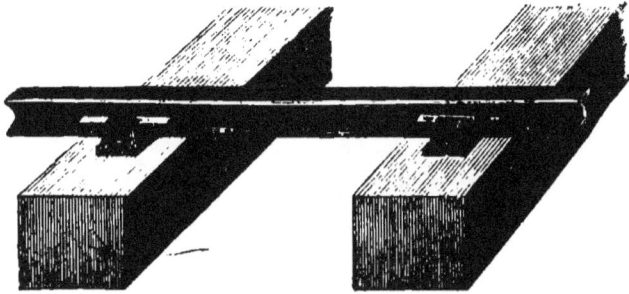

Fig. 262.

thin resistance ; but much difficulty arises, owing to the effects of atmospheric resistance being combined with those of friction. An extensive series of experiments was made by the author of this

Fig. 263.

volume, in the year 1838*, with a view to determine the amount of resistance to railway trains ; the result of which showed that this resistance is much more considerable than it had been previously supposed to be ; but that it depends in a great degree upon the velocity, and probably arises more from the resistance of the air, than from friction, properly so called.

582. **Brakes.** — Friction is generally resorted to as the most convenient method of retarding the motion of bodies, and bringing

---

* The details of these experiments will be found in the published reports of the meetings of the British Association in 1838 and 1841.

them to rest. Expedients of various forms, called *brakes*, have been contrived for this purpose.

Fig. 264.

The form of brake called a *shoe*, used in travelling carriages, is shown in *fig.* 264.

Fig. 265.

The brake used in diligences and other heavy vehicles on the continental roads, which is similar in principle to those used in railway carriages, is shown in *fig.* 265. Surfaces of wood are pressed against the tires of the wheels by a combination of levers pressed by the hand of the guard or conductor.

Fig. 266.

A brake used in machinery is shown in *fig.* 266., consisting of a strap tightly drawn upon the tire of the wheel intended to be retarded or stopped.

583. Friction is sometimes used as a point of resistance, as

T 2

where a cable is coiled round a post to arrest the progress of a vessel (*fig.* 267.).

Fig. 267.

584. **Imperfect flexibility of ropes.** — When ropes or cords form a part of machinery, the effects of their imperfect flexibility are in a certain degree counteracted by bending them over the grooves of wheels. But although this so far diminishes these effects as to render ropes practically useful, yet still, in calculating the power of machinery, it is necessary to take into account some consequences of the rigidity of cordage, which even by these means are not quite removed.

To explain the way in which the stiffness of a rope modifies the operation of a machine, we shall suppose it bent over a wheel, and stretched by weights A B, *fig.* 268., at its extremities. The weights A and B, being equal, and acting at C and D in opposite ways, balance the wheel. If the weight A receive an addition, it will overcome the resistance of B, and turn the wheel in the direction

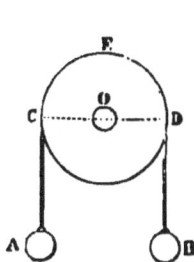

Fig. 268.

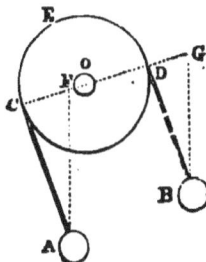

Fig. 269.

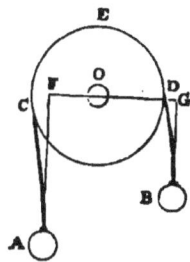

Fig. 270.

D E C. Now, for the present, let us suppose that the rope is perfectly inflexible, the wheel and weights will be turned into the position represented in *fig.* 269. The leverage by which A acts will be diminished, and will become O F, having been before O C;

and the leverage by which B acts will be increased to o G, having been before o D.

, But the rope, not being inflexible, will yield to the effect of the weights A and B, and the parts A C and B D will be bent into the forms represented in *fig.* 270. The preponderating weight A still has a less leverage than the weight B, and consequently, a proportionate part of the effect of the moving power is lost.

The extent to which the rigidity of cordage affects the motion of machinery has been ascertained by experiment in a still more imperfect manner than the results of friction. Many incidental circumstances vary the conditions, so as to throw great difficulties in the way of such an investigation. Different ropes, and the same ropes at different times, produce extremely different effects, influenced by the circumstances of their dryness or humidity, the quality of their material, the mode in which they are prepared and twisted, &c. These circumstances, it is evident, do not admit of being estimated or expressed with any degree of accuracy. It may, however, be stated generally that the resistance produced by the rigidity of a rope is directly proportional to the weight that acts upon it, and to its thickness. Other things being the same, it is also in the inverse proportion of the diameter of the wheel or axle upon which the rope is coiled; the greater the weights, therefore, which are moved, and the stronger the ropes, the greater will be the resistance proceeding from rigidity; and, on the other hand, the greater the diameter of the wheel or axle on which the rope runs, the less in proportion will be the force necessary to overcome the rigidity.

The resistance from new ropes is greater than from those of the same quality which have been some time in use. Ropes saturated with moisture offer increased resistance on that account.

585. **Resistance of fluids.** — Since all the motions which commonly take place on the surface of the earth are made in the atmosphere or in water, it is of great practical importance to ascertain the laws which govern the resistance offered by these fluids to the motion of bodies passing through them. A body moving through a fluid must displace as it proceeds as much of that fluid as fills the space which it occupies; and in thus imparting motion to the fluid, it loses by reaction an equivalent quantity of its momentum.

If a body thus moving were not impelled by a motive force in constant action, it would be gradually deprived of its momentum, and at length brought to rest. Hence it is, that all motions which take place on the surface of the earth, and which are not sustained by the constant action of an impelling power, are observed gradually to diminish, and ultimately cease.

T 3

Since the resistance produced by a fluid to the motion of a solid through it is equivalent to the momentum imparted by the solid to the fluid, which it thrusts out of its way in its motion, it follows . evidently that, other things being the same, this resistance will be proportional to the density or weight of the fluid. Thus the resistance produced by air is less than that produced by water, in the proportion of the weight of air to the weight of an equal bulk of water.

The resistances are proportioned to the quantity of the fluid which the moving body thrusts from the path; and this again depends upon the form and magnitude of the body, and more especially on its *frontage*.

586. The resistance which a body encounters in moving through a fluid is greater therefore, with a broad end foremost, than with a narrow end foremost. A ship would evidently encounter a much greater resistance if it were driven sideways, than if it move in the direction of the keel. It would also encounter a greater resistance if it moved stern foremost than in the usual direction.

The blade of a sword would be wielded with difficulty if moved with its flat side against the air, whereas it is easily flourished when moved edge foremost. Bodies whose foremost ends have the form of a wedge or a point, move through a fluid with less resistance than if the pointed ends were cut off, and they presented a flat surface to the fluid medium, because in their motion they act upon the principle of the wedge, and more easily cleave the fluid. Nature has formed birds and fishes in this manner to facilitate their passage through the air and through the water.

587. The resistance which a body moving through a fluid encounters, increases in a high ratio with its velocity. If the body move with the velocity of one foot per second, it will act in each second upon a column of the fluid, whose base is equal to its own transverse section, and whose length is one foot, and it will impel such a column with a velocity of one foot per second; but if the velocity of the moving body be doubled, it will not only drive before it in one second a column of fluid two feet long—that is, double the former length, but it will impel this column with double the former speed.

The resistance, therefore, will be doubled on account of the double quantity of the fluid, and again doubled on account of this quantity receiving double the velocity. The moving force, therefore, imparted by the body to the fluid when the velocity is doubled, will be increased in a fourfold proportion.

In the same manner it may be shown that if the velocity of the body be increased in a threefold proportion, it will drive from its path three times as much of the fluid per second, and impart to it three times as great a velocity; consequently, the moving force which it will impart to the fluid will be nine times that which it imparted moving at the rate of one foot per second.

In general, therefore, it follows that the moving force which the body imparts to the fluid in moving through it, will be increased in proportion to the square of the velocity; but as the resistance which the body suffers must be equal to the momentum which it imparts to the fluid, it follows that the resistance to a body moving through a fluid will be proportional to the square of its velocity.

588. **Ponderous missiles.** — Hence it follows that missiles lose a less proportion of their moving force in passing through the air, as their weight is increased; for, according to what has been stated, the resistance which they suffer at a given velocity will be proportional to their transverse section, which in this case is in the ratio of the squares of their diameters; but as their weight increases as the cubes of their diameters, and is proportional to their moving force, it follows that in increasing their magnitude their moving force is increased in a higher ratio than the resistance they encounter. For example, if two cannon-balls have diameters in the proportion of 2 to 3, the resistances which they will encounter at the same velocity of projection will be in the ratio of 4 to 9, while their weights will be in the ratio of 8 to 27. The resistance, therefore, of the smaller ball will bear to its weight a greater ratio than that of the larger.

589. **Resistance of the air to the motion of falling bodies.** — It has been shown that a body obedient to the action of gravity would descend in a vertical line with a uniformly accelerated motion. Its velocity would increase in proportion to the time of its fall, so that in ten seconds it would acquire ten times the velocity which it acquired in one second; but these conclusions have been obtained on the supposition that no mechanical agent acts upon the body save gravity itself. If, however, the body fall through the atmosphere, which in practice it must always do, it encounters a resistance which augments with the square of the velocity. Now, as the accelerating force of gravity does not increase, while the resistance continually increases, this resistance, if the motion be continued, must at length become equal to the gravitation of the falling body, and, when it does, the velocity of the falling body will cease to increase. It follows, therefore, that when a body falls through the atmosphere, its rate of acceleration is continually diminished; and there is a limit beyond which the velocity of its fall cannot increase, this limit being determined by that velocity at which the resisting force of the air will become equivalent to the gravity of the body.

As the resisting force of the air, other things being the same, increases with the magnitude of the surface presented in the direction of the motion, it is evidently possible so to adapt the shape of the falling body that any required limit may be impressed upon the velocity of its descent. It is upon this principle that parachutes have been constructed.

When a body attached to a parachute is disengaged from a balloon, its descent is at first accelerated, but very soon becomes uniform, and as it approaches the earth, the air becoming more and more dense, the resistance on that account increases, and the fall becomes still more retarded.

The theory of projectiles, which is founded upon the supposition of bodies moving in vacuo, is rendered almost inapplicable in practice, in consequence of the great effect produced by atmospheric resistance to bodies moving with such velocities as those which are generally imparted to missiles. According to experiments and calculations, it has been found that a 4-lb. cannon-ball, the range of which in vacuo would be 23226 feet, was reduced to 6437 feet by the resistance of the air. Hutton showed that a 6-lb. ball, projected with a velocity of 2000 feet per second, encountered a resistance a hundred times greater than its weight.

CHAP. VI.

STRENGTH OF MATERIALS.

590. THE solid materials of which structures, natural and artificial, are composed, are endued with certain powers, in virtue of which they are capable of resisting forces applied to bend or break them. These powers constitute an important class of resisting forces, and are technically called in mechanics the strength of materials.

Experimental inquiries into the conditions which determine the strength of solid bodies, and their power to resist forces tending to tear, break, or bend them, are obstructed by practical difficulties, the nature and extent of which have deterred many from encountering them.

591. These difficulties arise partly from the great force which must be employed in such experiments, but more from the peculiar nature of the bodies upon which such experiments are made.

The object of such an inquiry must necessarily be the establishment of a general law, or such a rule as would be strictly observed if the materials were perfectly uniform in their texture, and subject to no casual inequalities. In proportion, however, as such inequalities are frequent, experiments must be multiplied, so that they shall include cases varying in both extremes, so that the peculiar effects of each may be effaced from the general average result which shall be obtained. These inequalities of texture, however, are so great, that even when a general law has been established by a sufficiently extensive series of experiments, it can only be regarded as a mean result from which individual examples will be found to depart in so great a degree, that the greatest caution must be observed in its practical observation.

Although the details of this subject belong more properly to engineering than to an elementary treatise like the present, it may nevertheless be useful to give a general view of the most important principles which have been established.

592. A mass of solid matter may be submitted to the action of a force tending to separate its parts in several ways, of which the principal are —

    I. A direct pull; as when a weight is suspended to a wire or a rope, or when a tie-beam resists the separation of the walls of a structure.

    II. A direct pressure or thrust; as when a weight rests upon a pillar, or a roof upon walls.

    III. When a force is applied to twist or wrench a body asunder by turning a part of it round a point within it.

    IV. A transverse strain; as when a beam, being supported at its centre, weights are suspended from its ends, or, being supported at its ends, a weight is suspended from its centre.

593. When a rod, rope, or wire is extended between forces applied to its ends, and tending directly to stretch it, its strength to resist such force is, other things being the same, in proportion to the magnitude of its section.

Thus, suppose an iron wire stretched by a weight which it is just able to support without breaking. It is evident that a wire having twice the quantity of iron of the same quality in its thickness would support double the weight, because such wire would be in effect equivalent to two wires like the former combined. In the same manner, a wire having three times the quantity of iron of the same quality in its thickness, being equivalent to three wires like the first, would support three times the weight, and so on. Thus the power of bodies to resist a direct pull will be in general in proportion to the area of their transverse section. In practice it is found that when the length is much increased, the strength to resist a direct pull is diminished.

This departure, however, from the general law is explained by the increased probability of casual defects of structure in the increased length; and, subject to such qualification, it may be stated generally, that the strength of a body to resist tension, or a direct pull drawing from end to end, is in the direct ratio of the area of its section made at right angles to the direction in which it is stretched.

594. The strength of bodies to resist a direct pull is experimentally estimated by attaching the upper extremity securely to a point of support, and suspending weights to the lower extremity, which are increased gradually until the body under experiment is broken: the weight which breaks it is taken as the expression of its strength. The bodies which have been subjected to experiments of this kind are chiefly metals, woods, and ropes, these being most generally used in structures, in which their capacity for resisting tension is of great importance, as in suspension bridges, iron roofs, cables, cordage, &c.

595. In the following table are collected the results of the most recent and extensive experiments on this subject. The bodies subjected to experiment are supposed to be in the form of long rods, the cross section of which measures a square inch. In the second column is given the amount of the breaking weights, which are the measure of their strength. This table is selected from various recent works.

Table showing the strength with which prisms of the undermentioned substances resist a direct pull, expressed in lbs. per square inch of the area of their transverse section.

Name.	lbs.	lbs.	Name.	lbs.	lbs.
1st. Metals:—			2nd. Woods:—		
Steel, untempered	from 110690 to 127094		Teak	from 12915 to 15405	
— tempered	„ 114794 — 153741		Sycamore	„ 9630	
— cast	„ 134250		Beech	„ 12225	
Iron, bar	„ 53182 — 84611		Elm	„ 9720 — 15040	
— plate, rolled	„ 53920		Memel fir	„ 9540	
— wire	„ 58730 — 112905		Christiana deal	„ 12346	
— Swedish malle-			Larch	„ 12240	
able	„ 72064		Oak	„ 10367 — 25851	
— English do.	„ 55872		Alder	„ 11453 — 21730	
— cast	„ 16243 — 19464		Lime	„ 6991 — 20796	
Silver, cast	„ 40997		Box	„ 14210 — 24243	
Copper, do.	„ 20320 — 37380		Pinus sylv.	„ 17056 — 20395	
— hammered	„ 37770 — 39968		Ash	„ 13480 — 23455	
Brass, cast	„ 17947 — 19472		Pine	„ 10038 — 14955	
— wire	„ 47114 — 58931		Fir	„ 6991 — 12870	
— plate	„ 51240				
Gold	„ 20490 — 65237		3rd. Cords:—		
Tin	„ 3228 — 6666		Hemp twisted	lbs.	
— cast	„ 4736		¼ to 1 inch thick	8746	
Bismuth, cast	„ 3137		1 to 3 „	6800	
Zinc	„ 2820		3 to 5 „	5345	
Antimony, cast	„ 1062		5 to 7 „	4800	
Lead, molten	„ 887 — 1824				
— wire	„ 2543 — 3823				

596. From this table it appears that the strongest of the bodies for resisting tension is iron, and that the strongest condition of iron is that of tempered steel. In general, metals when cast are less strong than when hammered. Thus, cast iron has not one third of the tenacity of wrought iron. It is also found that metals

which are composed of two or more alloyed together are often stronger than any of their components. Thus, brass wire, which is composed of zinc and copper, has greater tenacity than copper wire, although the tenacity of zinc, as appears by the table, is extremely small.

597. It is also found that the strength of metals is affected by their temperature, being diminished in general as their temperature is raised. Sudden, frequent, and extreme changes of temperature impair tenacity.

598. The woods are subject to extreme variations, produced in general by the great inequalities which are incidental to them. Thus the strength of oak varies, as appears by the table, between the limits of 10000 and 25000 lbs. It is found that trees which grow in mountainous places have greater strength than those which grow on plains, and also that different parts of the same tree, such as the root, trunk, and branches, vary in strength within wide limits.

599. It is found that cords of equal thickness are strong in proportion to the fineness of their strands, and also to the fineness of the fibres of which these strands are composed. It is found also that their strength is diminished by being overtwisted. Ropes which are damp are stronger than ropes which are dry, those which are tarred than the untarred, the twisted than the spun, and the unbleached than the bleached. Other things being the same, silk ropes are three times stronger than those composed of flax.

The strength of many substances is increased by compressing them; this is the case with leather and paper, for example.

600. Animal and vegetable substances which, being originally liquid, are rendered solid by evaporation, change of temperature, or exposure, often possess extraordinary strength. Examples of this are presented in the gums, glue, varnish, &c. Count Rumford found that a copper plate having the thickness of 1-20th of an inch, rolled into the form of a cylinder, had its strength doubled when coated with well-sized paper of double its own thickness; and that a cylinder composed of sheets of paper glued together, having a sectional area of one square inch, was capable of supporting a weight of 15 tons for every square inch in its sectional area; and, in fine, that a cylinder composed of hempen fibres glued together had a strength greater than that of the best iron, being capable of supporting 46 tons per square inch of sectional area.

601. According to the theory of Euler, the strength of a column composed of any material and of any prismatic form to resist the crushing force of a weight placed upon it, increases as the square of the number expressing its sectional area divided by the square of the number expressing its height. This law, which is of great

practical importance on account of its application to architectural purposes, is found to be in very near accordance, within practical limits, with the experiments of Musschenbrock and others ; and more recently with the still more extensive experiments of Mr. Eaton Hodgskinson made on pillars of wrought iron and timber.

It is necessary to observe here, that when the height is reduced to a very small magnitude, the pillar is more easily crushed than would be indicated by this law. Exceptions are also presented in the case of pillars formed of particular materials ; such, for example, as cast iron, which Mr. Hodgskinson found to vary in rather a higher proportion in reference both to the sectional area and to the height.

According to Eytelwein's experiments, the strength of rectangular columns to resist compression is directly as the product of the larger side of their section multiplied by the cube of its shorter side, and inversely as the square of their height. This will coincide with that of Euler when the pillar is square.

602. Eytelwein gives the following table of the weights *necessary to crush pillars composed of the materials expressed in the first column, the numbers expressed in the second column being the total crushing weights in lbs. per square inch :* —

Name.	lbs.	lbs.	Name.	lbs.	lbs.
1. Metals : —			2. Woods : —		
Cast Iron	from 115813	to 177776	Oak	from 3860	to 5147
Brass, fine	,, 164864		Pine	,, 1028	
Copper, molten	,, 117088		Pinus sylv.	,, 1606	
— hammered	,, 103040		Elm	,, 1284	
Tin, molten	,, 15456		3. Stones : —		
Lead, molten	,, 7728		Gneiss	,, 4970	
			Sandstone, Rothenburg	,, 2556	
			Brick, well baked	,, 1092	

603. Mr. Hodgskinson gives the following results of his experiments as to the strength of timber pillars : —

TABLE.

Description of Wood.	Strength per Square Inch in Lbs.		Description of Wood.	Strength per Square Inch in Lbs.	
	Moderately dry.	After subjected to drying Process.		Moderately dry.	After subjected to drying Process.
Alder	6831	6960	Mahogany	8198	8198
Ash	8683	9363	Oak, Quebec	4231	5982
Bay	7518	7518	— English	6484	10058
Beech	7733	9363	Pine, Pitch	6790	6790
English Birch	3297	6402	— Red -	5395	7518
Cedar	5674	5863	Poplar	3107	5124
Red Deal	5748	6586	Plum, dry	8241	10493
White Deal	6781	7293	Teak	—	12101
Elder	7451	9973	Walnut	6063	7227
Elm	—	10331	Willow	2898	6128
Fir (spruce)	6499	6819			

The numbers in the first column are the number of lbs. per

square inch, deduced from experiments made on cylinders one inch in diameter and two inches in height, with flat sides, the wood being moderately dry. The second column gives the strength of similar pillars of the same woods, which had been subjected to a drying process in a warm place for two months and upwards. The great difference in the strength of the two cases, shows in a striking manner the effect of dryness upon the strength of timber.

604. Neither theory nor experiment has thrown much light on the laws which govern this mode of resistance to the separation of the constituent parts of bodies. The following results, showing the comparative strength of various metals, were obtained from a series of experiments made by Mr. Rennie.

Table showing the comparative Strength of various Metals to resist Torsion.

Lead	-	-	1000	English iron	-	- 10125
Tin	-	-	1438	Cast-iron	-	- 10600
Copper	-	-	4312	Blister-steel	-	- 16688
Brass	-	-	4688	Shear-steel	-	- 17063
Gun metal	-	-	5000	Cast-steel	-	- 19562
Swedish iron	-	-	9500			

It is stated by Mr. Bankes, that a square bar of cast-iron which would measure an inch, is wrenched asunder by an average force of 631 lbs. applied at the extremity of a lever two feet long.

605. Bodies submitted to a transverse strain are usually considered as having the form of prismatic beams, at right angles to which the strain is applied. A prismatic beam is one, all whose transverse sections are equal and similar figures, and the character of the beam is determined by the figure of this section. Thus, a cylindrical beam is one whose transverse section is a circle; a rectangular, one whose transverse section is a right-angled parallelogram; a square, whose transverse section is a square; and so on. The force producing the strain may be resisted by one or by two points of support, in the latter case the weight being placed between these two points.

In the first case, let B C, *fig.* 271., be a prismatic beam, fixed in a wall, or other vertical means of support, at B A, and let a weight W be supported from its end C. We shall, for the present, omit the consideration of the weight of the beam itself.

The effect of the weight W produced at B A will be to turn the beam round the point A, as if it were a hinge, as represented in *fig.* 272., and thus to tear asunder the fibres which unite the parts of the body forming its transverse section at B A. Let f be the force corresponding to the unit of surface of this section, let A be the area of the section; then $f \times A$ will express the total force of the body over the whole surface of the section. But if the beam tends to turn round the point A as a fulcrum, this force acts by a leverage at each point, corresponding to the distance of such point from A. The total force expressed by $f \times A$ may therefore be

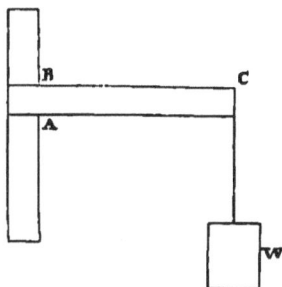
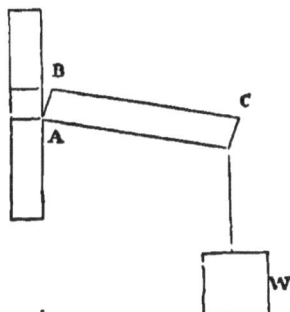

Fig. 271.   Fig. 272.

considered as having an average leverage, determined by the various distances of all the parts in the section from a horizontal line passing through A.

Now the point determined by these conditions is the *centre of gravity* of the section of the beam at A B. Let the distance of this centre of gravity from the horizontal line passing through the point A be c; we shall then have the effect of the forces of cohesion by which the body is united in the surface of the section expressed by $f \times A \times c$. Against this the weight w acts with the leverage c B, that is to say, the length of the beam.

Let this length be expressed by L, and we shall have, according to the properties of the lever,

$$w \times L = f \times A \times c,$$

and therefore

$$w = f \times \frac{A \times c}{L}.$$

It appears from this, therefore, that the weight necessary to fracture a beam placed in this manner, will be in the direct ratio of the area of the transverse section multiplied by the height of the centre of gravity above the lowest point of this section, and divided by the length of the beam.

For beams of the same length, their strength is proportional to the area of their section, multiplied by the distance of the centre of gravity above the lowest point; and for beams of the same section, their strength is inversely as their length.

By the length of the beam in this case is to be understood the distance of the part of the beam at which the weight is suspended from the point of support.

606. Let us now consider the case in which the body is supported at two points, the breaking weight being placed at some intermediate point. Let such a beam be represented in *fig.* 273., supported at the points D and C, the weight being placed at an intermediate point A. In this case, the tendency of the weight is to rupture the fibres in the vertical section

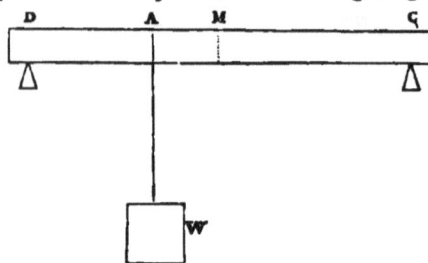

Fig. 273.

of the beam passing through A, and in doing so the beam would break as if a hinge were placed at A, on the upper surface of the section supporting the weight, as represented in *fig.* 274.

The fracture may in this case be considered to be produced by the reaction of the points of support D and C.

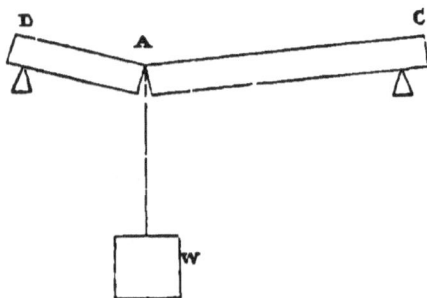

Fig. 274.

To determine the effects of this reaction, we are to consider that, by the properties of the lever, the part of the weight which acts it C as expressed by

$$w \times \frac{D\,A}{C\,D},$$

and the part of the weight which acts at D is expressed by

$$w \times \frac{C\,A}{D\,C}.$$

But the former of these acts upon the section at A with the leverage C A, and the latter with the leverage D A. If each be therefore multiplied by its respective leverage, we find that the effect of each of these actions in producing the fracture at A will be expressed by

$$w \times \frac{D\,A \times C\,A}{C\,D};$$

the total effect therefore will be

$$2 \text{ w} \times \frac{\text{D A} \times \text{C A}}{\text{C D}}.$$

Against this force the strength of the beam acts, and the strength of the beam at the section A is determined and expressed in exactly the same manner as in the former case, except that, in the present case, the average leverage by which the strength of the fibres resists rupture, is the distance of the centre of gravity below the highest point A of the section. Expressing this distance as before by c, we shall have the condition of equilibrium at the moment of fracture expressed by

$$2 \text{ w} \times \frac{\text{D A} \times \text{C A}}{\text{C D}} = f \times \text{A} \times c.$$

This may be simplified in the following manner: —

Let M, *fig.* 273., be the middle point of the beam between the two points of support, and let a express D M, half its length; let x express M A, the distance of the point where the breaking weight is placed from the middle point. We shall then have, by well-known principles of geometry,

$$\text{D A} \times \text{C A} = a^2 - x^2.$$

Hence we shall have

$$2 \text{ w} \times \frac{a^2 - x^2}{2\,a} = f \times \text{A} \times c,$$

and consequently we shall have

$$\text{w} = \frac{f \times \text{A} \times c \times a}{a^2 - x^2}.$$

Hence it follows that if the weight be placed at the middle point between the two points of support, we shall have $x = 0$; and consequently,

$$\text{w} = \frac{f \times \text{A} \times c}{a}.$$

It appears from this, that the weight placed at the centre of a beam, between two points of support necessary to break it, is double that which would be sufficient to break the same beam if it were supported only at one point, the weight being placed at the other point.

Such are the general practical principles for determining the transverse strength of beams as established by Galileo; and although other practical formulæ have been proposed by later writers, the above have been found in such near accordance with the average results of experiments made upon a large scale, that they have been generally adopted by mechanical writers as the basis of their investigations upon the strength of materials.

Let us now see how the preceding formulæ are modified for beams of particular forms.

607. If the beam be rectangular, let its breadth be b, its depth d, and its length $2\,a$. The distance of the centre of gravity of its section, therefore, from its upper or its lower surface will be $\frac{1}{2}\,d$. Now the area of its section will be $b \times d$: hence we have

$$A = b \times d,$$
$$c = \tfrac{1}{2}\,d,$$
$$L = 2\,a;$$

and consequently we have, if the beam be supported at one end only,

$$w = f \times \frac{b \times d^2}{4\,a}.$$

In like manner, if the beam be supported at both ends, the weight being placed at the middle, we shall have

$$w = f \times \frac{b \times d^2}{2\,a}.$$

If the beam be square, its breadth will be equal to its depth, and the breaking weight, where there is but one point of support, will be

$$w = f \times \frac{b^3}{4\,a};$$

and when there are two points of support,

$$w = f \times \frac{b^3}{2\,a}.$$

608. From the general principles here established it is evident that, so long as the quantity of matter composing the beam, and therefore its sectional area, remains the same, its strength will be augmented by any modification of form which will carry its centre of gravity to a greater distance from its lower surface if the beam have but one, and from its upper surface if it have two points of support. Some curious and instructive consequences ensue from this.

Thus, the strength of a rectangular beam, when its narrow side is horizontal, is greater than when its broad side is horizontal, in the same proportion as the width of its broad side is greater than the width of its narrow side. Hence, in all parts of structures

U

where beams are subject to transverse strain, as in the rafters of floors, roofs, &c., they are placed with their narrow sides horizontal, and their broad sides vertical.

If a beam, supported at two points, bear a strain at any intermediate point, a given quantity of matter composing it will have greater strength if it be so formed that its area shall be less in the upper part than in the lower.

Thus, a beam of triangular section, such as A B C (*fig. 275.*), will be stronger than a rectangular beam of equal section, such as A B C D (*fig. 276.*), because the centre of gravity of the triangle with its vertex upwards will be further from B than that of the parallelogram from B D.

But if the beam be supported at one point only, then the position of the triangle must be inverted.

Fig. 275.   Fig. 276.

609. If a solid and a hollow cylinder of equal lengths have the same quantity of matter, so that their sectional areas shall be equal, then their strength will be proportional to the distances of their centres of gravity from their external surface. But their centres of gravity being at their geometrical centres, it follows that the strength of the solid cylinder will be less than the strength of the hollow cylinder, in the ratio of the diameter of the solid cylinder to the diameter of the external surface of the hollow cylinder.

It appears, therefore, that the strength of a tube is always greater than the strength of the same quantity of matter made into a solid rod; the practical limitation of the application of this principle being, that the thinness of the tube should not be so great as to cause a local derangement of its form by the application of a strong force.

610. Innumerable striking and beautiful examples of this principle occur in the organised world. The bones of animals of every species are hollow cylinders, thereby combining strength with lightness. The stalks of numerous species of vegetables which have to bear a weight at their upper end are also tubes, whose lightness is remarkable when their strength is considered. These are intended to resist not only the crushing weight of the ear which they bear at their summit, but also the lateral strain produced by the movement of the air.

The quills and the plumage of birds, and especially of their wings, present a still more striking example of the application of this principle in the animal structure. The surface of the extended wing acting on the air produces a strong transverse strain upon the quill, which has a single point of support near the joint of the wing.

The number expressed by f in the preceding formulæ is the strength of a beam, whose sectional area is the square unit, which in this case is generally taken as the square inch. This number is always the same for the same species of material, but different for different materials. In tabulating the strength of materials obtained from experiment, this is accordingly the number which is taken to designate the strength of each species of substance. When this number f is known, the strength of a beam of any proposed dimensions can be immediately calculated by the formulæ; for it is only necessary to multiply this number f by the number of square inches in the sectional area, and the number of inches in the distance of the centre of gravity of this area from the upper or lower surface, as the case may be, and to divide the product by the length of the beam, and the result of this arithmetical process will give the breaking weight.

It must be understood that this weight, and the force expressed by the number f, are to be expressed in the same unit.

611. In ascertaining by experiment the average strength of each material, the number which it is important to determine is therefore that which is here expressed by f, and which may be considered the unit of strength; because, this being once determined, the strength of a beam of the proposed materials of any proposed denomination can be immediately computed.

This quantity f can be determined by direct experiment. If the amount of the breaking weight upon a beam of given dimensions be ascertained, we shall know the numbers severally expressed by w, b, d and a in the preceding formulæ; and in this case we shall have

$$f = \text{w} \times \frac{2\,a}{b \times d^2}.$$

In practice it will be found that the values obtained for the number f will be subject to considerable variations in different experiments, owing to casual inequalities and defects which affect each particular beam submitted to experiment.

The value of f must therefore, for each material, be obtained by taking an average of the results of numerous experiments, the casual inequalities being therein made to disappear from the result in proportion to the number and variety of trials.

612. In the following table is given the results of a series of experiments made by Tredgold upon the transverse strength of prismatic beams. The numbers in the second column represent the values of f in the preceding formulæ, while the numbers in the

third column express the number of lbs. weight in a cubic foot of
the material under experiment.

Table.

Name.	$f.$	lbs. Weight in Cubic Foot.	Name.	$f.$	Lbs. Weight in Cubic Foot.
1. Metals : —			2. Woods : —		
Malleable Iron -	17800	475	Oak - - -	3960	52
Hammered iron -	—	487	Fir (red or yellow) -	4290	34.8
Cast Iron - -	15300	450	Pine (American yel-		
Brass - - -	6700	586.25	low) - - -	3900	26.75
Zinc - - -	5700	439.25	Fir (white) - -	3630	29.3
Tin - - -	2880	455.7	Ash - - -	3540	47.5
Lead - - -	1500	709.5	Elm - - -	2340	34

613. Professor Barlow, of Woolwich, gives a table of the
strength of beams to resist transverse strain, from which the
following is extracted. The numbers in the second column in
this case represent the same number f as those in the second
column of the preceding table.

Table. •

Name.	$f.$	Name.	$f.$
Teak - - - -	9848	Elm - - -	4052
Poon - - - -	8884	Pitch pine - -	6528
English oak, 1st specimen -	4724	New England fir -	4408
„ 2nd specimen -	6688	Riga fir - -	4432
Canadian oak - -	7064	Mar Forest fir -	5048
Ash - - - -	8704	Larch - - -	4996
Beech - - -	6224	Norway spar - -	5896

614. If the weight producing the strain upon a beam, instead of
being concentrated at a single point, as supposed in the cases here
investigated, be equally distributed over the whole beam, the
power of suspension becomes twice as great as if it were applied at
the middle point of the beam.

It has been also shown that each point of the beam has a greater
supporting power the nearer it is to either point of support. It is
evidently, therefore, advantageous in all structures, in the distribution of the weight upon them, to throw a less quantity at the
centre, and to increase the quantity towards the points of support.
In this manner of loading, a far greater weight may be placed near
the walls than at the centre. In all cases, however, a concen-
tration of weight at a single point is to be avoided. A sheet of ice

---

• The numbers given by Tredgold are the values of

$$W \times \frac{L}{4\,B \times D^2}.$$

We have obtained the numbers given above by multiplying this by 4.

which would break under the weight of a skater would sustain the same skater if his weight were equally distributed over its surface. In allowing for the weight of the beam itself, this weight may be considered as uniformly spread over its surface.

615. According to Peschel, the transverse strength of a beam of timber may be greatly increased by sawing down from one third to one half of its depth, and driving in a wedge of metal or hard wood until the beam is forced at the middle out of the horizontal line, so as to form an angle presented upwards. It was found by such an experiment that the transverse strength of a beam thus cut to one third of its depth was increased one nineteenth; when cut to one half of its depth, it was increased one twenty-ninth; and when cut to three fourths of its depth, it was increased one eighty-seventh.

616. It follows from the principles which have been explained, that if any structure be increased in magnitude, the proportion of its dimensions being preserved, the strength will be augmented as the squares of the ratio in which it is increased. Thus, if its dimensions be increased in a two-fold proportion, its strength will be increased in a four-fold proportion; if they be increased in a three-fold proportion, its strength will be increased in a nine-fold proportion, and so on. But it is to be considered that, by increasing its strength in a two-fold proportion, its volume, and consequently its weight, will be increased in an eight-fold proportion; and by increasing its dimensions in a three-fold proportion, its volume and weight will be increased twenty-seven times, and so on. Thus it is apparent that the weight increases in a vastly more rapid proportion than the strength, and that consequently, in such increase of dimensions, a limit would speedily be attained at which the weight would become equal to the strength, and beyond this limit the structure would be crushed under its own weight. On the other hand, the more below this limit the dimensions of the structure are kept, the greater will be the proportion by which the strength will exceed the weight.

617. The strength of a structure of any kind is therefore not to be determined by its model, which will always be much stronger relatively to its size. All works, natural and artificial, have limits of magnitude which, while their materials remain the same, cannot be exceeded. Small animals are stronger in proportion than larger ones. We find insects and animalculæ capable of bodily activity, exceeding almost in an infinite degree the agility and muscular exertion manifested by the larger class. The young plant has more available strength in proportion than the forest tree.

An admirable instance of beneficence in the consequences of this principle is, that children, who are so much more exposed to accidents, are less liable to injury from them than grown persons.

CHAP. VII.

MOVING POWERS.

618. **Mechanical agents.** — The natural forces used for the production of mechanical effects are very various. Those applied to move machines, sometimes called *prime movers*, are animal power, water, wind, and steam. Besides these, however, there are several others which have been applied on a more limited scale to particular purposes, or under special circumstances. Some of these would be altogether inapplicable as movers of machinery, because their development cannot be rendered sufficiently constant and uniform. Others would be applicable; but their use has been hitherto excluded because of the expense which attends their production.

619. **The dynamical unit.** — Since all moving powers are applied to overcome resistances, and all resistances can be represented by equivalent weights, while the spaces through which they are urged can be represented by corresponding heights, the effect produced by a moving power is always expressed by a certain weight raised a certain height. Thus, for example, if a horse draw a loaded waggon with a force by which the traces are stretched with the same force as if 200 lbs. were suspended vertically from them, and if the horse thus acting draw the waggon over the space of 100 feet, the mechanical effect produced is said to be 200 lbs. raised 100 feet; or, what is the same, 20000 lbs. raised 1 foot.

In the expression, therefore, of all mechanical effects, it must be understood that the dynamical unit adopted is 1 lb. raised 1 foot, and that the effect produced by any moving power or mechanical agent is expressed by an equivalent number of such units. The great convenience attending this conventional mode of expression is, that the mechanical effects produced in different times, places, and circumstances by different mechanical agents can be immediately compared one with another, and their numerical ratio ascertained.

620. **Machines and tools.** — It happens rarely that a natural mechanical agent is applied immediately to the object on which it

is intended to act. In most cases some instrument or expedient is interposed which, when its form is simple and its magnitude small, is called a *tool*, and when greater and more complex, a *machine.* The line of distinction, however, between tools and machines, is not clearly defined; for we sometimes find large and complicated engines, such, for example, as those used for planing, punching, and boring metal, called tools.

The expedient by which an agent acts upon an object, not immediately, but by the intervention of a tool, is a mark of intelligence which has been considered by some as the most conspicuous distinction between man and the lower animals. Many of these display unmistakeable evidence of skill, sagacity, and reasoning under circumstances which will not allow of the supposition of mere instinct; but we believe that no example can be found among the lower animals of the production of any effect upon objects external to them otherwise than by the immediate use of those members or organs with which nature has furnished them, but never by the intervention of anything analogous to a tool.

621. **Animal power.** — The strength of animals may be employed as a moving power in various ways: the animal may remain without change of position, working by the action of its members, as when a man works at a windlass; or it may advance progressively, transferring its own body, and carrying, drawing, or pushing a load. The mechanical effect produced by such a power is subject to extreme variation, according to the various conditions under which it is exerted; and in an economical point of view it is therefore of extreme importance to determine, with respect to each animal, the circumstances and conditions under which the greatest amount of useful effect can be obtained.

In considering this question, two extreme cases present themselves in which the useful effect altogether vanishes, and between which it continually varies, increasing gradually from nothing to a maximum quantity, and then decreasing again to nothing.

622. There is a certain amount of load or resistance which requires the entire strength of the animal to sustain, no force remaining for motion. Thus, we can conceive a load which a man or horse is just able to support, but unable to move; in this case, therefore, the useful effect is nothing.

We can conceive, on the other hand, a speed of motion so great that with it the animal is unable to carry or draw any load, however small. At all intermediate speeds, however, some load more or less can be carried or drawn, the amount of such load being obviously greater as the speed is less. A horse can draw a greater load at three miles an hour than at ten miles an hour. The useful effect of the agent being, however, measured by multiplying the

load by the speed, it is evident that if either the load or speed be
unduly diminished, the useful effect will consequently be diminished, and there is a certain relation or proportion between the
load and speed which will give a maximum effect. This relation
is not only different with different animals, but varies even with
the same animal, according to the different ways in which it acts.

623. If the force of a man be applied, for example, to the mere
elevation of a weight, without any reference to the continuance of
its action, it is found that the most advantageous mode of exerting
the force is when he endeavours to raise a weight placed between
his legs. The greatest weight which can be raised in this case
varies from 400 lbs. to 600 lbs., according to the strength of the
individual; its average amount, however, is much less, not exceeding 250 lbs.

The force which can be exerted by human strength varies, as
has been already observed, extremely, according to the way in
which it is applied. Thus, if men work at a pile-engine, such as
represented in *fig.* 288., it is found that each can pull with a force
sufficient to raise about 144 lbs. weight of the ram 1 foot high,
and that working thus they can make about 20 strokes per minute,
which, being continued for 4 or 5 minutes, they require to recruit
their strength by an interval of repose.

624. When men work at a capstan, it is found that each can
exercise a force of about 28 lbs. on the extremity of the lever
which he pushes, and that with this force they can move at the rate
of about 10 feet per minute.

When a man works at a winch, of which the arm is from 12 to
14 inches from the centre on which it revolves, it is found that he
can exercise a force of about 16 lbs. on the handle, and can make
from 20 to 25 turns per minute.

In general, however, human labour produces, in the long run,
the greatest effect when it is exercised with frequent intervals of
rest; and accordingly the greatest effect is produced when an
animal ascends, raising nothing but its own weight, and produces
the mechanical effect by that weight itself in descending, so that
the animal actually reposes in the intervals during which the
mechanical effect is produced.

625. Thus, for example, if two baskets be connected by a rope
which passes over a pulley at the level to which heavy matter is to
be raised, the length of the rope being such that when one of the
baskets is brought up to that level the other shall have descended
to the level from which the heavy matter is to be elevated, the
lower basket being charged, let one or more men get into the
upper basket, so as to give it, by their weight, a sufficient preponderance to make it descend, drawing up the lower basket with its

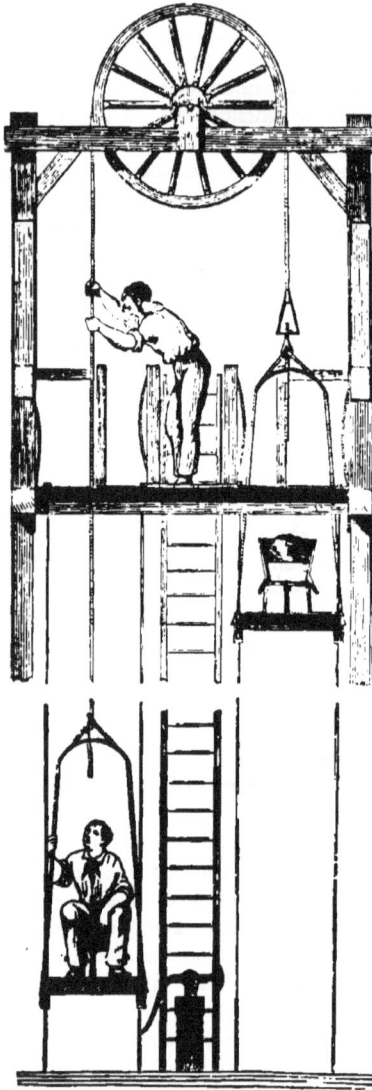

load. When the latter has
been brought up and dis-
charged at the upper level,
the men leave the lower
basket and charge it, when
men at the upper level get
into the upper basket, and
again bring up the lower
as before. Meanwhile, the
men who had previously
gone down again ascend to
the upper level by a ladder
or stair-case. In this manner the work may be con-
tinued without any inter-
mission, the labour of the
men consisting exclusively
in elevating the weights of
their own bodies while ascending the ladder or stair-
case.

It is found by experi-
ments, conducted upon a
large scale, that men working in this manner for eight
hours a day can produce in
each day an effect equivalent to 2,000000 lbs. raised
1 foot.

The great advantage ob-
tained by this mode of applying animal force will be
apparent, when it is stated
that the man who can thus
produce an effect of two
millions of pounds could
not produce, in working at
a windlass, a greater effect
than a million and a quarter;
while, if he acted upon a
pile-engine, such as that re-

Fig. 277.

presented in *fig.* 288., he would produce an effect of only three
quarters of a million.

626. An apparatus such as that here described is represented
in *fig.* 277 It was employed in making the earth-works at Vin-

, cennes, near Paris, and was found to be attended with great economy.

Analogous, though not quite identical with this mode of action, are the tread mill and the ladder wheel which are represented in *figs*. 148. and 150. In these cases the labour is expended upon the continual elevation of the body, which, however, descends as fast as it is elevated. The daily effect produced by men of the average strength working in this way for eight hours is found to be equivalent to 1,875000 lbs. raised 1 foot. Thus it appears that this method of employing human force is a little less advantageous than the method of ascent and descent, described above, which produces a daily effect of 2,000000.

The comparative disadvantage, slight as it is, is explained by the fact that in the one case the labour is intermitting, frequent intervals of repose alternating with those of labour, the most advantageous conditions under which animal force can be employed; while in the other case the labour is continuous.

627. Spade labour is one of the most disadvantageous forms in

Fig. 278.

which human force can be applied. The spade or shovel is a lever of the first or third kind, according as the one hand or the other applied to its handle is regarded as the fulcrum, the other being the power, and the earth taken up upon it being the weight. And since that part of the handle included between the two hands is always less than that between the lower hand and the earth, the force exerted is always greater than the weight of the earth raised.

In all extensive works, such, for example, as the formation of embankments on railways, the labour is executed by contract, the quantity of work being determined by the number of cubic yards of earth excavated and raised to a given height, or transported to a given distance. The number of cubic yards is computed by multiplying the superficial extent of the excavation, measured horizontally, by its average depth.

628. In the formation of railways, the surveying engineer endeavours, as far as possible, to render the quantity of earth-work in the cuttings, by which the road is carried through elevations, equal to the quantity necessary to form the embankments by which it is carried across the adjacent and intermediate valleys. In this case the quantity of labour is computed by taking the average distance between all the points of the cutting and all the points of the embankment, and assuming that the entire quantity of earth excavated in the cutting must be transported over this distance.

It rarely happens, however, that this exact equilibrium is attainable between the cuttings and embankments. If the earth of the cutting be in excess, the surplus is deposited in the nearest convenient place, forming what is called a *spoil-bank*; and if the earth of the cutting fall short of the quantity necessary for the embankment, the deficiency is made up by excavations made in the ground beside the embankment, called *side cuttings*.

629. **Horse-power.** — Before the invention and improvement of the steam-engine, the force of horses was very extensively used as a moving power; and although its application to machinery is now much less frequent, it is still resorted to, especially in places where fuel is expensive. Much of what has been said of the relative advantage of so applying the force of men is also applicable to that of other animals; and accordingly one of the most advantageous methods of applying such force is that represented in *fig.* 149., which is upon the principle of the tread mill, and in which the animal uses its strength in continually endeavouring to ascend, while, by the nature of the machine, those efforts are continually frustrated by its descent.

It is found, however, in practice, that the most convenient method of applying horse-power to machinery is by means of a large toothed wheel fixed on a strong vertical axis, and therefore turning horizontally; to the arms of this wheel two or more horses, called machiners, are yoked in such a manner as, by travelling constantly in a circle of which the axle of the wheel is the centre, and thus pushing against their collars, to make the wheel revolve. The wheel in this case may have the form called a crown wheel, as represented in *fig.* 280., in which case it gives motion to a horizontal shaft by means of a spur-pinion or basket

Fig. 279.

fixed on that shaft on which it acts; or the same object may be attained, perhaps preferably, by a bevelled wheel acting on a bevelled pinion, as explained in 450.

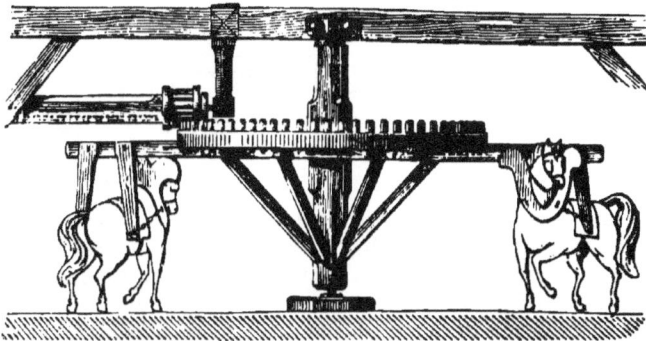

Fig. 280.

The greatest average maximum force which a horse can exert in drawing is about 900 lbs.; but when he works continuously, the force exerted must be much less than this. A good draught-horse working 6 days a week at 16 miles per day, and travelling 5 miles an hour, will bear a tractive force of about 110 lbs., and his daily labour will be equivalent to about 10,000000 lbs. raised 1 foot. A horse driving a machine, as represented in *fig.* 280.

produces a somewhat less effect than a draught-horse. To prevent the animal from needless fatigue by travelling in too small a circle, the arm to which he is yoked should not have less than 20 feet in length.

630. The effect produced by a horse of average strength so working is found to be equal to that of 7 average men working at a windlass.

The average effect produced per second of time by a machiner is equivalent to about 300 lbs. raised 1 foot.

An ox harnessed to a vehicle can exercise a tractive force equal to that of a horse, but he produces less than half the effect, owing to his natural slowness; but as a machiner, where speed is not required, the efficiency of an ox is nearly equal to that of a horse.

An ass of average force acting as a machiner has about one fourth of the efficiency of a horse.

631. When it is considered how very variable animal power is, whether we compare one animal of the same species with another, or even the same animal with itself, at different times and under different circumstances, it will be easily understood, that the estimates of different observers of the average force exercised by each species of animal differ considerably; it will, therefore, be useful here to give a few of the most generally received of these estimates.

In the following table is shown the estimates made by the observers named in the first column, of the average speed with which a man can walk without a load, for any continuance.

Observer.	Feet per Second.	Miles per Hour.
Schulze	5.37	3.66
Bernoulli	6.56	4.47
Guenyveau	from 6.56 to 9.84	4.47 6.70

The average powers exerted by a man employed in working a pump, turning a winch, ringing a bell, and rowing a boat are, according to Mr. Buchanan, in the proportion of the numbers 100, 167, 227, and 248. It follows from this that the act of rowing is that in which the force is exerted with most advantage.

632. According to Mr. Peron, the following is the average relative power exerted by men employed in continuous labour in the countries indicated:—

England - 71.4
France - 69.2
Timor - 58.7
Van Diemen's Land - 51.8
New Holland - 50.6

633. According to Tredgold, the average speed with which a

horse can travel without a load will vary according to the number of hours per day he works, as shown in the following table : —

Time of March in Hours.	Greatest Velocity per Hour in Miles.	Time of March in Hours.	Greatest Velocity per Hour in Miles.
1	14·7	6	6.0
2	10·4	7	5·5
3	8.5	8	5.2
4	7·3	9	4·9
5	6·6	10	4·4

According to observations made upon the performance of draught horses in Scotland, it appears that the average load, which a single horse can draw at the rate of 22 miles per day in a cart weighing 7 cwt. is a ton.

634. The following estimates of the relative forces of a man and other animals have been given by the authorities whose names are indicated, — Coulomb's estimate of the labour of a man being, in each case, taken as the unit : —

Carrying loads on the back on a level road : —
Horse, according to Brunacci - 4.8
Do., according to Wesermann - 6.1
Mule, according to Brunacci - 7.6
In drawing loads on a level road with a wheeled vehicle : —
Man with wheelbarrow, according to Coulomb - - 10.0

In drawing loads on a level road with a wheeled vehicle : —
Horse in four-wheeled waggon, according to Wesermann - 175.0
Do. in two-wheeled cart, according to Brunacci - - 243.0
Mule in do. do. do. - 233.0
Ox in do. do. do. - 122.0

Hassenfratz gives the following comparative estimates : —

In carrying loads on a level road : —
Man - - - - 1.0
Horse - - - 8.0
Mule - - - 8.0
Ass - - - 4.0
Camel - - - 31.0
Dromedary - - 25.0
Elephant - - 147.0
Dog - - - 1.0
Rein-deer - - 3.0

In drawing a load on a level road : —
Man - - - - 1.0
Horse - - - 7.0
Mule - - - 7.0
Ass - - - 2.0
Ox - - - 4 to 7.0
Dog - - - 0.6
Rein-deer - - 0.2

635. **Steam-horse.** — Every one is familiar with the term horse-power as applied generally to steam-engines, as well as to water mills and other machines, but few have a definite notion of its import ; horse-power thus applied, is a term merely conventional, having no reference whatever to the actual work of the animal from which its name is taken. A steam-engine which is capable of a mechanical effect per minute equivalent to 33000 lbs. raised 1 foot, is a steam-engine of 1 horse power ; and in general, if the effect produced per minute by any machine, whatever be the power which impels it, be expressed as usual by an equivalent number of pounds' weight, the number of horse-power which characterises it will be found by dividing the latter number by 33000.

636. **Water-power.** — This power is rendered available for
the motion of machinery by intercepting its fall or progress by
wheels having buckets or leaves at their circumference, upon which
the water acts either by its impact or weight so as to keep them in
revolution with a corresponding force; the form and effect of these
wheels, however, will be explained more fully under the head of
hydraulics.

637. **Wind-power.** — The force of the atmosphere in motion
is rendered available as a mover for machinery by means of arms,
called sails, which are driven round in a plane nearly vertical,
giving revolution to a horizontal axis. These machines will be
explained more fully under the head of pneumatics.

638. **Steam-power.** — When heat is applied in sufficient quantity to water, the liquid is converted into vapour which has all the
mechanical qualities of air, being elastic, expansive, and compressible. Its expansive force cannot be said to have any practical
limit, but the space through which it acts will decrease in nearly
the same proportion as that in which its intensity increases; thus,
if it be confined by a resistance equivalent to a pressure of 15 lbs.
per square inch, the steam will move the resistance until it has
swelled to about 1800 times the bulk it occupied as water. If
the resistance is 30 lbs. per square inch, it will move it, until it
has obtained a space equal to about 900 times its bulk as water,
and so on.

It follows from this, that when water contained in a vessel such
as a cylinder, confined by a movable piston pressing on it with a
force of 15 lbs. per square inch, is converted into steam, the piston
will be raised through about 1800 inches for every inch of depth
of the water evaporated. A moving force would, therefore, be
developed equivalent to a weight of 15 lbs. raised 1800 inches, or
150 feet; this would be in effect the same as 15 times 150 lbs.,
or 2250 lbs. raised 1 foot. It may, therefore, be stated in round
numbers, that when a cubic inch of water is evaporated a mechanical force is produced equivalent to a ton weight raised 1 foot
high. It does not matter under what pressure the evaporation is
produced, since any increase of pressure will be attended with a
proportionate decrease of the space through which the resistance
is moved.

The water, however, must in this case be regarded merely as a
medium, by which the mechanical effects of heat are evolved. The
real moving power is therefore, not the water, but the combustible
by which heat is produced, and that combustible being usually pit
coal, it becomes a question of the highest economical importance
to determine the average quantity of coal which is consumed in

the evaporation of a given quantity of water, and in the consequent evolution of a given amount of moving force.

639. The consumption of coal in the evaporation of water in steam boilers, varies considerably according to the circumstances under which the fuel is applied. In Cornwall, however, where the conditions most favourable to economy are strictly observed, it has been found that in experiments conducted under circumstances of the greatest precision, a bushel of coal, that is, 84 lbs. weight, of good quality, has produced a mechanical effect equivalent to 120,000000 lbs. raised 1 foot. This must, however, be regarded as an extreme experimental result. We may take, perhaps, 100,000000 pounds as the maximum mechanical effect attainable in regular work by the combustion of a bushel of coals.

Since it has been shown that the average maximum daily labour of a man, working under the most favourable circumstances, is 2,000000, and that of a horse 10,000000, it follows that 1 bushel of coals consumed daily can perform the work of 50 men or 10 horses.

640. It has been computed, on data such as these, that the materials of which the great pyramid of Egypt is formed, could have been raised from the ground to their actual position by the combustion of something less than 700 tons of coal. Herodotus states that 100000 men were employed for 20 years in raising this structure.

The Menai bridge consists of about 2000 tons of iron, placed at 120 perpendicular feet above the water level; its entire weight could have been raised to that height by the combustion of 400 lbs. weight of coals.

A train of coaches weighing 80 tons, and conveying 240 passengers, is drawn from Liverpool to Birmingham and back by the combustion of 4 tons of coke, the cost of which is 5*l.* To carry the same number of passengers daily, in stage coaches, on a common road, would require an establishment of 20 coaches and 3800 horses.

The circumference of the earth measures about 25000 miles; if it were begirt with an iron railway, such a train would be drawn round it in five weeks, by the combustion of about 500 tons of coke.

641. **Springs and weights.** — In a certain qualified sense, springs and weights may be considered as moving powers; they are, in fact, the movers in all watch and clock work. It must be observed, however, that the title of movers cannot be given to them in the same absolute sense as that in which it is applied to the several powers which we have described above. It has been already explained that when a watch or clock is wound up, the force which is exerted by the hand is expended in coiling up the

main spring or elevating the weight, and that the motion of the watch or clock from that moment until it ceases is merely produced by the spring or weight giving back by slow degrees and during a comparatively long interval the force which was exerted by the hand in winding them up. Strictly speaking, therefore, the moving power is in this case the force of the hand, and the spring or weight is merely the depository in which that force is collected, and from which it is given out until it is completely exhausted, and the clock ceases to move.

642. **Forces produced by heat.**—It is a physical law of high generality that when heat is imparted to or withdrawn from a body, expansion or contraction is produced. The force with which such expansion or contraction takes place has been sometimes used as a moving power or mechanical agent.

Air being eminently susceptible of such expansion and contraction by variation of temperature, attempts have been made, with more or less success, to apply the force of such expansion and contraction as a moving power. One of the most recent and ingenious of these projects is the caloric-engine invented by Captain Ericcson as a substitute for the steam-engine. This engine has been constructed on a large scale, and tried experimentally at New York in the propulsion of a vessel. The piston is urged, as in the steam-engine, on one side by the expansive force of heated air, the resistance of the air on the other side being diminished by a great reduction of temperature.

Hitherto this experiment has not been successful; the engine is reported to have operated with a certain efficiency, but under conditions not favourable to economy.

The expansion and contraction of metals has been used with great success in many places as a mechanical agent where great force has been required to be exerted through small spaces; examples of this have already been given.

643. **Mechanical force of congelation.**—Water at a certain point of the thermal scale exhibits a striking exception to the general law of contraction by cold. Just before congelation commences, the reduction of its temperature is attended, not with contraction, but with expansion; and this expansion takes place with irresistible force. When water, having percolated into the fissures and clefts of rocks, is frozen there, the force with which it expands in the process of congelation is such that the rocks are split asunder, and fragments of enormous weight and magnitude detached from them.

This force has been sometimes applied artificially for the fracture of large masses of stone.

644. **Electro-magnetic force.**—By expedients which will be

x

fully explained in another part of this Hand-book, a mass of soft iron can be rendered instantaneously magnetic, and as instantaneously deprived of its magnetism ; and these alternations in its magnetic condition can be made to succeed each other at any desired rate, however rapid. The magnetism imparted to it may also have any desired intensity. It is, therefore, easy to imagine that such a piece of iron will attract another situate at a limited distance from it, with a certain definite force while it is magnetic, and that the moment it loses its magnetism, the iron which it had thus attracted may be made to recoil from it either by its weight, by the reaction of a spring attached to it, or by any other convenient means. A force, acting alternately in one direction and another, of definite intensity, will thus be produced, which may be applied as a moving power in the same manner exactly as that in which the alternate motion of the piston of a steam-engine is applied.

Various attempts have been made to render this available as a general moving power for machinery; although the desired effect has been in many cases attained, the expense has hitherto so much exceeded that of steam power, that no successful practical result has been attained, except in the case of the workshop of M. Froment, the well known philosophical instrument maker of Paris, who has for some years used electro-magnetism as a moving power for the mechanism by which the divisions are engraved upon the limbs of circles, quadrants, and other similar instruments of precision. We shall describe this machinery more fully in another volume of this Hand-book.

645. **Chemical agency.** — A great variety of powerful mechanical effects are produced by the chemical combination or decomposition of bodies. These phenomena, however, are generally developed under circumstances and conditions which render their application as mechanical agents often difficult, and sometimes impracticable. The circumstances under which the mechanical force is thus developed are generally those in which solid or liquid bodies of comparatively small dimensions are suddenly converted into gaseous bodies of great volume, the change being usually attended with a large and intense development of heat. The gases thus evolved at a high temperature, expanding with prodigious force, drive before them whatever resistances may be opposed to their dilatation.

One of the most familiar examples of this class of phenomena, and that which has been most extensively applied, is gunpowder. This substance is a mixture of nitre, charcoal, and sulphur, all in a very pure state, and in certain definite proportions; thus, in 100 lbs. weight of gunpowder there are generally 75 lbs. of nitre, 15 lbs. of charcoal, and 10 lbs. of sulphur These proportions are subject to a small variation in different qualities of powder, which need not be noticed more particularly here.

When a spark is applied, the charcoal and the sulphur heated by it, attract the oxygen, which is one of the constituents of the nitre; and, combining with it, form gases, called carbonic oxide, carbonic acid, and sulphurous gas; while the constituent of the nitre, disengaged from the oxygen, is the gas called nitrogen, or azote. The result, therefore, of the phenomenon is the instantaneous evolution of a mixture of the four gases above named; this evolution being attended with such intense heat, that the gases are incandescent, or luminous. It is found that the gunpowder, in this process of explosion, swells into 2000 times its volume. Count Rumford found that 28 grains of gunpowder, screwed into a cylindrical space within a piece of iron, tore it asunder with a force of 400000 lbs.

646. **Gun cotton.**—If a piece of common raw cotton, usually called cotton wool, be steeped in a mixture composed of equal measures of sulphuric and nitric acids, and be then pressed and dried, it will, to all external appearance, be the same as before; but if it be weighed it will be found to be nearly one half heavier. The change which it has undergone is this: the cotton has lost a quantity of water which was combined with it, equal to about one third of its weight, but it has entered into combination with such a quantity of nitric acid as to give the whole a weight 50 per cent. greater than that of the original weight of the cotton.

The cotton thus prepared is highly explosive, and the effects of its explosion are explained on the same principle as those of gunpowder. It is considered by chemists that, weight for weight, the force of this substance is greater than that of gunpowder. It explodes also at a lower temperature, and, when of good quality, leaves no perceptible residuum. It is, therefore, probable that when the process of making it has undergone those improvements which it is likely to receive, it will replace gunpowder in fire-arms, as it has already done to a certain extent in the blasting of rocks.

647. **Capillary attraction.**— The mechanical agency supplied by this principle in the arts, and its use in the vegetable economy, will be explained in another volume of the Hand-book.

648. **Perpetual motion.**—There is no mechanical problem on which a greater amount of intellectual ingenuity has been wasted than that which has for its object the discovery of the perpetual motion. Since this term, however, is not always rightly understood, it will be useful here to explain what the perpetual motion is not, as well as what it is.

The perpetual motion, then, which has been the subject of such anxious and laborious research, is not a mere motion which is continued indefinitely. If it were, the diurnal and annual motion of the earth, and the corresponding motions of the other planets and satellites of the solar system, as well as the rotation of the sun upon its axis, would be all perpetual motions.

To understand the object of this celebrated problem, it is necessary to remember that, in considering the construction and performance of a machine, there are three things involved : 1st, the object to which the machine gives motion ; 2ndly, the structure of the mechanism ; and 3rdly, the moving power, the effect of which is transmitted by the machine to the object to be moved. In consequence of the inertia of matter, the machine cannot transmit to the object more force than it receives from the moving power; strictly speaking, indeed, it must transmit less force, since more or less of the moving force must be intercepted by friction and atmospheric resistance. If, therefore, it were proposed to invent a machine which would transmit to the object to be moved the whole amount of force imparted by the moving power, such a problem would be at once pronounced impossible of solution, inasmuch as it would involve two impracticable conditions; first, the absence of atmospheric resistance, which would oblige the machine to be worked in a vacuum ; and secondly, the absence of all friction between those parts of the machine which would move in contact with one another.

But suppose that it were proposed to invent a machine which would transmit to the object to be moved a greater amount of force than that imparted by the moving power, the impossibility of the problem would in this case be still more glaring, for even though the machine were to work in a vacuum, and all friction were removed, it could do no more than convey to the object the force it receives. To suppose that it could convey more force, it would be necessary to admit that the surplus must be produced by the machine itself, and that, consequently, the matter composing it would not be endowed with the quality of inertia. Such a supposition would be equivalent to ascribing to the machine the qualities of an animated being.

But the absurdity would be still greater, if possible, if the problem were to invent a machine which would impart a certain motion to an object without receiving any force whatever from a moving power ; yet such is precisely the celebrated problem of the perpetual motion.

In short, a perpetual motion would be, for example, a watch or clock which would go so long as its mechanism would endure without being wound up ; it would be a mill which would grind corn or work machinery without the action upon it of water, wind, steam, animal power, or any other moving force external to it.

It is not only true that such a machine never has been invented, but it is demonstrable that so long as the laws of nature remain unaltered, and so long as matter continues to possess that quality of inertia which is proved to be inseparable from it, not only in all places and under all circumstances on the earth, but throughout the vast regions of space to which the observations of astronomers have extended, the invention of such a machine is an impossibility the most absolute.

BOOK THE FOURTH.

ILLUSTRATIONS OF THE APPLICATION OF MECHANICAL PRINCIPLES IN THE INDUSTRIAL ARTS.

~~~~~~~~~

CHAPTER I.

CRANES.

649. **Cranes** are pieces of mechanism usually consisting of combinations of toothed wheels and pulleys, by means of which mate-

Fig. 281.

rials for building are raised to the stage where the builders are engaged — goods elevated from vessels to the quays or wharves, and loaded on the waggons by which they are to be transported, or, on the contrary, transferred from the quays or wharves into the vessels—and for all similar purposes.

650. **Movable cranes.**— One of the most simple forms of movable cranes commonly used for building in France is shown in *fig.* 281.

It consists of a sort of strong triangular ladder, at the top of which is a fixed sheave c, over which the rope attached to the object to be elevated passing is carried down to the cylindrical axle T, upon which it is rolled by means of bars inserted in holes, as shown in *fig.* 145.

This ladder is inclined more or less from the vertical by means of a rope o D, which is carried to some fixed object at a distance, to which it is attached.

651. Another form of movable crane, also used for building purposes, is shown in *fig.* 282., which consists of a shaft driven

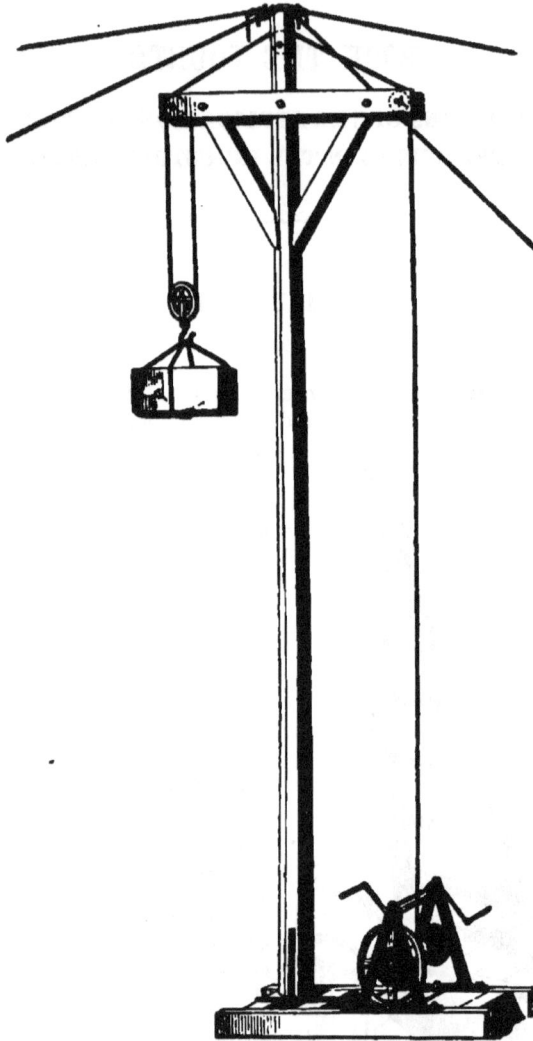

Fig. 282.

by a combination of toothed wheels and a pair of winches. Upon the shaft a rope is coiled, which, being carried up, is passed over three sheaves, and then, passing downwards, is attached to the object to be raised.

652. Movable cranes for manufactories and railway stations. — Another form of movable crane, much used in large

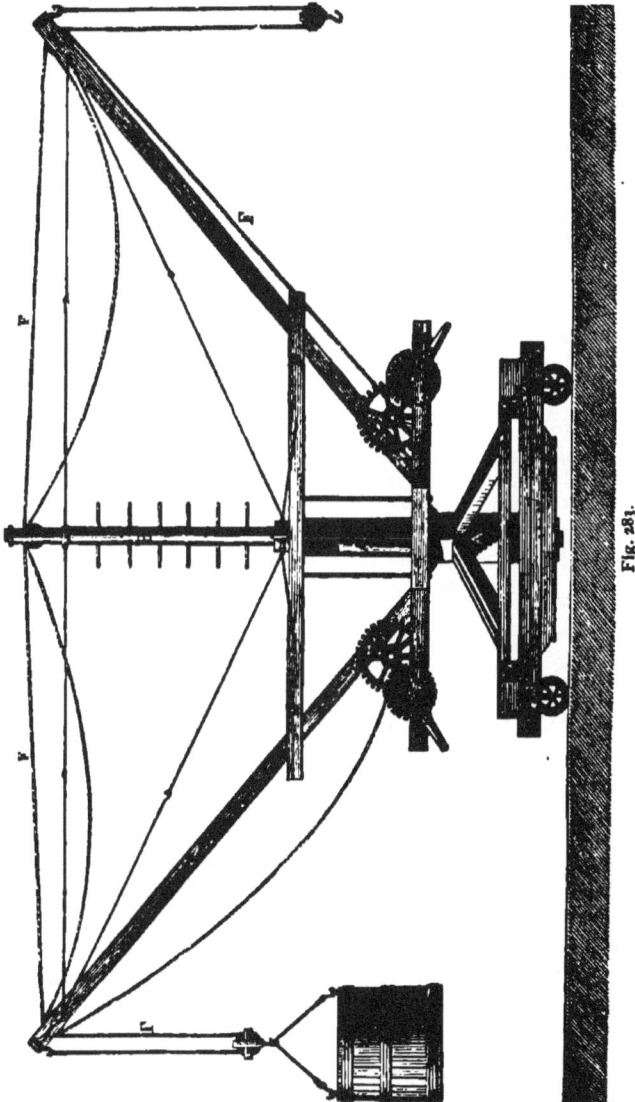

Fig. 283.

engineering establishments and railway stations, is shown in *fig.* 283. The figure is drawn upon a scale of 1 inch to 12 feet.

The stand upon which the apparatus is fixed is supported on four small wheels, which, in some cases, are made to run upon rails like those of a railway waggon. It is evident, from what has been explained in Chap. V. Book II., that the apparatus must be so constructed that the centre of gravity of it and its load must be in a vertical line which shall fall within the quadrangle formed by the lines which join the four wheels.

A strong pillar of wood, A A, is secured upon the base of the apparatus by diagonal ties. In this a circular vertical hole is perforated, in which the round vertical beam B which supports the crane is inserted, so that it can revolve horizontally without much resistance from friction. The horizontal beams C C and D D are attached to collars, which also turn freely round A, while they are supported by it. The diagonal pieces E E are attached to these horizontal pieces, and also tied by iron rods to the central axis, both in horizontal and diagonal directions. Two winders with systems of wheelwork attached are fixed at C and C, by which the axles upon which the ropes are coiled are turned. In the position represented in the figure, the right-hand winch alone is worked. The rope F coiled upon its axle passes over a sheave at the top of the right-hand diagonal beam E, from which it is carried horizontally over a sheave at the top of the vertical pillar B, and thence to a sheave in the top of the left-hand diagonal beam E, from whence it passes downwards and under the movable pulley which supports the weight to be elevated, and is finally attached to a hook or ring at the top of the left-hand diagonal beam E.

The winch C and its apparatus of wheelwork acts in a similar manner over another series of sheaves in juxtaposition with the former upon the movable pulley suspended from the right-hand diagonal beam E.

653. **Fixed cranes.** — When cranes are stationary, as is usually the case upon wharves and landing-places, they are constructed in

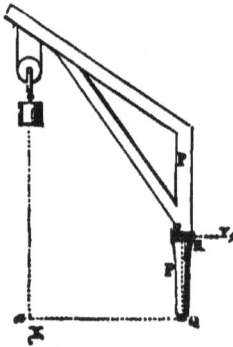

a more efficient manner, giving greater efficacy to the power by the interposition of a greater number of wheels, axles, and pulleys. They are usually made to revolve upon a centre, so that the goods raised from the vessel may, by lowering the crane, be brought over the waggon on which they are to be deposited. This is accomplished by inserting a round piece of metal, P Q (*fig.* 284.), at the lower part of the apparatus, into a hole of corresponding magnitude made in the wharf. The level of the wharf being R Y, the crane, when it turns in the hole, will carry the object

Fig. 284.

elevated in a circle, of which Q X is the radius, round P Q, the centre or axis.

. A crane thus constructed is represented, with all its accessories, in *fig.* 285., upon a scale of 1 to 100, or an inch to 8 ft. 4 in. The wheelwork is shown upon a larger scale and more evidently in *fig.* 286.

The weight is attached to a single movable pulley suspended from the upper end of the crane. The rope which is carried over the fixed sheave, and along the inclined arm of the crane to the axle A (*fig.* 285.), is stretched by a force equal to half the weight raised. Upon the axle A is fixed a toothed wheel B, which is driven by a pinion C (*fig.* 286.). Upon the axle of this pinion is fixed another large toothed wheel D, which is driven by a

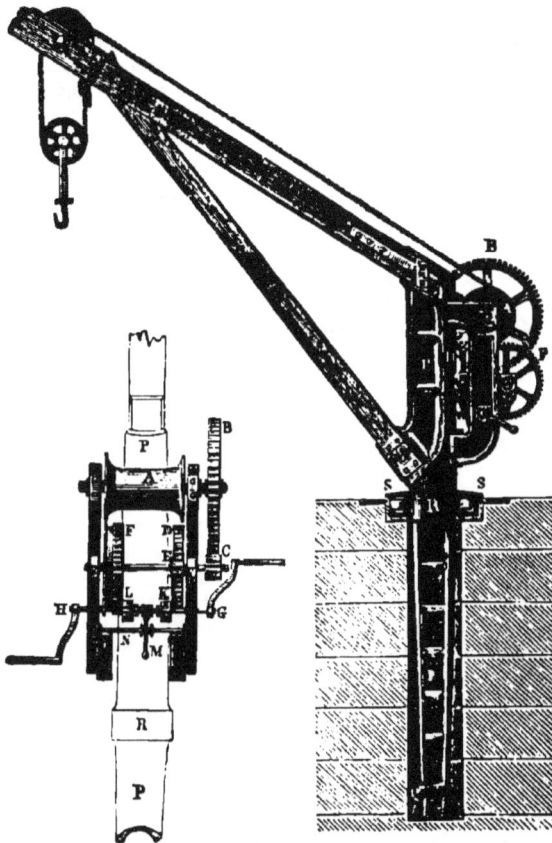

Fig. 286.   Fig. 285.

pinion E. Upon the axle of this pinion E is fixed another large toothed wheel F. The two axes of the wheels D and F being upon the same level, as shown in the figure, one conceals the other.

Two winches G and H work an axle upon which two pinions L and K are fixed, which are capable of being moved through a small space right and left horizontally by a lever M. These pinions are formed to engage in the wheels D and F, and can be thrown into or out of connection with their wheels by the lever M. As shown in the figure, both are out of connection with D and F, and the crane is inoperative. If we suppose them both in connection with the wheels, the pinion L, driven by the winch H, will drive F. and consequently E, which will drive D. At the same time the pinion K,

driven by the other winch G, will drive D. Thus, the wheel D will receive
at once the effect of the two powers applied to the winches G and H. This
wheel D will drive the pinion C fixed upon its axle, which will drive B, and
consequently will cause the axle A on which the rope or chain is coiled to
revolve, and thus to raise the object to be elevated.

After what has been explained respecting the relation of the power and
weight, it will be evident that the efficiency of such an apparatus will altogether depend on the relative dimensions of the wheels and pinions. But in
all cases it is apparent that the weight must bear a very large ratio to the
power. A crane constructed upon this principle by M. Cavé, the well-known
engine maker at Paris, has been erected in the government dockyard at
Brest, in which the wheels have the following numbers of teeth :—

The wheel B, 66 teeth.
The pinion C, 11 teeth.
The wheel D, 54 teeth.
The pinion E, 9 teeth.

The wheel F, 54 teeth.
The pinion K, 9 teeth.
The pinion L, 9 teeth.

By this crane a power of 20 lbs. applied to each of the winders G and H will
be sufficient to raise 25 tons.

Either of the pinions K or L may be put into connection with the wheels
without the other, so that, when the weights to be elevated are not too great,
the crane may be worked by one winch only.

**654. Drops for loading and unloading vessels at quays
and wharves.** — To facilitate the process of loading and unloading vessels, an ingenious contrivance, which is very nearly self-
acting, has been adopted, which is represented in *fig.* 287.

An iron railway, properly supported on the level of the quay or wharf, is
continued a few feet beyond its edge, so as to overhang the water. To the
end of a long lever which turns on a centre attached to the quay wall, nearly
level with the deck of the vessel, is suspended a platform, upon which the
waggon containing a portion of the matter with which the vessel is to be
freighted stands. The end of this lever is connected by a rope G with the
axle of a windlass C, established on the wharf. Coiled upon the same axle,
but in the contrary direction, is another rope F, attached to the end of
another lever E, to which a counter-weight D is suspended. The waggon B
loaded preponderates a little over the weight D, and by this preponderance
it descends upon the deck of the vessel, suitably placed to receive it. When
the contents of the waggon are discharged in the vessel, the weight D preponderates, and draws the waggon up.

In this manner the weight itself of the goods to be loaded becomes the
moving power by which the process is executed.

If, on the contrary, it be required to unload a vessel, the windlass C must
be worked by some power adequate to raise the weight deposited in the
waggon to the level of the wharf. No advantage would be gained in that
case by increasing the counter-weight D, so as to give it a preponderance
over the loaded waggon, since a power should in that case be applied to let
the empty waggon down. But if it could be so managed that while one
cargo was being unloaded another could be loaded, the work would be executed with a very small power, by adjusting the counter-weight D so as to
balance, as nearly as possible, the weight of the loaded waggon both in
ascending and descending. In that case power need only be applied sufficient
to overcome the friction and other resistances of the machinery, and any

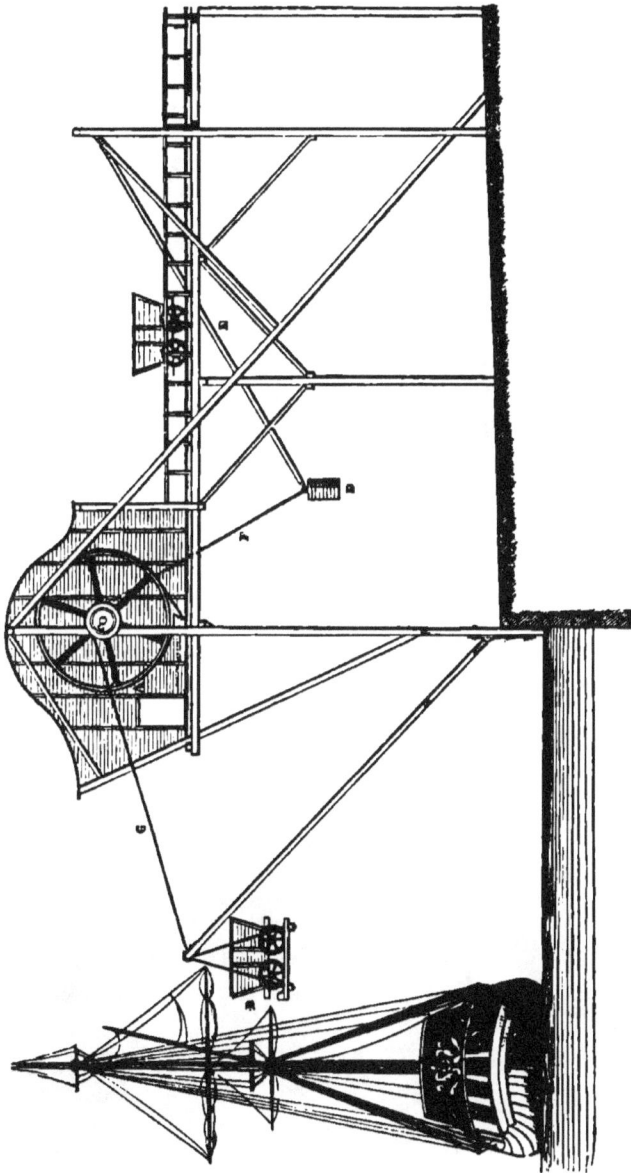

Fig. 287.

small difference that might exist between the effect of the loaded waggon
and the counterpoise.

CHAP. II.

ENGINES WORKING BY IMPACT.

655. **Pile-engines.** — This class of machine is an example of the application of the momentum acquired by a heavy mass falling from a certain height as a mechanical agent.

Heavy weights are raised by the moving power to a greater or

Fig. 283.

less height, from which they are let fall on the object upon which they are intended to act. The forms and magnitudes of such weights, and the mechanism by which they are elevated and detached, vary according to the purposes to which the machines are applied.

The form of machine properly called a Pile-engine is applied to drive piles into the ground, upon which bridges or other structures are intended to be erected.

One of the most simple and inartificial forms of such a machine is represented in *fig.* 288., where D is the pile to be driven into the earth, the top of which is bound by a ring of iron to prevent it from being split or flattened by the blows; A is the weight let fall upon it, which is usually a large block of cast iron called a ram, having two ears cast upon its sides, which play in vertical grooves formed in the two uprights C C, by which the ram is guided in its fall, and prevented from deviating from the vertical line. In this case the ram is elevated by human labour, the extremity of the rope being connected with several independent cords each of which is pulled by a man; the rope passes at the top of the framing over a fixed sheave B, from which it descends, and is tied to a ring at the top of the ram.

When the ram is elevated to the top of the framing, the men let go the ropes simultaneously by a signal. It is very important that the ropes should be all let go at the same instant, since if some men held on after the others had let go they would be dragged up, and grave accidents would ensue. Where such machines are used, the men are accustomed to accompany their work by a sort of chaunt which regulates the moment of dismissal.

A less inartificial pile-driver is shown in *fig.* 289., where the rope which raises the ram, instead of being directly drawn by human power, is worked by a windlass having two winches, A A, by means of which two or four men can raise a ram of enormous weight.

The axle B, turned by the winches, has upon it a pinion E which works in the teeth of a large wheel attached to the drum upon which the rope is coiled. When the ram is raised by the windlass to the top of the framing, the pinion E is disengaged from the teeth of the large wheel by means of a lever C D E, which turns horizontally on a centre D, and is terminated in a fork at E, which, laying hold of the axle B, moves it horizontally so as to make it slide in its bearings, and thus to withdraw the teeth of the pinion E from those of the great wheel. The wheel being thus detached, the ram would be free to fall upon the pile, but in that case it would draw the rope after it with a force and velocity which would soon wear it out; this is prevented by an expedient which is represented in *fig.* 290.

The rope, instead of being immediately attached to the head of the ram O, is attached to a piece of mechanism consisting of a sort of pincers H O K, the two arms of which play upon centres and are pressed outwards by the springs I I, which keep the points of the hooks K in close contact. The lower surfaces of these hooks are curved, so that, when they descend upon the ram, the ring forces itself between them, the springs I I yielding to the pressure. When the upper part of the ring has passed above the edges of the hooks, the latter are forced into the hole in the centre of the ring by the reaction of

Fig. 287.

the springs, the ring then resting upon the upper surface of the hooks. The apparatus is thus firmly connected with the ram G, so that the whole can be raised by the windlass. When it is elevated to the top of the framing, the

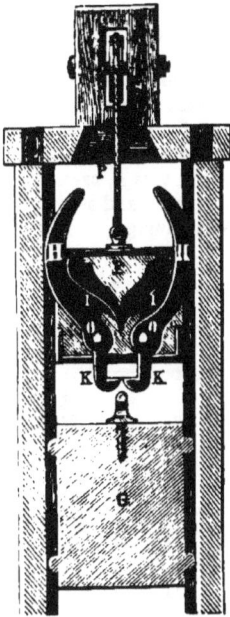

Fig. 290.

tops of the curved handles H H pass into an opening with oblique sides, which, pressing upon them,
force the handles together, the springs I I yielding to the pressure; the hooks K K are thus
separated, and the ring attached to the ram
being no longer supported by them, the latter
falls upon the pile. Meanwhile, the block to
which the apparatus H O K is attached, slightly
preponderating over the resistance of the rope,
slowly descends, and the hooks K K again
engage themselves in the ring so that the ram
can be again elevated by the windlass.

656. **Stamping-engine.** — One of the
most extensively useful forms which engines constructed upon this principle assume, is that which is applied to the
pounding of ore raised from mines. The
raw ore consists, as is well known, of
parts which are more or less metallic, mechanically mixed with earthy, stony, and
non-metallic matter, from which it is separated by reducing it to dust by pounding. The pieces of mechanism by which
this is accomplished, called " stamping
mills," consist of a combination of pile-
engines, identical in principle with those described above, but
worked in somewhat a different
manner.

A front view of a set of four
stampers, with the apparatus which
drives them, is shown in *fig.* 291.,
and a side view of part of a stamping mill is shown in *fig.* 292.

The stampers are mounted in sets of
four or more in immediate juxtaposition,
and set in framing, as shown in *fig.* 291,
To each stamper o o (*fig.* 292.) a projecting piece B is attached, upon which four
curved projecting pieces A A A A, called
cams, attached to an axis, made to revolve by the moving power, act. When
one of these cams A comes against B it
throws up the stamper c; and after it
passes from under B, the stamper c falls,
and the pounder, at its lower end, strikes
upon the metal placed under it. The
moment that this takes place the next
cam A comes under the piece B, and again

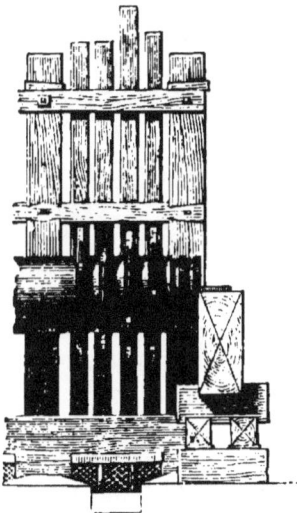

Fig. 291.

raises and lets it fall. In this manner each stamper is raised and let fall four times in each revolution of the axis. There is a distinct series of four cams for each stamper, and each series of four cams holds a different position on the axis, so that no two of them will act at the same moment upon the stamper. It may be easily conceived that the 16 cams which thus work a system of four stampers, let fall those stampers at 16 equal intervals in the time of a single revolution of the axis. Thus the ore will receive 16 blows in each revolution. A stream of water flowing through the pipe D, passes through a spout, along E, to the ore, which it washes, the earthy and stony being separated from the metallic matter by the process of straining.

Fig. 292.

A fly wheel is represented in the figure as attached to the apparatus, to equalise the action of the moving power; but such an appendage is not necessary when the cams divide a revolution of

Y

the axis into so many parts as is here supposed, the resistance being in that case nearly continuous.

657. **The sledge hammer** is another example of accumulated force obtained by the descent of a heavy mass. This apparatus,

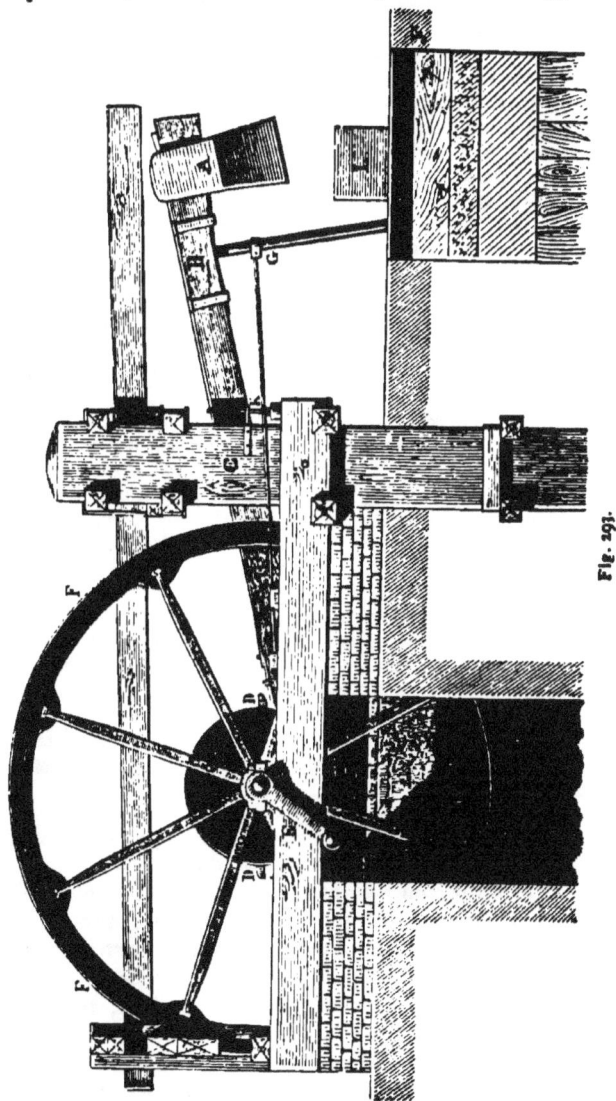

Fig. 293.

in its largest and most efficient form, is represented in *fig.* 293., upon a scale of 1 in. to 7 feet.

The head of the hammer A, is a mass of cast iron, weighing from 1½ to 2 tons, pierced with a large opening, in which the head of the handle B is inserted and secured by a wedge. This handle turns upon a centre c, and is acted upon at its extremity by cams D D, projecting from an axle, driven by an arm E, which receives its motion from some moving power, such as steam or water. When the cams descend, and strike the end of the handle H, the hammer A is raised, and as the motion is rapid, it is jerked upwards against a horizontal beam of wood above it, by the elasticity of which it is repelled, and descending falls upon the mass of red hot iron to be forged, which is previously laid upon the anvil L. Since the axle driven by E encounters no resistance, except while the cams act upon the handle of the hammer, the machine would be unduly accelerated in these intervals of suspension of the resistance, were it not that a heavy fly wheel F F is mounted on the axle; which, receiving the force of the moving power, becomes a depository thereof, and combines with the moving power in acting upon the hammer when the next cam encounters the handle.

A workman holds a long iron rod, to the end of which the mass of red hot iron to be forged is attached, and turns it between stroke and stroke of the hammer, from one position to another, so that it can be beaten in any desired manner.

A rod of iron, G, is governed by a lever H G, by which it can be brought at will under the handle of the hammer, so as to prevent it from falling on the anvil when it is desired to remove one piece of iron already forged and to replace it by another. During this process the axis continues to be driven, and the cams still act upon the hammer, but the hammer only plays between the rod G and the beam above it. When it is desired to recommence the forging on another piece of iron, the rod G is withdrawn and the hammer again allowed to fall.

The horizontal beam, which reacts by its elasticity on the heel of the hammer, adds nothing to its force, for if it were not present, the hammer would rise to a greater height, and would acquire an increase of force in its fall exactly equal to that which it receives from the reaction of the beam, or, more strictly speaking, a little more, the beam not being perfectly elastic. The advantage of the beam is that it makes the hammer return to the anvil in shorter intervals, and consequently to give a greater number of blows to the iron in a given time.

658. **The coining press.** — The first process in the fabrication of coin consists in the production of the proper composition of metal of which the coin is to be made, for the silver and gold used for coining, when absolutely pure, do not possess the requisite hardness, — a quality which is, however, easily imparted to them, by alloying them with a small proportion of baser metal. When thus properly alloyed, and formed into ingots or bars, they are submitted to the process of rolling, so as to be reduced to the form of plates of a regulated thickness,—an object which is attained with great precision by forcing them between rollers of tempered steel, placed at a distance one from the other, equal to the thickness of the intended plate of precious metal.

The plates thus produced are then submitted to the action of

circular punches of hard steel, by which circular discs are punched
from them of the exact size of the coins to be made.   These discs,
which are called blanks, are then submitted to the coining press.

This press has been constructed in various forms at different
mints, but is generally constructed upon the principle of the screw.

The characters and figures to be raised upon the coin are first
engraved with the greatest precision in sunk characters, or "*inta-
glio,*" upon circular surfaces of hardened steel.   These are called
dies, and two of them are of course required for each coin.   These
dies are placed in the press, one at its base, with its engraved face
upwards, and the other attached to the end of a rod driven by the
screw, with its engraved face downwards.   The blank is then placed
upon the lower die, and the upper die is driven down upon it by
the screw with a force sufficient to drive the metal of the blank
into all the sunk characters of the two dies, in the same manner
exactly as wax is forced into the sunk characters of a seal by the
pressure of the hand.

Such being the general principle of the coining press, the details
of its mechanism, which vary much in different places, may be illustrated by those of the press represented in elevation in *fig.* 294.,
the parts being shown on a larger scale in *figs.* 295. and 296.

Fig. 294.

The lever by which the screw is driven is a long metal bar c c, having
heavy knobs at the ends, and strong bunches of ribbon or cord, which the
men who drive the press lay hold of, to impart the necessary motion to the
screw.   The masses c c, receiving thus from the hand a considerable velocity,
acquire a great momentum, the effect of which, transmitted to the upper die
attached to the screw, is augmented in the proportion of the circumference

described by the knobs c c, to the intervals between the threads; a proportion which, it is obvious, is always very great.

The base plate of the press, with the circular cavity to receive the lower

Fig. 295.

die, is shown in *fig.* 296. Upon this plate is attached a piece H, which is moved horizontally by the mechanism M N L (*fig.* 295.), so that when the

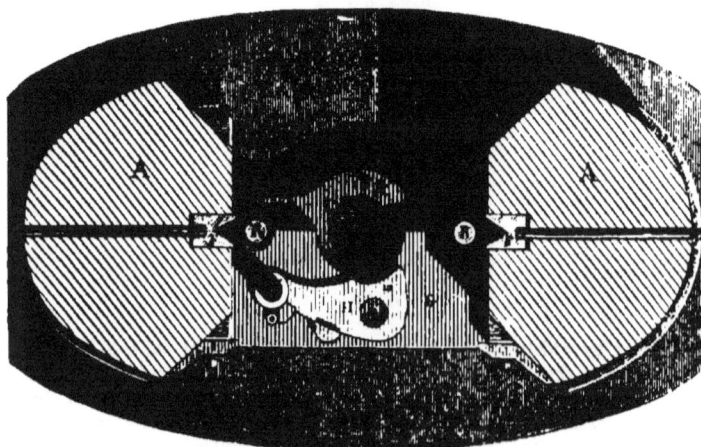

Fig. 296.
Y 3

screw rises after making the impression upon the blank, the piece H turns upon a joint at its extremity, moves towards the blank with a rapid jerk, throwing out the coined piece, and dropping into its place another blank to be coined by the next stroke of the press.

There are various other practical details in this mechanism, which it will not be necessary here to discuss, the general principle being rendered sufficiently clear from what has been stated above.

In some still more recent coining presses the screw action has been replaced by a mechanism constructed upon the principle of the knee-lever explained in 427. One of these presses, of the

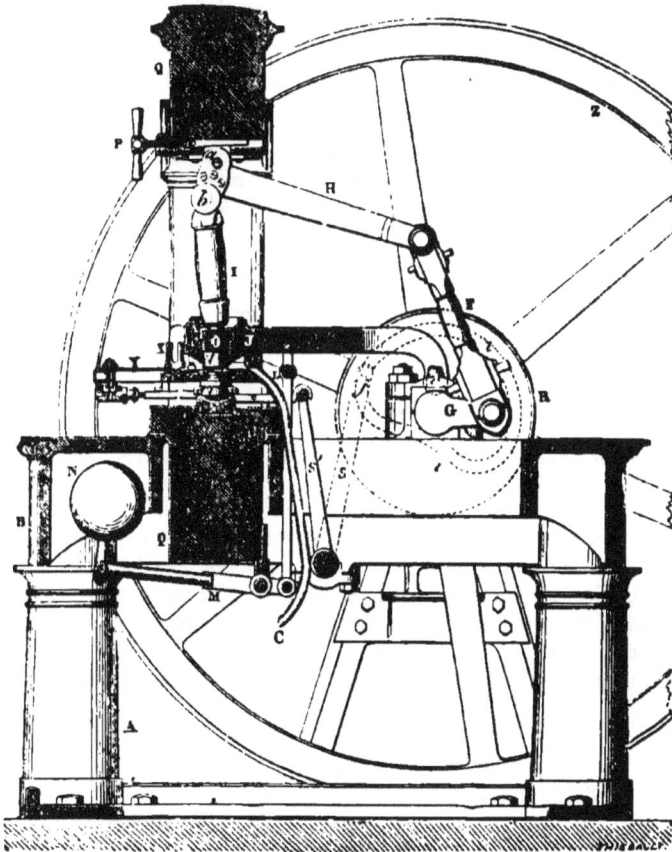

Fig. 297.

most recent construction and improved form, is represented in *fig.* 297., being that which was constructed by M. Thonnelier for the mint at Paris, where it is now worked.

The working shaft which imparts motion to the machine carries a large fly wheel z to equalise its motion. Upon this shaft a short crank ᴏ is connected with a lever ʜ by a rod ꜰ. The lever ʜ is made to vibrate on the centre *a* by the motion of the crank, and a corresponding oscillatory motion is imparted to the short arm *a b*, which is at right angles to the long arm ʜ. This short arm is connected with a piece ᴏ, which carries the upper die, by the rod ɪ, and the knee-joint at *b* transmits to the upper die an intense pressure, upon the principle already explained in 427.

This action differs essentially by substituting pressure for a blow, and is consequently more favourable to the preservation and durability of the mechanism.

The rod ɪ acts upon the piece ᴏ by a ball resting in a corresponding socket, and this piece ᴏ is pressed upwards against the ball by heavy counter-weights ɴ acting upon the lever ᴏ ᴊ *c* through the intervention of the lever ᴍ and the vertical rod ʟ. The lower die is placed in a cavity in the upper part of the fixed block ǫ, and the blanks are placed upon it and removed, after being coined, by mechanism which does not differ much in principle from that already described, and which is worked by the levers ʙ' and ʙ, the latter of which receives an oscillating motion by a pin *f*, which is moved along a sinuous groove formed in a central plate fixed upon the main shaft, and turning with the fly wheel. This groove is indicated in the figure by a dotted line.

CHAP. III.

BORING, PLANING, SAWING, AND SCREW AND TOOTH CUTTING ENGINES.

659. **Boring tools.** — All instruments of this class, great and small, present examples of continued rectilinear, produced by continued circular motion. The point of the boring tool is driven continually in one direction by means of a circular motion imparted to a disc or lever attached to it.

The common bow-drill (*fig.* 298.) presents a familiar example of this.

Fig. 298.

The bit and brace (*fig.* 299.) presents another example. The handle is turned rapidly, while the bit is pressed with its point upon the object to be bored.

The form of the point of the bit varies with the material to be bored, and the magnitude and shape of the hole. Different forms of bits are shown in *figs*. 300—306.

Fig. 299.   Fig. 300.   Fig. 301.

Fig. 302.   Fig. 303.   Fig. 304.   Fig. 305.   Fig. 306.

For heavier work, drilling and boring machines, with self-acting mechanism, and driven by steam or water power, are constructed on a large scale. Machines of this kind are used for drilling the holes in the end plates of large tubular boilers, in the heads and flanges of steam cylinders, and other heavy work. The principle is the same as that of the ordinary hand-drill, the details of the mechanism for imparting the motion, for changing the position of the drill or of the work, or both, and the scale of the machine, only being different.

660. **Planing machine.** — This machine presents an example of the production of a continuous rectilinear by a continuous circular motion. The purpose to which it is applied is to render the surfaces of pieces of metal perfectly plane.

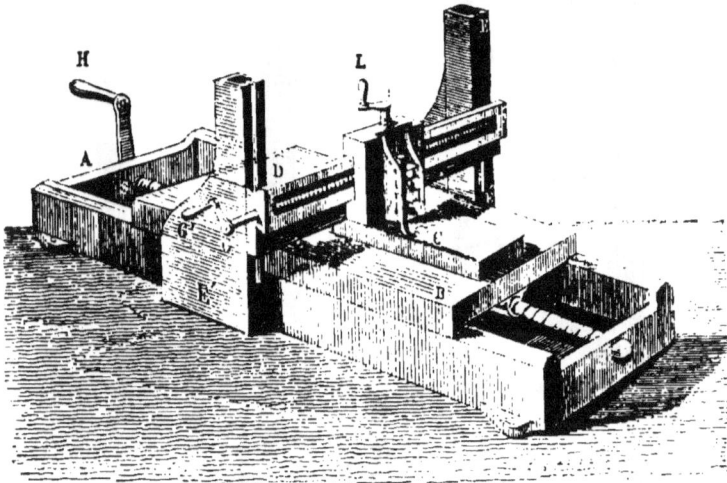

Fig. 307.

The block or plate of metal to be planed is firmly attached to a bed B (*fig.* 307.), which is connected with a screw s s', which extends longitudinally from end to end of the machine, and which is turned by a winch A H, or other similar contrivance. The screw plays in nuts attached to the bottom of the bed B; and as it turns in sockets in the ends s s' of the frame, and has therefore no progressive motion, it must impart a progressive motion to the frame B, and consequently to the plate of metal to be planed. By each revolution of the screw the metal to be planed is thus advanced through a space equal to the interval between the threads. This motion may be rendered as slow as is necessary by regulating the motion of the winch A H. A frame D is fixed horizontally in grooves in the upright pieces E E'. In this frame is inserted a screw F', worked by a winch G G'. This screw carries a piece I, in which is inserted a cutter C, formed of hardened steel, presenting its point in a direction contrary to that in which the metal to be planed is moved. This cutter is raised and lowered at pleasure by a winch L. The piece I which carries the cutter is moved transversely, by means of the winch G G' turning the screw F'.

When, by means of the winches G' and L, the position of the cutter C is properly adjusted, the surface of the metal is moved slowly against it, by turning the winch H. The cutter thus acting detaches from the metal a narrow shaving, and forms in it a very narrow and shallow groove.

When the entire length of the metal has thus been drawn under the cutter, the action of the machine is suspended; the cutter is raised by the winch L, and the metal advanced towards the end s' of the frame by turning the winch H in the proper direction.

The cutter c is now advanced transversely through a very small space towards E, by turning the winch o'. It will be understood that this motion is susceptible of any required degree of minuteness, since an entire revolution of the winch o' would advance the cutter through no greater space than the interval between the threads of the screw F', the threads of which may be made fine, and the winch o' may be turned only through a small fraction of an entire revolution. The machine is now prepared to recommence its operation; and, by turning H continuously, another groove will be cut in the surface of the metal parallel to the former, and equal to it in breadth and depth.

In fine, by continuing this process, the surface of the metal will be rendered not plane but grooved; and by making the grooves finer and finer, they will at length, for all practical purposes, disappear, and the surface will be plane.

All planing engines, whatever be their form or magnitude, are constructed upon the general principle which has been illustrated in that here described. They differ, however, in their details. It will be observed that in the form of machine here described the operation is intermitting, being suspended while the bed supporting the object to be planed is carried back from s' to s. It is, therefore, desirable that this returning motion should be more rapid than the slow motion which takes p lace from s' to s. In the best engines a provision is accordingly made for a quick return motion; but in some a contrivance is introduced for reversing the cutting tool, so that the operation can be executed while the bed D is moved backwards as well as forwards.

In some machines a circular motion can be given to the bed, so that the cutter, instead of making longitudinal grooves, makes concentric circular or spiral grooves.

By thus varying the adjustments and motions, as well of the bed or table supporting the object operated on, as of the cutting tool, these machines can be applied to the formation of all sorts of work, such as shaping and planing levers, cranks, straps, and crossheads, and, in short, all the parts of the larger class of machines. They are so constructed as to be self-acting, being driven by steam or water power, and are sometimes constructed on a scale of great magnitude.

661. **Saw-mills.** — The mechanical action of a saw, whatever be the substance it is applied to divide, is so regular and uniform, that it is obvious that machinery may easily be adapted to move it instead of the hand. It was, therefore, one of the earliest forms of labour, in which the human arm was relieved by inanimate power. Saws driven mechanically by wind mills, have been worked from

7

time immemorial in Holland. The application of water power to this industry is very general in mountainous and woody countries, where the frequent falls of water supply means of working the machinery, by which the timber⬤ divided and adapted for transport at almost a nominal cost

Fig. 308.

The forms of saws thus driven is either circular or straight. A circular saw is, as the name implies, a thin round disc of steel, having the teeth of a saw formed upon its edge, which is fixed upon a shaft to which a very rapid motion of revolution is imparted by the moving power. The saw in this case remaining stationary in its position, the object to be divided by it is urged against its teeth, with a regulated pressure, and constantly advanced, as the incision is made. When it is applied to divide wood, for example, the blade of the saw, being vertical, rises through a slit in a level table, at which the operator stands, and upon which he places the block of wood to be cut, pressing it gently against the teeth of the saw, and advancing it continually until it is completely divided.

Such saws are used for squaring the ends of the iron rails used upon railways. The saws adapted for this purpose are usually about 40 inches in diameter and 1-10th of an inch thick, turning at the rate of about 900 revolutions per minute, or 15 per second. The iron rail is then applied to them red hot, and the lower part of the saw is immersed in a trough of water to keep it cool.

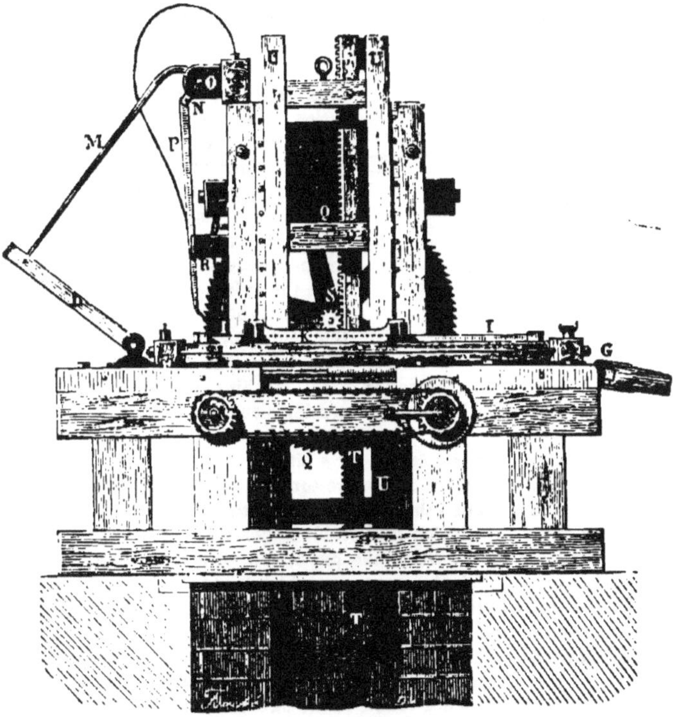

Fig. 309.

Straight saws worked by machinery consist of an oblong rectangular blade, like that of a common knife mounted upon a rectangular frame, with screw adjustments, for the purpose of giving the necessary tension to the blade. When these are applied to divide timber into boards or planks, they are placed vertically, and are moved with an up and down motion, similar exactly to that given to them when worked by hand in a saw pit. Instead, however, of advancing against the timber, as in the latter case, the timber being stationary, the timber is advanced against the saw, the saw being stationary. Saws worked in this way make from 120 to 150 strokes per minute, and cut at the rate of from one tenth to half an inch per stroke, according to the softness or hardness of the wood.

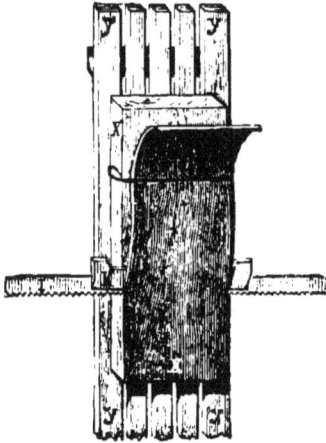

To render intelligible this important application of the saw, we shall here describe one of the most improved forms of the saw mill adapted to cut veneering from mahogany and other precious ornamental woods.

The moving power, as is customary in mills for all purposes, imparts revolution to a shaft or shafts, which are carried along the rooms under the ceilings in which the machines are established. Upon these shafts over each machine a drum A (*fig.* 308.) is fixed, which revolves with the shaft. This drum is connected by an endless band with the axle B of a fly wheel E E, and is kept tight by a bar D C fixed upon the centre D, and pressing against it with its weight by a roller C. When it is desired to suspend the motion of the machine, it is only necessary to raise the bar D C, when the band becoming slack, the motion of A will not be imparted to B.

Fig. 310.

A connecting rod G F is attached by a pin F to one of the arms of the fly wheel E E, at a distance from the centre equal to half the stroke intended to be given to the saw. The other end G is attached in like manner by a pin to the frame of the saw, which is movable in horizontal grooves, so that by each revolution of the fly wheel the pin G, and with it the saw, is moved back and forward through the space *m n*, which is equal to twice the distance of F from the centre of the fly wheel.

A view of the sawing machine, driven by the rod F G, is given in *fig.* 309., where the end G of that rod appears broken off. One side of the horizontal frame which carries the saw is shown at H H, the blade of the saw being behind this, and therefore not visible in the figure. A rod of strong wire is seen in front of H H, which passes through blocks at each end of the frame, and is secured by nuts screwed upon it, by means of which the rod is shortened so as to increase the tension of the blade of the saw. This blade is kept exactly vertical, the teeth being downwards, by a plate of iron, K, the lower edge of which is bevelled towards the saw.

According to what has been explained, the saw receives an alternate motion right and left through a space equal to *m n* (*fig.* 308.). If, therefore, the wood to be cut be pressed upwards against its teeth in a suitable manner, the desired separation will be effected. We shall now explain how this is accomplished.

In *fig.* 309. there is a vertical frame U U fixed behind the saw, to which a rack T T is attached. In the teeth of this rack a pinion s, fixed upon the axle of a ratchet wheel Q Q, works. In the teeth of this ratchet wheel are engaged two catches: one R, playing on a centre attached to the frame, and urged against the teeth by a spring; and the other fixed upon the lower end of a bar P, which turns on the centre N, and which is urged against the teeth by a spring which appears in the figure. A lever L works upon a centre attached to the sliding frame of the saw, and which, therefore, moves right and left with it. This lever imparts a corresponding motion to M, and the latter to N. By this means the short lever N moves P alternately upwards and downwards with each stroke of the saw. When P is raised, it draws the hook at its lower end upwards from the tooth of the ratchet wheel in which it was engaged; and when it has risen to the limit of its play, the spring presses it over the next tooth. When P descends, the hook presses the tooth of the wheel down; and, causing the wheel to move through a small space, the other catch R passes to the next tooth. In this manner with each stroke of the saw the ratchet wheel is advanced one tooth ; and the pinion s receiving a corresponding motion, the rack T T, and the frame U U to which it is attached, are raised through a corresponding small space.

Now, between this frame U U and the blade of the saw a space sufficiently wide is left for the introduction of the block of wood from which the sheets of veneering are to be cut. That block, therefore, being so introduced, is glued to the frame U U, its upper edge at the commencement of the operation being a little below the teeth of the saw.

The framing which carries the saw is traversed by two screws V V, connected by an endless chain, both of which receive a common motion from one handle which appears in the figure, attached to the right-hand screw. By means of these screws the frame U U, with a block of wood attached to it, can be moved through any desired space to or from the saw. In the commencement of the operation the upper edge of the block is by this means brought under the teeth of the saw, in such a position that a sheet of veneering of the requisite thickness shall be cut from it.

These arrangements being made, and the machine being put in motion by throwing the bar D O (*fig.* 308.) on the band, the work proceeds, and the saw detaches a sheet of veneering. The manner in which the saw thus operates is shown more evidently in *fig.* 310., where X X is the block of wood, and *y y* some bars, by means of which it is fastened to the frame U U (*fig.* 309.). It will be remembered that by each stroke of the saw, the frame U U, and with it the block X X, is raised through a small height by the action of the levers L M N P. The mechanism is so adjusted that this height shall be equal to the depth of the wood through which the saw can make an incision at each stroke. The bevelled bar K K separates the sheet of veneering from the block, so as to prevent the saw from being locked between them; and as this sheet is too thin to support its own weight, it is sustained by a loop of wire, a little below the top of the block, as shown in the figure.

In the practical operation of the machine, the drum A revolves at the rate of about one revolution per second, and if its diameter be 5 times that of the axle B, the latter will revolve 5 times per second, and the same number of strokes will be made per second by the saw. If the wood from which the veneering is cut be mahogany, the depth cut by each stroke will be about the fiftieth of an inch ; and such a saw will be capable of detaching about 400 square feet of veneering per day.

After each sheet of veneering is cut off, the action of the machine is suspended by raising the bar D C, and the frame U U is then lowered by disengaging the two catches from the ratchet wheel, until the upper edge of the block X X is brought below the teeth of the saw, and the screws V V are then turned until the upper edge of the block is brought sufficiently within the teeth of the saw to detach another sheet of veneering of the necessary thickness.

It is found that the thickness of wood converted into saw dust by this process is about the fiftieth of an inch, and if the thickness of the sheet of veneering be also the fiftieth of an inch, which it may be, and generally is, it is evident that 50 per cent. of the wood will be wasted in the process.

It is obvious that in the case of highly valuable woods, which are precisely those that are cut in the thinnest sheets, it is highly desirable to avoid a waste so extensive, and various attempts have been accordingly made from time to time to substitute the plane for the saw in the cutting of veneering, so that the wood would be detached in thin shavings without any waste at all. I am not aware, however, that as yet these attempts have been attended with any practical success.

662. **Screw-cutting engine.** — This engine affords another example of the reciprocal production of circular and rectilinear motion. It is represented in *fig.* 311.

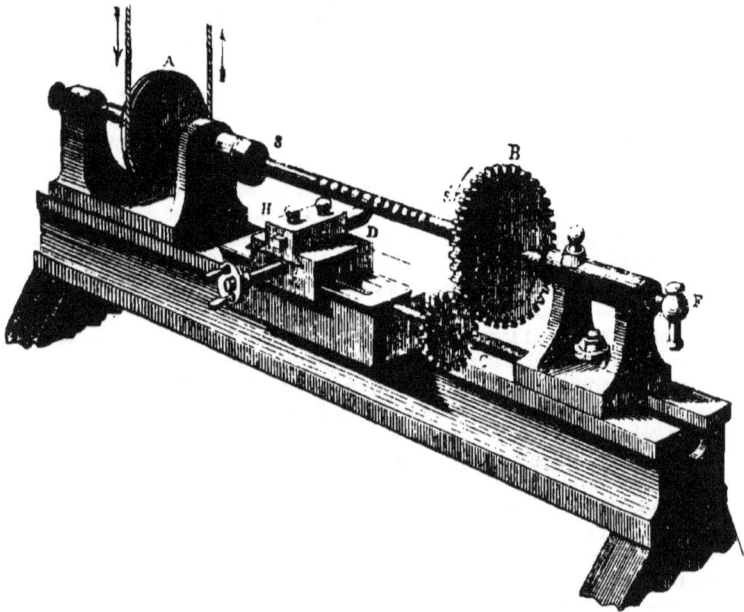

Fig 311.

The wheel A imparts rotation to the cylindrical rod s s', upon which it is desired to cut a thread. A toothed wheel B is fixed upon the rod s s', and revolves with it. This wheel imparts revolution to a smaller toothed wheel c, which is fixed upon the end of a fine screw by which a progressive motion is imparted to the piece H, in which a cutter D is inserted. It is evident that the proportion between the progressive motion imparted to the cutter D, and the rotatory motion of the rod s s′, is regulated by the proportion of the wheels B and C, and the thread of the screw driven by C.

According to this combination, while the cutter D moves slowly from s to s′, the rod s s′ revolves, and it is evident that the incision made by the cutter in the rod, will be of a spiral or helical form, and that the interval between the threads of the screw thus produced will be equal to the space through which the cutter D advances during one revolution of the rod s s′.

The depth of the incision made by the cutter is regulated by a screw E, which advances or withdraws the cutter at pleasure.

663. **Engine for cutting the teeth of wheels.** — This engine furnishes an example of the combination of a continuous with an intermitting circular motion. It is represented in *fig.* 312.

Fig. 312.

The disc A, upon which the teeth are to be cut, is fixed upon a rod C, which is turned by a disc B, the circumference of which is divided into equal intervals by holes, into which a point projecting from the end of a lever D, can be inserted so as to render the disc B, and consequently A, stationary. The teeth are cut in A, by means of a circular cutter E, driven by a band F, receiving its motion from any moving power. The piece carrying the cutter is moved longitudinally by a screw driven by a winch G, and it is moved to and from the wheel to be cut by a screw driven by the winch H.

The form of the teeth is determined by the shape of the cutter E, and their depth by the extent to which the incisions are continued by acting on the winch H.

The wheel A is placed upon the shaft and removed from it by means of the lever I, by which the shaft c can be removed from the frame at pleasure.

•

CHAP. IV.

FLOUR MILLS.

664. THE process of converting grain into flour is executed by passing the grain between two cylindrical stones, having a common vertical axis, smooth horizontal surfaces, and mounted so that the distance between these surfaces is less than the thickness of the grain. The lower stone is stationary, and the upper kept in revolution round its axis, with a regulated velocity, by some moving power, such as wind, water, or steam. The grain is let fall in small and regulated quantities through an opening which surrounds the axis in the upper stone ; it descends thus into the space between the two stones, and while it is broken between their surfaces, its particles are carried round by the rotation of the upper stone, and being slightly affected by the centrifugal force produced by the rotatory motion, it recedes gradually from the centre as it revolves, each particle describing a sort of spiral. The particles, at length arriving at the edge, fall into a groove made to receive them, from which they are discharged through openings provided for the purpose into a proper receptacle.

Such being the general principle upon which flour mills are constructed, their details and mode of operation will be better understood by reference to *figs.* 313. and 314.

The horizontal shaft A (*fig.* 314.) receives a motion of rotation from the moving power, whatever it may be, outside the mill; it imparts by the bevelled wheels a motion of rotation to the vertical shaft B, upon which a large toothed wheel C is fixed; this wheel works into small toothed wheels D and E, fixed upon two vertical shafts N and F. These wheels D and E are capable of moving through a limited space with a sliding motion of the shafts N and F, so as to be thrown into or out of connection with the great wheel C at pleasure. The moving power may thus be made to impart revolution to either or both of the shafts N and F. These shafts pass respectively through the centres of two pairs of mill-stones, that through which F passes being shown in section in the figure, and that through which N passes being enveloped in its octagonal wooden casing, on the top of which is placed the apparatus by which grain is supplied to the stones.

z

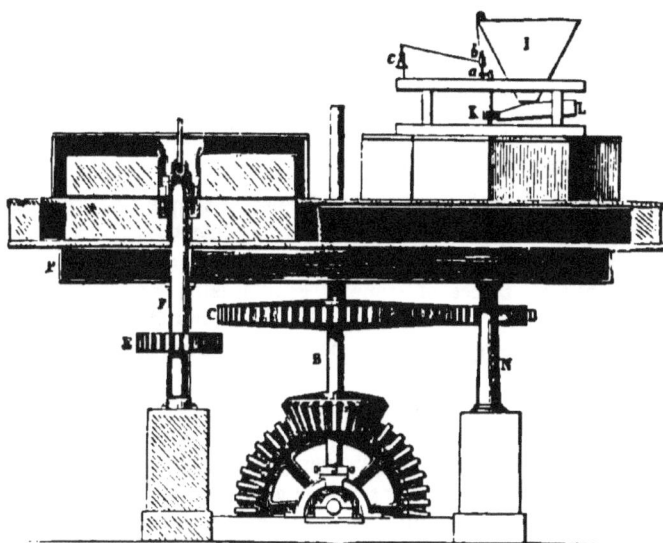

Fig. 313.

The shafts pass through holes in the centres of the lower stones, in which they turn freely, without moving these stones, and the space around them is

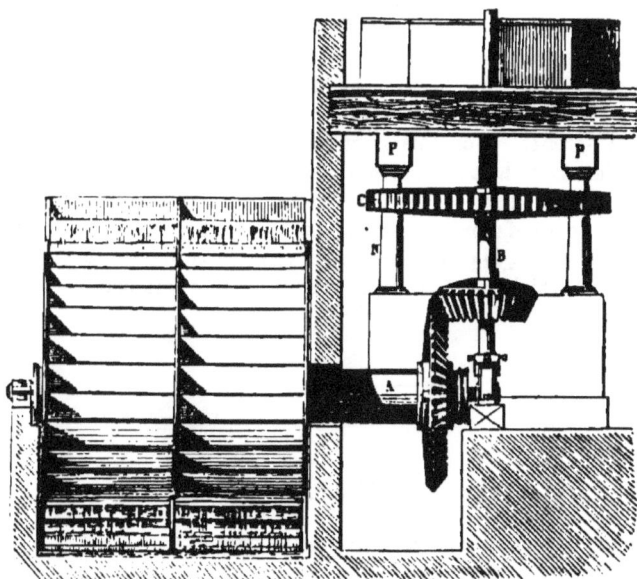

Fig. 314.

stopped up with leather so as to prevent the grain from falling between. The upper stone is fixed upon the top of the shaft and turns with it; a space, however, being left open between the shaft and the stone to let in the grain. The grain is supplied from a funnel-shaped box I, with an opening in the bottom, of regulated magnitude, through which it falls into the inclined spout L, which receives a slight oscillating motion right and left, from pieces which project from the rod K. By this motion the grain is shaken out, and falling little by little through the central opening in the upper stone, passes between the stones, where it is ground.

Mill-stones have been generally formed of simple blocks of stone, but these having been found to be often defective, a better class of stones have been produced, by cementing together, by means of plaster, chosen pieces of stone of good quality, the whole being bound together by hoops of iron. The diameters of the mill-stones, which were formerly from 5 to 6 feet are now generally reduced to 3½ feet.

Each pair of stones is capable of grinding about forty-two imperial bushels in twenty-four hours, and it is found that the best velocity for the stones is about seventy revolutions per minute.

The mere grinding of the grain is not, however, all that is accomplished in the flour mill. The grain, as is well known, consists of the material which constitutes the flour, properly so called, and the husky envelope, which forms the bran, so that the immediate product of the mill stones is a mixture of flour and bran, generally called "*whole meal.*" To produce fine flour from this, it is necessary to separate the bran from the flour, and this process, which may be executed with a greater or less degree of perfection, according to the fineness of the flour intended to be produced, is accomplished in the mill by machinery driven by the same power which moves the mill-stones. For this purpose, proper shafts are connected with the main shaft driven by the power, and their motion is conveyed to those parts of the mill in which is an apparatus consisting of wire sieves of more or less fineness. The whole meal, as it comes from the stones, is passed over a series of these sieves, through the meshes of which the flour falls, the bran being retained.

CHAP. V.

CLOCK AND WATCH WORK.

665. **Time measurer a necessity of civilised life.** — After the supply of the absolute necessaries of physical existence — food, clothing, and lodging — one of the first wants of a society emerg

z 2

ing from barbarism, is the means of measuring and registering time. In civilised society, all contracts for labour, and for all kinds of service, are based upon time. Even in the cases of the highest public functionaries, and where the service rendered is purely social and intellectual, still it is regulated, limited, and compensated with relation to time. Time measurers or chronometers were therefore among the earliest mechanical and physical inventions.

Although nature has supplied visible signs to measure and to mark the larger chronometric units, such as days, months, and years, she has not furnished any corresponding measures of the lesser units of hours, minutes, and seconds. There are no visible marks on the firmament by passing from one to another of which the sun can note the hours, still less are there any signs for minutes or seconds. These subdivisions are therefore merely artificial and conventional, and, to measure and mark them, artificial motions must be contrived.

666. **Sun-dials.** — Rough approximations were first made to the chief divisions of the day, by observing the apparent motion of the sun from rising to setting. Thus the direction of the meridian, or of the south, being once known, and marked by some fixed and visible object, the time of noon was known by observing when the sun had this direction. The hours before and after noon were roughly estimated by the position of the sun between noon and the times of its rising and setting. Greater precision was given to this method, by erecting a wand or gnomon, the shadow of which would fall upon a level surface, in a direction always opposite to that of the sun. Thus, after sunrise, the shadow would be inclined towards the west, the sun being then towards the east. From the moment of sunrise until noon, the shadow would move continually nearer and nearer to the direction of the north, and at noon it would have exactly that direction. From noon to sunset the shadow would be more and more inclined towards the east.

It is evident, however, that such a dial would not afford uniform indications at all seasons of the year, so that the hour-lines of the shadow determined in spring, for example, would not show the same hours in winter as in summer. Without much astronomical knowledge, it is easy to be convinced of this. At the equinoxes, the sun rises and sets at six o'clock, and at the east and west points precisely ; and, therefore, at these seasons, the six o'clock hour-lines of such a dial would be for the morning due west, and for the evening due east. But on the first day of summer (21st June), the sun rises and sets at points of the horizon very much north of the east and west points, and at six o'clock in the forenoon and afternoon its bearing is north of the east and west points.

A dial so constructed at any given place would be useless as a time indicator. To render it useful, it would be necessary that the shadow of the style should fall in the same directions at the same hours at all seasons of the year. Now, to attain this object, the style must be not vertical, but must be directed to the celestial pole. It is easy to comprehend that in that case a plane passing through the style and the sun would always be carried round the style with an uniform motion by the diurnal motion of the sun, and that at all seasons this plane would at the same hours have the same position.

It is for this reason that the gnomon of sun-dials is placed at such an inclination with the plate of the dial, that when the dial is properly set the gnomon will be directed to the north pole of the heavens, and being so placed, its shadow will fall upon the same lines of the dial at the same hours, whatever be the season of the year.

667. **Position of gnomon.** — It is evident, therefore, that dials must be differently constructed for places which have different latitudes. It is shown in astronomy that the elevation of the celestial pole is equal to the latitude of the place, and consequently the inclination of the gnomon of a sun-dial must be also equal to the latitude of the place where the dial is intended to be set. It follows, therefore, that a dial constructed for London would not be suitable for York, Newcastle, or Edinburgh.

The position of the plate of the dial upon which the shadow of the gnomon is projected is quite unimportant. All that is really important is the direction of the gnomon, which must always be that of the celestial pole, whatever be the position of the plate of the dial. Thus the plate of the dial may be either horizontal, vertical, or oblique. Its position will depend upon the place where it is to be erected. If it be in an open space, as in a garden or field, having a clear exposure on all sides, it will be generally most convenient to make it horizontal; and, hence, in such cases, it is usual to fix it upon the top of a column of three or four feet high, so that it may be easily observed by a person of ordinary height standing near it. Sometimes it is convenient to place it upon the wall of a building, such as a church. A wall with a southern exposure is in that case the most convenient; but to indicate the hours of the early morning in the spring and summer, an eastern exposure would be required, and to indicate those of the late evening, a western exposure would be necessary.

Where these vertical dials are erected, it is therefore frequently the practice to establish them at the same time on different walls of the same building.

Whatever be the position of the plate of the dial, the position of the hour-lines upon it is a matter of mere technical calculation, for which the formulæ and principles of spherical trigonometry are necessary, but which is not attended with any difficulty.

It must, however, be observed, that generally the hour-lines are inclined to each other at unequal angles, as may be seen by inspecting any ordinary sun-dial. There is one, and one only, position which could be assigned to the plate of the dial, such that the hour-lines would make equal angles with each other. That position would be at right angles to the gnomon, and a dial so constructed would be suitable to any place, whatever be its latitude. All that would be necessary would be to set it so that the gnomon would be directed to the celestial pole. The sun, however, would shine upon the upper or north side of it during the spring and summer, and on the lower or south side during the autumn and winter. It would, therefore, be necessary that it should be marked on both sides with hour-lines, and that a gnomon should be fixed on both sides.

The name dial is derived from the Latin word *dies*, a day, and the invention and use of the instrument as a time indicator is very ancient. According to Herodotus, the invention came to Greece from Chaldæa. The first dial recorded in history is the hemisphere of Berosus, who is supposed to have lived 540 B.C.

668. **Clepsydra.**— The first attempts to measure time by motions artificially produced, consisted in arrangements by which a fluid was let fall in a continuous stream through a small aperture in the pipe of a funnel, the time being measured by the quantity of the fluid discharged. The CLEPSYDRA, or water-clock, of the ancients, was constructed upon this principle. This and the sundial were the only instruments contrived or used by the ancients for the measurement of time.

Clepsydras were contrived by the Egyptians, and were in common use under the reign of the Ptolemys. In Rome, sun-dials were used in summer and clepsydras in winter. These instruments, though subject to very obvious defects, were, nevertheless, when skilfully used, susceptible of considerable accuracy, as may be easily conceived when it is stated that, before the invention of clocks and watches, they were the only chronometric instruments used by astronomers. The chief sources of their irregularities were the unequal celerity with which the fluid is discharged, owing to its varying depth in the funnel and its change of temperature.

669. **Sand-glass.**— The common hour-glass (*fig.* 315.) comes

Fig. 315.

under this class of chronometric instruments, but is the most imperfect of them. Nevertheless, for certain purposes, it is even now, advanced as we are in the application of science to the arts, still found the most convenient chronometer. The process of ascertaining a ship's rate of sailing or steaming by means of the log affords an example of its use. One man holds the reel from which the line runs off, while another holds the sand-glass, and gives the signal when the sand has run out. The number of knots run off from the reel is then the number of miles per hour in the rate of the vessel. The intervals between the knots, the quantity of sand in the glass, and the aperture through which it falls, are so adapted to each other as to give this result.

670. **Mercurial time measurer.**—Notwithstanding the great perfection to which the art of constructing chronometers has attained, an apparatus was not long since proposed by the late Captain Kater for the measurement of very small intervals of time, fractions of a second for example, which is a modification of the clepsydra. A quantity of pure and clean mercury is poured into a funnel with a small aperture at its apex, so that a stream of the quicksilver shall fall through it. The flow is rendered uniform by keeping the mercury in the funnel at a constant level. The apparatus is intended in scientific researches to note the exact duration of phenomena, and it is so managed, that the stream issuing from the funnel is turned over a small receiver at the instant the phenomenon to be observed commences, and is turned away from it the instant the phenomenon ceases. The mercury discharged into the receiver is then accurately weighed, and the number of grains and parts of a grain it contains being divided by the number of grains which would be discharged in a second, the number of seconds and the parts of a second which elapsed during the continuance of the phenomenon is found.

671. **Clocks and watches.**—For the purposes of civil life, as well as for the more precise objects of scientific research, all these contrivances have been superseded by clocks and watches, which are now so universal as to constitute a necessary article of furniture in the most humble dwellings, and a necessary appendage of the person in all civilised countries.

All varieties of this most useful mechanical contrivance include five essential parts.

I. A moving power.

II. An indicator, by whose uniform motion time is measured.

III. An accurately divided scale, upon which the indicator moves and by which its motion is measured.

IV. Mechanism, by which the motion proceeding from the moving power is imparted to the indicator.

V. A regulator, which renders the motion imparted to the indicator uniform, and which fixes its celerity at the required rate.

Thus, for example, in a common clock, the moving power is the weight suspended by cords over a pulley fixed upon the axle of a wheel, to which the weight in descending imparts a motion of rotation. The indicator is the hand. The scale is the dial-plate upon which the hours, minutes, and sometimes the seconds, are marked by equal divisions, over which the point of the hand moves. The mechanism is a train of wheelwork, so constructed that the rate of rotation of the last wheel, upon the axle of which the hand is fixed, shall have a certain proportion to the rate of rotation of the first wheel, upon the axle of which the weight is suspended. And if, as is generally the case, there be two or three hands, then the wheelwork is so constructed that while one of the hands makes one revolution, another shall make twelve revolutions, and the third shall make sixty revolutions during a single revolution of the latter, and therefore seven hundred and twenty during a single revolution of the former.

672. **Pendulum.** — If no other appendage were provided, the weight would, in such an apparatus, descend with a continually increasing velocity, and would therefore impart to the hands a motion of rotation more and more rapid, which would not consequently serve as a measure of time. This defect is removed by the addition of a pendulum, combined with a wheel upon which it acts called the escapement. It is the property of the pendulum that its oscillations are necessarily made always in equal times, and its connection with the escapement wheel is such, that one tooth of that wheel, and no more, is allowed to pass the upper part of the pendulum during each oscillation right and left. But this escapement wheel itself forms part of the train of wheelwork by which the first wheel, moved by the descending weight, is connected with the wheels which move the hands, and consequently, by regulating and rendering uniform the motion of this escapement wheel, the pendulum necessarily regulates and renders uniform the motion of the entire apparatus.

The instrument thus arranged, therefore, imparts an uniform motion of rotation to each of the hands, but this is not enough to render it a convenient time measurer. It is necessary that the motion of the hands should have some definite and simple relation to the natural and conventional division of time into days, hours, minutes, and seconds. For this purpose it is required not only that the hands should move uniformly, but that the first, or slowest of them, should make two complete revolutions in a day, or a single revolution in twelve hours; and, as a necessary consequence of this, that the second should make a single revolution in an hour, and the third in a minute.

From what has been stated, it will be apparent that the actual rate of motion imparted to the hands will be determined by the rate of oscillation of the pendulum. It has been shown that for each oscillation, right and left, of the pendulum, one tooth of the escapement wheel passes, and if the escapement wheel have thirty teeth, and if the pendulum take one second to make a single swing, it will allow the escapement wheel to make a complete revolution while it makes thirty swings from right to left, and thirty from left to right, that is, in sixty seconds, or one minute; so that, if the axis of the third hand were in this case fixed upon the axle of the escapement wheel, that hand would make one complete revolution in a minute, and consequently the second would make one complete revolution in one hour, and the third in twelve hours. The required conditions would therefore be in this case fulfilled.

To render this explanation of the regulating property of the pendulum complete it will be sufficient to show—1st, that the time of vibration must be always rigorously the same with the same pendulum; 2nd, that this time can be made shorter or longer by varying the length of the pendulum, so that a pendulum can always be constructed which will vibrate in one second, or in half a second, or, in short, in any desired time; and 3rd, that the connection of the pendulum with the escapement wheel can be so constructed, that the motion of the latter shall be governed by the vibrations of the former, in the manner already described.

The isochronism of the pendulum, and the manner in which its rate of oscillation may be made to vary by varying its length having been already explained, it only remains to show how it is connected with the escapement wheel, so as to regulate the motion of the hands.

The pendulum acts upon the escapement wheel by means of a fork o (*fig.* 316.) between the prongs of which its rod passes, so that as the pendulum swings alternately from side to side, it carries the fork with it. This fork o projects from a vertical rod r, which is connected with a horizontal shaft D. This shaft is connected with the escapement, to the anchor of which it gives an alternate motion that will be explained hereafter.

673. **Compensation pendulums.** — It has been explained that the centre of oscillation of a pendulum will vary with every

change in the relative positions of its point of suspension and its centre of gravity. Now, since all bodies are susceptible of dilatation by increase, and contraction by decrease of temperature, pendulums will necessarily be subject to a change in the position of their centres of oscillation, and therefore of the rate of vibration, with every change of their temperature. To ensure the uniformity of the rate of vibration, a class of contrivances has been invented by which every variation of temperature is made to produce two effects on different parts of the pendulum, which counteract each other, and keep the centre of oscillation at an invariable distance from the point of suspension.

One of the means of accomplishing this, is, by connecting the bob of the pendulum with the point of suspension by rods composed of materials expansible in different degrees, so arranged that the dilatation of one shall augment the distance of the centre of oscillation from the point of suspension, while the expansion of the other diminishes it.

Let s (*fig.* 317.) be the point of suspension, and o the centre of oscillation, and let s be supposed to be connected with o by means of two rods of metal, s A and A o, which are united at A, but independent of each other at every other point.

Fig. 316.

If such a pendulum be affected by an increase of temperature, the rod s A will suffer an increment of length, by which the point A and the rod A o attached to it will be lowered ; but, at the same time the rod A o, being subject to the same increase of temperature, will receive an increment of length, in consequence of which the point o will be raised to an increased distance above the point A, at which the rods are united. If the increment of the length of the rod A o be in this case equal to the increment of the rod s A, then the point o will be raised as much by the increase of the length of A o as it is lowered by the increase of the length of s A, and, consequently, its distance from the point s will remain the same as before the change of temperature takes place.

To fulfil these conditions, it is only necessary that the length of the rod A o shall be less than the length of the rod s A in exactly the same proportion as the expansibility of the metal composing A o exceeds the

Fig. 317.

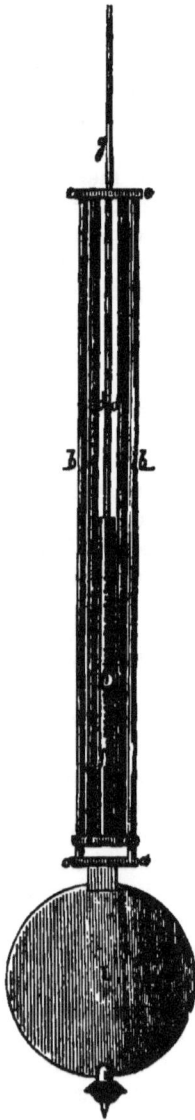

expansibility of the metal composing s A. If the lengths of s A and A o were equal, their increments of length would be proportional to their dilatations; but the length of the more dilatable rod A o, being less than that of the less dilatable s A, in the same proportion as the dilatability of the former is greater than that of the latter, the absolute increments of their length wili necessarily be equal, the greater dilatability of A o being compensated by its lesser length.

674. **Harrison's gridiron pendulum.** — This principle is variously applied in different pendulums; that which is best known is Harrison's gridiron pendulum, represented in *fig.* 318. A heavy disc of metal L, is suspended from a cross piece of brass *a a*, fixed to the lower extremities of two iron rods *b b*. These rods themselves are attached at the upper extremities to another brass bar *c c*, which is supported upon the upper ends of two zinc rods *d d*. These zinc rods are supported below by a third cross bar *e e*, which is fixed to the lower end of a brass tube *f*, into which the iron rod *g* passes, and is held by a linch-pin *h*. This rod *g* is continued upwards to the point of suspension. When the temperature rises, the rod *g* and tube *f*, being elongated by expansion, the cross piece *e e* is removed farther from the point of suspension. If the rods *d d* suffered no change of dimension, the cross piece *c c* would be lowered just as much as *e e*, and would be removed farther from the point of suspension. The dilatation which the iron rods *b b* undergo causes the cross piece *a a* to be removed farther from *c c*, and in consequence of these several dilatations of the rod *g*, the tube *f*, and the rods *b b*, the lens L would descend to a lower position than that which it previously had, but the zinc rods *d d*, which we have here supposed to remain unchanged, not only dilate like the other rods,

Fig. 318.

but dilate in a much greater degree in proportion to their length, and, by dilating, cause the cross pieces *a a* and *c c* and the rods *b b* which connect them, and consequently the lens L, to rise and therefore to come nearer to the point of suspension.

It appears, therefore, that while the dilatation of g, f, and b b cause the lense L to descend, that of d d causes it to ascend ; and if the length of these be properly proportioned, the ascent produced by the latter dilatation will be exactly equal to the descent produced by the former, and consequently, the centre of oscillation and the rate of the pendulum will remain unaltered.

675. **Mercurial pendulum.** — This pendulum, which depends on the same principle, consists of two cylindrical glass vessels b b, (*fig.* 319.) containing mercury, which supply the place of the bob or lens, the rod of the pendulum a passing between them and supporting a brass plate upon which they rest. An increase of temperature, causing an expansion of the rod a and its appendages, would cause the vessels of mercury to descend ; but the same increase of temperature producing a still greater expansion in the mercury, causes it to rise in the glass cylinders, and consequently to produce a corresponding elevation in the centre of gravity of the mercury. These two dilatations, like the former, may be so arranged that they shall precisely counteract each other, and so keep the centre of oscillation, and consequently the rate of the pendulum, unvaried.

Fig. 319.

676. **Another compensation pendulum.** — If two straight bars of differently dilatable metals be soldered together, every change of temperature will bend the combined bar into the form of a curve, the more dilatable metal being on the convex side of the curve when the temperature is raised, and on the concave side of it when it is lowered.

Let the more dilatable metal be called A, and the less dilatable B. Now, if the temperature be raised, A will become longer than B, and, as they cannot separate, they must assume such a form, being still in contact, as is consistent with the inequality of their lengths. This is a condition which will be satisfied by a curve in which the bar A is on the convex and the bar B on the concave side.

If the temperature, on the other hand, be lowered, the more dilatable metal being also the more contractible, the bar A will be more diminished in length than the bar B, and being, therefore, the shorter, will necessarily be on the concave side of the curve.

This principle has been ingeniously applied as a compensator.

Such a compound bar as we have just described is placed at right angles to the rod of the pendulum, and has, at its extremities, two bobs. When the temperature rises, and the centre of oscillation is, by expansion of the pendulum, removed to a greater distance from the point of suspension, this compensating bar is bent into the form of a curve concave towards the point of suspension, as represented in *fig.* 320.; and the bobs which it carries at its extremities, being brought closer to the point of suspension, compensate for the increased distance of the bob of the pendulum from that point.

If, on the other hand, the temperature falls, and the rod of the pendulum contracting brings the bob and the centre of oscillation nearer to the point of suspension, the compensating bar is bent into a curve, which is concave downwards, as represented in *fig.* 321.; and the bobs which it carries being removed to an increased distance from the point of suspension, compensate for the diminished distance of the bob of the pendulum.

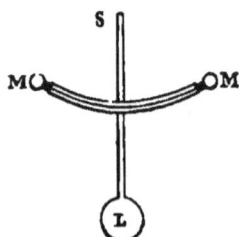 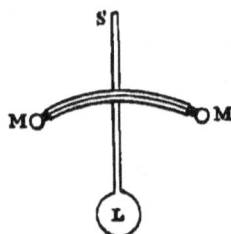

Fig. 320.   Fig. 321.

677. Connection between pendulum and escapement. — Supposing, then, the pendulum to be so adjusted that it shall make its vibrations at any required rate, one per second for example, let us see how the motion of the indicating hands is governed by such vibrations.

Upon the axis on which the pendulum oscillates is fixed a piece of metal in the form of an anchor, such as D B A C (*fig.* 322.), so that this piece shall swing alternately right and left with the pendulum. Two short pieces, *m* and *m'*, called pallets, project inwards at right angles to it from its extremities A and C.

The form and dimensions of the anchor A B C are accommodated to those of the escapement wheel, W W', which is part of the clockwork, and which, in common with the other wheels forming the train, is moved in the direction indicated by the arrow by the weight or mainspring. When the anchor swings to the right the pallet *m* enters between two teeth of the wheel, the lower of which coming against it, the motion of the wheel is for the moment arrested. When it swings to the left, the pallet *m* is withdrawn from between the teeth, and the wheel is allowed to move, but only for a moment, for the other pallet *m'* enters between two teeth at the other side, the upper of which coming against it the motion of the wheel is again arrested.

The wheel, therefore, is thus made to revolve on its axis E, not with a continuous motion, as would be the case if it were impelled by the weight or mainspring, without the interference of any obstacle, but with an intermitting motion. It moves by starts, being stopped alternately by one pallet or the other coming in the way of its teeth.

When the pendulum, and therefore the anchor, is at the extreme right of its play, the pallet m, having entered between two teeth, a tooth rests against its lower side, the wheel is arrested, and the pallet m', is quite disengaged from, and clear of, the teeth of the wheel. When in swinging to the left the arm D B becomes vertical, the tooth of the wheel on the left has just *escaped* from the pallet m, and the wheel, being liberated, has just commenced to be moved by the force of the weight or mainspring. But at the same moment the pallet m', enters between the teeth of the wheel on the right, and when the anchor has arrived at the extreme left of its play, the tooth of the wheel which is above the pallet m', will have fallen upon 'it, so that the motion will again be arrested. Thus it appears that during the first half of the swing from right to left, the motion of the wheel is arrested by the pallet m, and during the remaining half of the swing the wheel moves, but is arrested the moment the swing is completed.

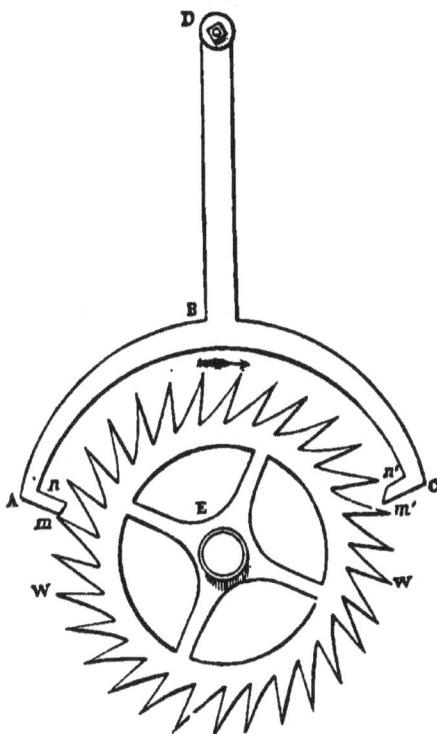

Fig. 322.

In like manner it may be shown that during the first half of the swing from left to right, the motion of the wheel is arrested by the pallet m', that it is liberated and moves during the latter half-swing, and is again arrested when the swing is completed.

678. **Motion of hand intermitting.** — The motion which is imparted to the hands upon the dial necessarily corresponds with this intermitting motion of the escapement wheel. If the clock be

provided with a seconds-hand, the circumference of the dial being divided into sixty equal parts by dots, the point of the seconds-hand moves from dot to dot during the second half of each swing of the pendulum, having rested upon the dot during the first half swing. The whole train of wheelwork being affected with the same intermitting motion, the minute and hour hands must move, like the seconds-hand, by intervals, being alternately moved and stopped for half a second. This intermission, however, is not so observable in them as in the seconds-hand, owing to their comparatively slow motion. Thus, the minute-hand, moving sixty times slower than the seconds-hand, moves during each half swing of the pendulum through only the sixtieth part of the space between the dots, and the hour-hand, moving twelve times slower than the minute-hand, moves in each half swing of the pendulum through the 360th part of the space between the dots. It is easy, therefore, to comprehend how changes of position so minute are not perceptible.

679. **How the motion of the pendulum is sustained.** — If the pendulum vibrated upon its axis of suspension unconnected with the clockwork, the range of its oscillation would be gradually diminished by the combined effects of the friction upon its axis and the resistance of the air, and this range thus becoming less and less, the oscillation would at length cease altogether, and the pendulum would come to rest. Now this not being the case when the pendulum is in connection with the wheelwork, but on the contrary, its oscillations having always the same range, it is evident that it must receive from the escapement wheel some force of lateral impulsion, by which the loss of force caused by friction and the resistance of the air is repaired.

It is easy to show how the effect is produced. It has been shown that during the first half of each swing, a tooth of the escapement wheel rests upon one or other pallet of the anchor. The pallet reacts upon it with a certain force, arresting the motion of the wheelwork, and receives from it a corresponding pressure. This pressure has a tendency to accelerate the motion of the pendulum, and this continues until the tooth slips off, and is liberated from the pallet. It is this force which repairs the loss of motion sustained by the pendulum by friction and atmospheric resistance.

Thus we see that while on the one hand the pendulum regulates and equalises the motion imparted to the wheelwork by the weight or mainspring, its own range is equalised by the reaction of the weight or mainspring upon it.

680. **Its action upon the escapement.** — If the action of the anchor of the pendulum upon the escapement wheel be attentively considered, it will be perceived that one tooth only of the escapement passes the anchor for each double vibration made by the pen-

dulum. Thus, if we suppose that when the pendulum is at the extreme left of its range, the right-hand pallet is between the teeth m' and n', the tooth n' will escape from the pallet c when the pendulum, swinging from left to right, comes to the vertical position, which is the middle of its swing. While it rises to the extreme right of its range, the tooth n' advances to the place which m' previously occupied, and at the same time the tooth m advances to the place which n previously occupied; but, at the same time, the pallet A, carried to the right, enters between m and the succeeding tooth, and arrests the further progress of the wheel. When the pendulum then swings to the left, the wheel continues to be arrested until it arrives at the middle of its swing, when the tooth below m escapes from the pallet A, but at the same moment the pallet c enters below the tooth which is above n', and, receiving it at the end of the swing, stops the motion of the wheel. Thus it appears that tooth after tooth, in regular succession, falls upon the pallet c upon the arrival of the pendulum at the extreme left of its play after each double oscillation.

If the pendulum be so constructed that it shall vibrate in a second, and that it be desired that the escapement wheel shall make a complete revolution in a minute, that is, during sixty vibrations of the pendulum, the wheel must have thirty teeth. In that case, one tooth passing the anchor during each double oscillation from right to left, and back from left to right, thirty teeth, that is the whole circumference of the wheel, will pass the anchor in thirty double oscillations, or in sixty single swings of the pendulum, the time of each swing being one second.

681. **Motion of hands, how proportioned.** — The manner in which different rates of revolution can be imparted to the different hands of a clock or watch, by tooth and pinion work, is easily rendered intelligible.

The wheels commonly used in watch and clock work are formed from thin sheets of metal, usually brass, which are cut into circular plates of suitable magnitude, upon the edges of which the teeth are formed. The edges of the wheels thus serrated are brought together, the teeth of each being inserted between those of the other, so that if one be made to revolve upon its axle, its teeth, pressing upon those of the other, will impart a motion of revolution to the other.

When a large wheel works in the teeth of a much smaller one, which is a very frequent case in all species of wheelwork, the smaller wheel is called for distinction a *pinion*, and its teeth are called *leaves*.

682. **Method of cutting wheels and pinions.** — The method of manufacturing the pinions and smaller wheels used in watch and clock work is very ingenious. A rod of wire, the diameter of

which a little exceeds that of the wheel or pinion to be made, is drawn through an aperture cut in a steel plate, having the exact form and magnitude of the wheel or pinion to be formed. After being forced through this aperture by the ordinary process of wire-drawing, it is converted into a *fluted* wire, the ridges of the fluting corresponding exactly in form and magnitude to the edge of the aperture, and therefore to the teeth or leaves of the pinion or wheel.

This fluted wire, called *pinion wire*, is then cut by a cutter, adapted to the purpose, into thin slices, at right angles to its length. Each slice is a perfect wheel, or pinion ; and it is evident that all of them must be absolutely identical in form and magnitude.

Such a wire-drawing plate, with apertures of different forms and sizes, is represented in *fig.* 323.

Fig. 323.

683. **Moving power. — Weight.** — It has been already stated that the moving power applied to clock or watch work is either a weight or a mainspring.

If a weight be the moving power, it is suspended to a cord which is coiled upon a drum fixed upon an horizontal axis, the first wheel of the train which gives motion to the hands being fixed on the same axis, so that it shall turn when the drum turns.

Such an arrangement is represented in *fig.* 324., where A B is the drum, C D the wheel attached to and moved by it, W the weight which is the moving power suspended to the cord E, which is coiled upon the drum A B. The end, F, of the axis of the drum projecting beyond it, is made square, so as to receive a key made to fit it, by which it is turned, so as to coil the cord upon the axis, when it has been uncoiled by the descending motion of the weight.

The direction in which the wheel C D is turned by the force of the descending weight is indicated by the arrow, and in that direction it will continue to turn so long as the weight acts upon the coil of the cord upon the drum. But as soon as the cord, by the continued descent of the weight, shall have been discharged from the drum, the rotation imparted to C D must cease. It is then that the key must be applied to the square end, F, of the axis of the drum, and turned continually in the direction contrary to that in which the weight would turn the drum in descending .

A A

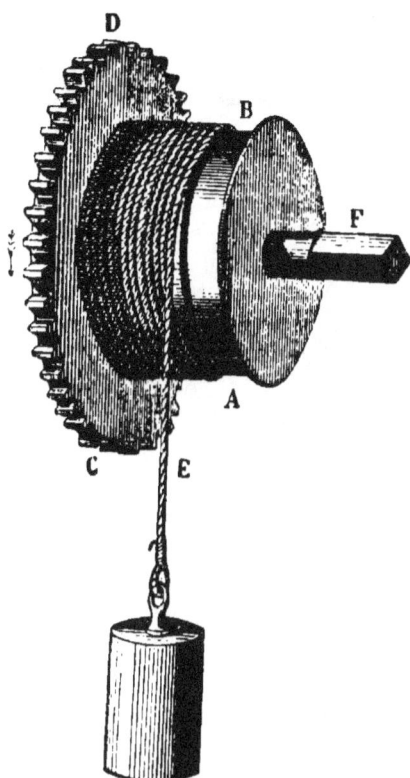

Fig. 324.

It will be perceived that, in this case, the hands of the clock would . be always turned backwards while the clock is being wound up, unless some special provision were made against such an effect; for it is evident that if the wheel c D, when turned by the descent of the weight w, in the direction of the arrow, give a progressive motion to the hands, the motion imparted to c D, by the ascent of the weight w, while the clock is being wound up, must necessarily impart to the hands a motion in the contrary direction, that is, a backward motion.

In all clocks this is prevented by an expedient called a ratchet wheel and catch, the one being attached to the barrel A B, and the other to the face of the wheel c D, the effect of which is to allow the barrel A B to be turned while the clock is being wound up in the direction contrary to that indicated by the arrow without turning the wheel c D; but when the barrel A B is turned by the descent of the weight w, in the direction of the arrow, the catch acting in the teeth of the ratchet wheel, the motion of A B is imparted by the action of the catch on the ratchet wheel to the wheel c D, and by it to the hands.

The form and mode of application of the ratchet wheel will be presently more clearly explained.

684. **Mainspring.** — The moving power of a weight can only be applied to time-pieces where the space necessary for the play of the weight in its descent and ascent can be conveniently obtained. This condition is obviously incompatible with the circumstances attending pocket-watches, all portable and movable time-pieces,

chimney, table, and console clocks, and in general all time-pieces constructed on a small scale.

The moving power applied to these universally is a mainspring, which is a ribbon of highly tempered steel bent into a spiral form, as represented in *fig.* 325. At one end, A, an eye is provided, by

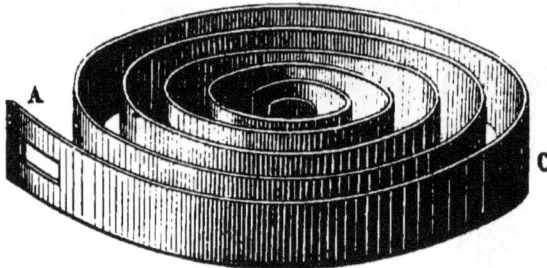

Fig. 325.

which that extremity may be attached either to a fixed point or to the side of the barrel to which the spring is intended to impart motion. In the centre of the spiral an arbor, or axle, is introduced, to which the inner extremity of the spring is attached. Supposing the extremity A to be attached to a fixed point, let the arbor B (*fig.* 326.), be turned in the direction indicated by

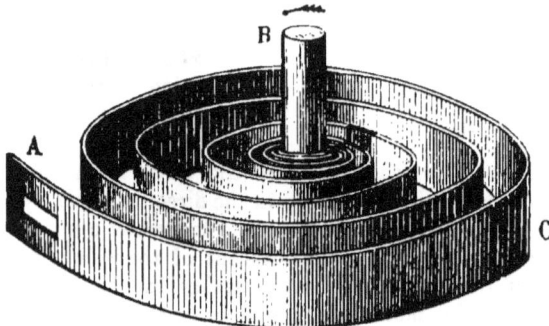

Fig. 326

the arrow. ' The spring will then be coiled closer and closer round the arbor A B, while its exterior coils will be separated one from another by wider and wider spaces.

After the spring has been thus coiled up by turning the arbor, it will have a tendency to uncoil itself and recover its former state, and if the arbor B be abandoned to its action and be free to revolve, it will receive from the reaction of the spring a motion of revolution contrary in direction to that which was given to the arbor in coiling up the spring, and such motion would be imparted to a wheel fixed upon the axle, and might from it be transmitted to the hands in the same manner as if the arbor wheel received its motion from the power of a weight.

But between such a moving force and that of a weight there is an obvious difference. The tension of the cord by which the weight is suspended, and consequently its effect in giving revolution to the barrel upon which the cord is coiled, is always the same until the clock altogether goes down. The moving force of the spring, on the contrary, is subject to a continual decrease of intensity. At first, when completely coiled up, its intensity is greatest, but as it turns the arbor B it becomes gradually relaxed, and its intensity is continually less and less. It exerts, therefore, a continually decreasing power upon the wheel fixed upon the arbor, and therefore upon the hands to which the motion is transmitted.

685. **Fusee.** — As a varying power would be incompatible with that uniformity and regularity which are the most essential and characteristic conditions of all forms of clockwork, such a spring would be quite unsuitable if some expedient were not found by which its variation could be equalised.

This has been accordingly accomplished, by a very beautiful mechanical contrivance, consisting of the combination of a flexible chain and a conical barrel arranged to receive its coils, called a *fusee.*

This arrangement is represented in *fig.* 327. The mainspring is

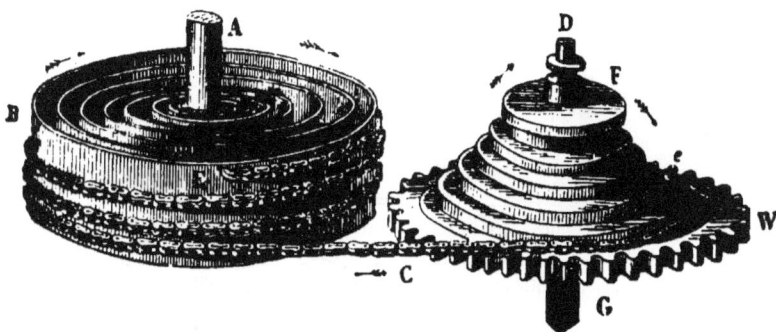

Fig. 327.

attached by its inner extremity to the fixed arbor A, and by its outside end at E, to a barrel B, which is capable of being turned round the fixed axis A. A jointed chain is attached by one extremity to the barrel at E, and, being coiled several times round it, is extended in the direction c, to the lowest groove of the fusee ʀ, to which its other end, e, is attached. This fusee is a conical-shaped barrel, upon which a spiral groove is formed, continued from the base to the summit, to receive the chain. The base is a toothed wheel, by which the motion imparted by the mainspring and chain to the fusee is transmitted to the hands through the wheelwork. The fusee is fixed upon an arbor, ᴅ ɢ, the lower end of which, projecting outside the case containing the works, is formed square, to receive a key made to fit it, by which the clock or watch is wound up.

The action of the spring transmitted by the barrel to the chain, and by the chain to the fusee, has a tendency to impart to the fusee a motion of rotation in the direction of the arrows. The fusee is connected with the wheel, w, by means of a ratchet wheel and catch similar to that described in the case in which the moving power is a weight, by means of which the fusee ꜰ imparts rotation to the wheel when it turns in the direction of the arrows, but does not move it when turned in the opposite direction.

These arrangements being understood, let us suppose the key applied upon the square end ɢ of the arbor ᴅ ɢ of the fusee, and let it be turned round in the direction contrary to that indicated by the arrows. The fusee will then be turned, but will not carry the wheel w with it ; the chain c will give to the barrel ʙ a motion of revolution contrary to the direction of the arrows, the chain will be gradually uncoiled from the barrel ʙ, and will be coiled upon the spiral groove of the fusee, winding itself from groove to groove, ascending on the spiral until the entire length of the chain has been uncoiled from the barrel ʙ, and coiled upon the fusee ꜰ, as represented in *fig*. 328.

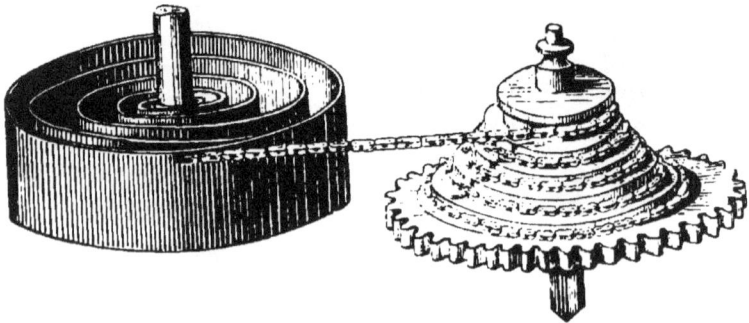

Fig. 328.

During this process, the external extremity of the mainspring, attached to the barrel at ᴇ, is carried round with the barrel, while

A A 3

the internal extremity is fixed to the arbor A, which does not turn with the barrel. By this means the spring is more and more closely coiled round the arbor A, until the entire chain has been discharged from the barrel to the fusee, when the spring will be coiled into·the form represented in *fig.* 328., and in this state the intensity of its force of recoil, and the consequent tension of the chain c, extended from the barrel B to the fusee F, is greatest.

The clock being thus wound up and left to the action of the spring, the tension of the chain c, directed from the fusee to the spiral, will make the fusee revolve in the direction indicated by the arrows. This tension at the commencement acts upon the highest and smallest groove of the fusee. As the chain is gradually discharged from.the fusee to the barrel, the tension is gradually decreased by reason of the relaxation of the spring, and at the same time the chain acts upon a larger and larger groove of the fusee. In this way the tension of the chain is continually decreased, and the radius of the groove on which it acts is continually increased, until the entire action has passed from the fusee to the barrel, and the clock goes down.

Now the power of the chain to impart a motion of revolution to the fusee depends on two conditions; first, the force of its tension, and secondly, the leverage by which this tension acts upon the fusee. This leverage is the semi-diameter of the groove, upon which the chain is coiled at the point where it passes from the fusee to the barrel. It will be easy to perceive that it requires less force to turn a wheel or barrel if the force be applied at a great distance from the axle than if it be applied at a small distance from it. Upon this principle generalised, it follows that the power of the tension of the chain to impart revolution to the fusee is augmented in exactly the same proportion as the magnitude of the groove on which it acts is increased.

The form given to the fusee is such that as the chain is gradually discharged from it, the diameter of the groove on which it acts increases in exactly the same proportion as that in which the tension of the chain decreases. It follows, therefore, that the power of the chain upon the fusee gains exactly as much by the increase of its leverage as it loses by the decrease of its tension, and consequently it remains invariable.

Complete compensation is therefore obtained by this beautiful and simple expedient, and a variable force is thus made to produce an invariable effect. It may be useful to state that this is only a particular application of a mechanical principle of great generality. In all cases whatever, the varying energy of a moving power may be equalised by interposing between it and the object to be moved some mechanism by which the leverage, whether simple or complex, through which its force is transmitted, shall vary in the exact inverse proportion of the variation of the power, — increasing as the intensity of the power is decreased, and decreasing as the intensity of the power is increased.

686. **Balance wheel.** — Whatever be the moving power, whether it be a weight or mainspring, it would, if not controlled and regulated, impart to the hands a motion more or less accelerated, and therefore unsuitable to the measurement of time, which requires a motion rigorously uniform. It is on that account that the moving power must be controlled and governed by some expedient by which it shall be rendered uniform.

How the combination of a pendulum and escapement wheel accomplishes this has been already explained. But this expedient requires that the time-piece to which it is applied shall be stationary; the slightest disturbance of its position would derange the mutual action of the pendulum and the escapement wheel, and would either stop the movement or permanently derange the mechanism. It is evident that a pendulum is not only inapplicable to all forms of pocket time-piece, but that it cannot even be used for marine purposes, the disturbances incidental to which would be quite incompatible with the regularity of its action.

The expedient which has been substituted for it with complete success in all such cases is the *balance wheel.*

This is a wheel, like a small fly wheel, having a heavy rim connected with the centre by three or more light arms, as shown at A B C, in *fig.* 329. Under, and parallel to it, is placed a spring resembling in form the mainspring, but much finer and lighter,

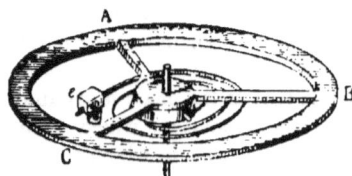

Fig. 329.

and having much less force. This spring is formed of extremely fine and highly tempered steel wire, so fine that it is sometimes called a hairspring. One extremity of this spring is attached to the axis of the balance wheel, and the other to any convenient fixed point in the watch. The spring is so constructed that when at rest it has a certain spiral form, to which it has a tendency to return when drawn from it on the one side or the other. If we suppose it, therefore, to be drawn aside from this position of rest and disengaged, it will return to it, but on arriving at it, having acquired by the elasticity a certain velocity, it will swing past it to the other side, to a distance nearly as far from its position of rest as that to which it had been originally drawn on the other side.

It will then swing back, and will thus oscillate on the one side and the other of the position of rest, in the same manner exactly as that in which a pendulum swings on the one side or the other of the vertical line which is its position of rest.

687. **Isochronism.**—The balance wheel thus connected with a spiral spring, like the pendulum, is isochronous; that is, it performs all vibrations—long and short—in the same time. It will be recollected that this property of the pendulum depends on the fact that the wider is the range of its vibrations the more intense is the force with which it descends to the vertical direction, and consequently wide vibrations are performed in as short a time as more contracted ones Now the vibrations of the balance wheel are subject to like conditions. The wider the range of its vibrations, the more intense is the force with which the recoil of this spring carries it back to its position of rest, and consequently it swings through these wide vibrations in the same time as through more contracted ones, in which the force of the spring is proportionally less intense.

688. **Compensation balances.** — The balance wheel, like the pendulum, being subject to expansion and contraction by variation of temperature, will suffer a corresponding change in its diameter, and a consequent change in the time of its oscillation, and the rate of the time-piece which it regulates.

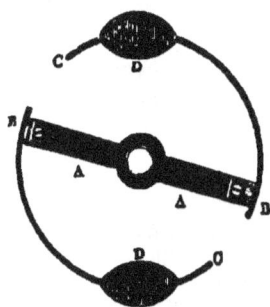

This irregularity has been compensated by attaching to the rim of the wheel a compound metallic arc, such as that already described. When the temperature rises, and the diameter of the wheel is augmented, this arc, (B c, *fig.* 330.) with its concavity towards the centre of the wheel, becomes more concave, and a weight D which it carries is brought nearer to the centre of the wheel, and this compensates for the increased magnitude of the diameter A A. If, on the other hand, the temperature is lowered, and the diameter A A of the wheel diminished by contraction, this arc B c becomes less concave, and the weight which it carries is removed to a greater distance from the centre, and this compensates for the diminished diameter of the wheel.

Fig. 330.

689. **Movement imparted to the hands of a watch.**—The oscillation of the balance wheel regulates the motion of watchwork in the same manner by means of an escapement wheel, as that in which the pendulum regulates the motion of clockwork. The

pallets and the escapement wheel are, however, very variously
formed in different watches.

Having thus explained generally the powers by which clocks
and watches are moved and regulated, it now remains to show
how the necessary motions are conveyed to the hands by suitable
combinations of wheels and pinions.

In *fig.* 331. are represented the works of a common watch, moved by a
mainspring A, and regulated by a balance wheel H; the wheels and pinions,

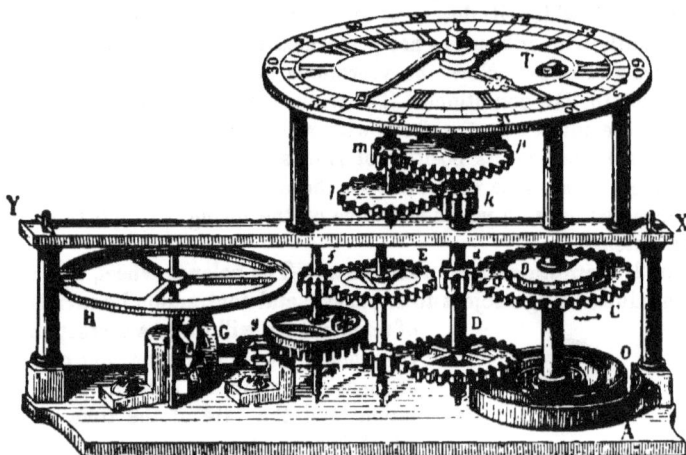

Fig. 331.

however, being changed in their relative positions, and the fusee being
omitted, so as to show more visibly the connections and mutual dependency
of the many parts. The external extremity of the mainspring is attached
to the base, O, of a column of the frame. Its internal extremity is attached
to the lower end of an axle, of which the square end, T, at the top enters a
hole in the dial-plate into which the key is inserted when the watch is to be
wound up. The ratchet wheel B is fixed upon this axis so as to turn with it,
but the other wheel C under the ratchet wheel is not fixed upon it, the axis
being free to turn in the hole in the centre of C, through which it passes. A
catch *n o* is attached by a pin on which it plays to the face of the wheel C,
and its point *o* is pressed against the teeth of the ratchet wheel B, by a
spring provided for that purpose. When the key is applied upon the end
T, and turned in the direction in which the hands move, the ratchet wheel is
turned with it, and the point *o* of the catch — pressed constantly against the
teeth while it turns — falls from tooth to tooth with an audible click, and
thus produces the peculiar sound, with which every ear is familiar, while the
watch is being wound up. During this process the wheel C does not turn
with the axle, which only passes through the hole in its centre without being
fixed upon it, but the mainspring A, being attached to the axle, is coiled more
and more closely round it, and reacts against the fixed point o with greater .
and greater force.

If the fusee, which is omitted in this figure, were introduced, it would occupy the place of the spring, and would be turned by the axle imparting a like revolution to the axis of the spring by means of the chain.

When the watch is wound up, the reaction of the spring, rendered uniform in its force by the fusee, imparts a motion of revolution to the ratchet wheel B, in the direction of the arrow. By this motion the tooth of the ratchet wheel in which the point o of the catch is engaged, presses against the catch so as to carry it round with it in the direction of the arrow; but the catch being attached to the face of the wheel c, at n, this wheel is carried round also in the same direction, and with a common motion.

The teeth of the wheel c act in those of the pinion d, which is fixed upon d D. Upon the same axle is fixed the wheel D, so that the wheel D and the pinion d receive a common motion of revolution from the wheel c.

The wheel D, in precisely the same manner, imparts a common motion of revolution to the pinion e, and the wheel E; and the wheel E imparts a common motion of revolution to the pinion f and the wheel F.

This last wheel F is of the form called a crown wheel, and acts upon the pinion g, imparting to it, and to the escapement wheel G, a common motion of revolution. This escapement wheel is acted upon and controlled by the pallets or other contrivances attached to the axis of the balance wheel H, so as to regulate its motion by the oscillations of that wheel in the same manner as the escapement wheel of a clock is regulated by the anchor of the pendulum.

It may be asked why so long a series of wheels and pinions are interposed between the mainspring and the balance wheel, and why the first pinion d may not act directly upon the escapement wheel. The object attained by the multiplication of the wheels and pinions is to cause the mainspring, by acting through a small space, to produce a considerable number of revolutions of the escapement wheel, for without that the spring would be speedily relaxed, and the watch would require more frequent winding up. Thus by the arrangement here shown, while the mainspring causes the wheel c to revolve once, it causes the pinion d and the wheel D to revolve as many times as the number of teeth in c is greater than the number in d. Thus, if there are ten times as many teeth in c as in d, one revolution of c will produce ten of d and D. In like manner if D have ten times as many teeth as e, one revolution of D will produce ten of e and E, and so on. In this way it is evident that one revolution of the first wheel c, which is on the axis of the fusee, can be made, by the mutual adaptation of the intermediate wheels and pinions, to impart as many revolutions as may be desired to the escapement wheel G.

The wheels which govern the motion of the hands are those which appear in the figure between the watch face and the frame X Y. The relative power of the mainspring and balance wheel must be so regulated that the wheel D shall make one revolution in an hour. The axis upon which this wheel is fixed, passing through the centre of the dial, carries the minute-hand, which therefore revolves with it, making one complete revolution on the dial in an hour.

Upon this axle of the minute-hand is fixed a pinion k, which drives the wheel l, on the axle of which is fixed the pinion m, which drives a wheel p, through the centre of which the axle of the minute-hand passes without being fixed upon it. Upon the axle of the minute-hand a small tube is placed, within which it can turn. Upon this tube the hour-hand, as well as the wheel p, is fixed. The pinion k, therefore, fixed upon the axis of the minute-hand, imparts motion to the hour-hand by the intervention of the wheel l, the pinion m, and the wheel p. Since the hour-hand must make one revolution while the minute-hand makes twelve, it is necessary that the relative numbers of the teeth of these intermediate wheels shall be such as to produce that relation between the motions of the hands. An unlimited variety of combinations would accomplish this, one of the most usual being the following : —

Pinion k	8 teeth.
Wheel l	24 „
Pinion m	8 „
Wheel p	32 „

By this arrangement p will make eight revolutions, while m and l make thirty-two; or, what is the same, p will make one revolution, while m and l make four. In like manner, l will make eight revolutions, while k, and therefore the minute-hand, makes twenty-four; or, what is the same, l will make four revolutions, while k and the minute-hand make twelve. It follows, therefore, that p, and therefore the hour-hand, makes one revolution, while k, and therefore the minute-hand, makes twelve, which is the necessary proportion.

In this case there is no seconds-hand : but, if there were, its motion would be regulated in like manner by additional wheels and pinions.

690. **Movement of the hands of a clock.** — The manner in which the moving power of a weight, and the regulating power of a pendulum are applied in a clock to produce the motion of the hands, does not differ in any important respect from the arrangement explained above. Nevertheless, it may be satisfactory to show the details of the mechanism. The train of wheels connecting the weight with the anchor of the pendulum is shown in *fig.* 332.

A side view of the mechanism, showing the wheels which more immediately govern the motion of the hands, and also the pendulum, with its appendages, is given in *fig.* 333.

The weight w acts by a cord on a barrel, as already explained. This barrel and the ratchet wheel, with its catch, are mounted upon the axis of the great wheel c, and are behind it, as represented in *fig.* 332., their form and position being shown by the white lines. The catch is attached to the wheel c by the screw n, and its point o acts on the teeth of the ratchet wheel, which is attached to the barrel on which the rope is coiled. The spring which presses the catch against the teeth of the ratchet is also shown. When the clock is wound up by the key applied to the square end т (*fig.* 333.) of the axis of the barrel, the barrel is turned in the direction opposite to that indicated by the arrows, and the catch falls from tooth to tooth of the ratchet wheel, making the clicking noise which attends the process of winding up. When the clock has been wound up, the weight acting on the barrel presses the tooth of the ratchet wheel against the catch, and thereby carries round with it the wheel c. This wheel transmits the motion to the escapement wheel G, *fig.* 332., through the series of wheels and pinions, d, D, e, E, f, F, and g, in the same manner exactly as has been already described; and the pendulum, by means of the anchor N N, regulates the motion.

The wheels which more immediately govern and regulate the motion of the hands are those which appear in *fig.* 333. in front of the plate X Y.

The pendulum consists of a heavy disc of metal, seen edgeways at V in *fig.* 333., attached to the end of a metal rod, B R, represented broken, to bring it

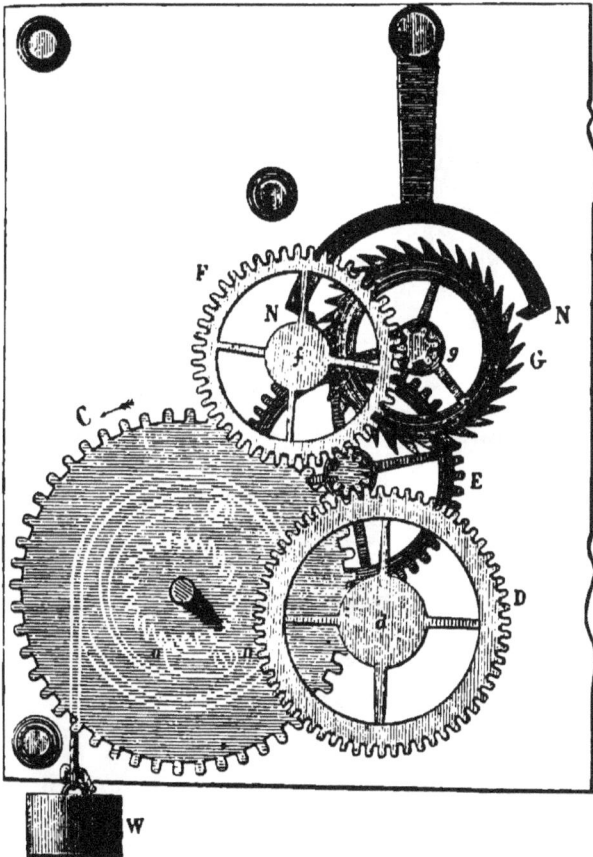

Fig. 332.

within the limits of the figure. This rod is suspended by various means, but often, as in the figure by two elastic ribbons of steel, s s, which permit its swing right and left. It passes between the prongs of a fork U, by which a rod r r is terminated, so that this rod swings right and left with the pendulum. Upon the axis of this rod, and over the escapement wheel G, is fixed the anchor N of the escapement.

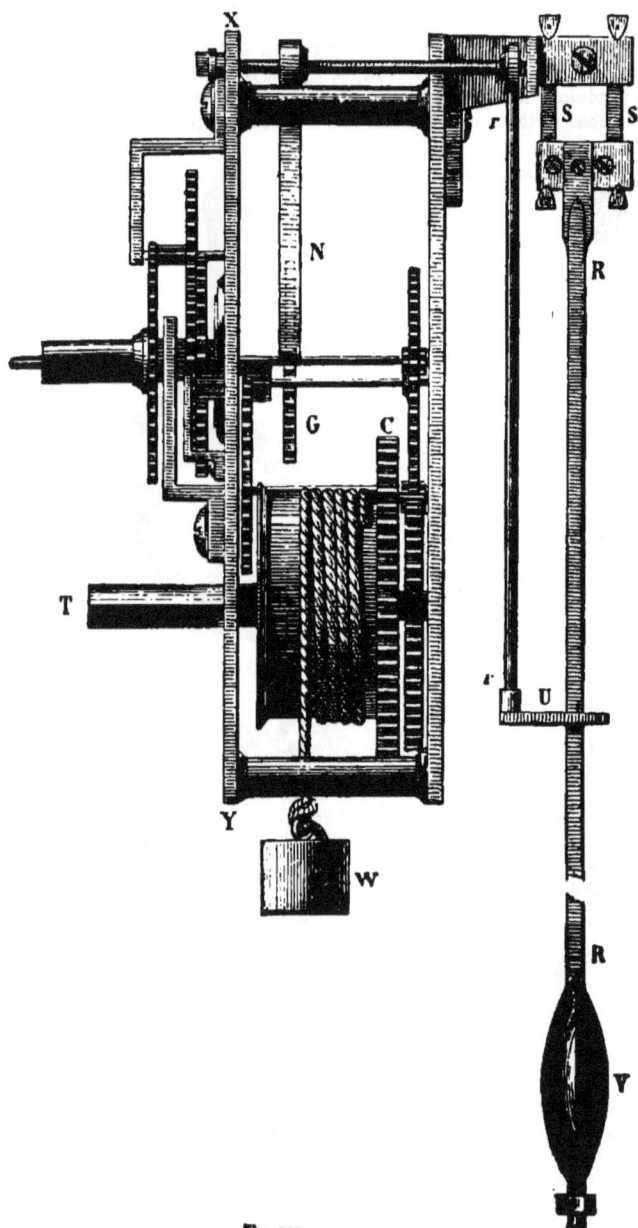

Fig. 333.

691. **To regulate the rate.** — Whether the movement be regulated by a pendulum or a balance wheel, it is necessary to provide some means of adjustment by which the rate of vibration may be increased or diminished at pleasure within certain limits, for although in its original construction the regulator may be made so as to oscillate *nearly* at the required rate, it cannot be made to do so *exactly*. Besides, even though it should vibrate exactly at the required rate, it will be subject, from time to time, to lose that degree of precision, and to vibrate too fast or too slowly, from the operation of various disturbing causes.

It has been already shown that the rate of vibration of the pendulum is rendered more or less rapid by transferring the centre of gravity nearer to, or farther from, the point of suspension. Upon this principle, therefore, the adjustment of the rate of vibration depends. The heavy disc v, *fig.* 333., is made to slide upon the rod B B, and can be moved upon it, upwards or downwards, by a fine screw attached to it, which works in a thread cut in the rod. In this manner the centre of gravity of the disc v may be transferred nearer to the suspension s s, so as to shorten the time of its vibrations, or removed farther from s s, so as to lengthen the time. If the clock is found to lose or go too slowly, it is screwed *up*, and if it gain or go too fast, it is screwed *down*.

In chimney and table time-pieces, the pendulum is regulated in a different manner. It is usually suspended upon a loop of silken thread, which can be drawn up or let down through a certain limited space, by means of a rod, upon which one end of the thread which forms the loop is coiled. This rod, passing through the dial-plate, has a square end, upon which a key can be applied, by turning which in the one direction or the other, the loop is drawn up or let down.

The rate of oscillation of the balance wheel cannot in the same manner be so easily regulated by modifying its form; but, on the other hand, while the force which moves the pendulum, being that of gravity, is beyond our control, that which moves the balance wheel, being the force of the spiral spring, is at our absolute disposition. It is accordingly by modifying this spring that we are enabled to regulate the time of oscillation of this regulator.

One of the most common expedients by which this is accomplished is represented in *fig.* 334.

Near the fixed point G, at the external extremity of the spiral, is placed a small bar E, near the end F, of which is a notch, or hole, through which the wire of the spiral passes. This arrests the action of the spiral, so that the only part of it which oscillates is that which is included between F and its internal extremity. In a word the point F is the virtual external extremity

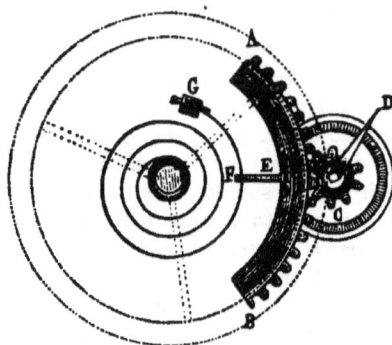

Fig 334.

of the spiral. Now this point F can be moved in the one direction or the other, so as to increase or diminish the virtual length of the spiral at pleasure, by means of the toothed arc A B and the pinion C, which latter is turned by the index D. If the index D be turned to the left, the bar E, and the point F, is moved towards G, and the length of the spiral is increased. If it be turned to the right, the point F is moved from G, and the length of the spiral is diminished.

In this manner the rate of vibration of the balance wheel may be adjusted by varying at will the vertical length of the spiral spring.

692. **Recoil escapement.** — The precision of the movement of all forms of time-pieces depends in a great degree on the mechanism of the escapement, and accordingly much mechanical skill and ingenuity have been directed to its improvement, and several varieties of form have been adopted and applied.

The recoil escapement, represented in *fig.* 331., consists of two pallets, which project from the axis of the balance wheel at right angles with each other, one of which acts at the top, and the other at the bottom of the escapement wheel G, the axis of which is horizontal and the wheel vertical. These pallets, as the balance wheel oscillates, engage alternately in the teeth of the escapement wheel, exactly in the same manner as do the pallets of the anchor of the escapement attached to the pendulum already described. This form of escapement was long the only one used, and is still continued in the more ordinary sorts of watches.

It has, however, been superseded, in watches and chronometers where greater precision is required, by others of more improved construction.

Cylindrical escapement. — In pocket watches, where great flatness is required, the cylindrical escapement is used, in which the axis of the balance wheel, instead of having pallets attached to it, is formed into a semicylinder, having a sort of notch in it, as represented on an enlarged scale in *fig.* 335. The semi-cylinder *a b* is cut away at *c*, through about half its height. The axis A B, *fig.* 336., of the escapement wheel is vertical, and the

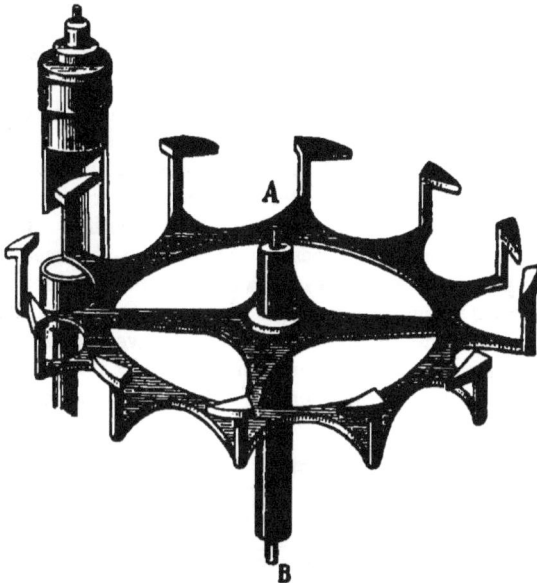

Fig. 335.　　　　　　　　　　Fig. 336.

teeth, raised at right angles to its plane, and therefore parallel to
its axis, have the peculiar form represented in the figure. As the
balance wheel oscillates on the one side and the other, the semicylinder upon its axis interposes itself alternately between the
teeth of the escapement wheel, stopping them and letting them
escape in the usual way. The manner in which the action takes

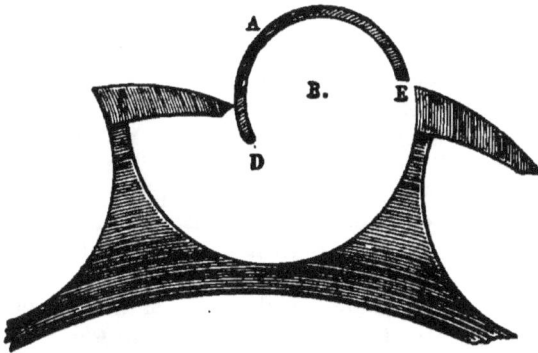

Fig. 337.

place will be more clearly understood by *figs.* 337. and 338., in
which a view in plan upon an enlarged scale is given of the

position of the semicylinder and the teeth of the escapement wheel after each successive oscillation.

In *fig.* 337., the balance wheel swinging from right to left, throws the convex side A D of the semicylinder before the tooth c of the escapement wheel, and thus for the moment arrests it, while the side A E of the semicylinder has turned out of the way of the preceding tooth, and has let it pass. The balance wheel then swings from left to right, and the convex side A D of the semicylinder slides against the point of the tooth c. When the edge D of the semicylinder passes the point of the tooth, the latter, in slipping over it, gives to it a slight impulse, which restores to the balance wheel the small quantity of force which it lost by the previous reaction of the tooth upon its convex surface.

The side A E of the semicylinder is now thrown before the tooth c. the point of which having advanced through a space equal to

Fig. 338.

the diameter of the semicylinder, is thrown against the concave surface of A E, as shown in *fig.* 338.

The balance wheel now swinging again from right to left, the point of the tooth c slides upon the concave surface of the semicylinder A E, until the edge E comes to it. The tooth then slips over the inclined face of E, and in doing so gives the semicylinder, and consequently the balance wheel, another slight impulse, restoring to it the force of which it deprived it while previously sliding upon its concave side.

The explanation here given of the action of this form of escapement is well calculated to render the conditions which all escapements should fulfil intelligible. These arrangements are primarily directed to the regulation of the movement of the wheelwork, so as to secure its uniformity. This will obviously be accomplished provided that the escapement, whatever be its form, lets a tooth

B B

of the wheel pass for each oscillation of the balance wheel. But owing to the friction of the axis of the balance wheel, and of the pallets on the teeth of the escapement wheel, and the resistance of the air, the range of its oscillations would be gradually diminished, so that at last it would not be sufficient to allow the successive passage of the teeth of the escapement wheel, and the watch would stop unless some adequate means be provided by which the balance wheel shall receive from the mainspring through the escapement wheel as much force as it thus loses. All escapements accomplish this by the peculiar forms given to the edges of the pallets and the teeth of the escapement wheel. In the present case, the object is attained by making the edge D round and the edge B inclined, and by giving to the teeth the form shown in the figure.

This form of escapement supplies a sufficiently good regulating power for the best sorts of pocket watches, and is attended with the advantages of allowing the works to be compressed within a very small thickness. It is the form most commonly used in the French and Swiss watches.

693. **Duplex escapement.** — The form of escapement used in the best English-made watches consists of an escapement wheel, which partakes at once of the double characters of a spur and crown wheel, and is hence called the *duplex escapement.*

The spur teeth, A B C, &c. (*fig.* 339.), are similar in their form

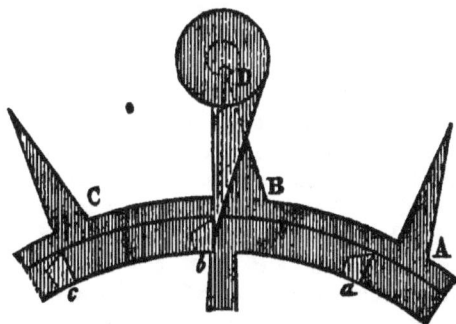

Fig. 339.

and arrangement to those of the cylindrical escapement described above. The crown teeth, *a b c*, &c., project from the face of the wheel, and have a position intermediate between the spur teeth. Upon the axle of the balance wheel just above the plane of the escapement wheel is fixed a claw pallet, called the *impulse pallet,* which, by the combination of the oscillations of the balance and the progressive motion of the escapement wheel, fall successively between the crown teeth of the latter, receiving from their reaction, as they escape from it, the restoring action which maintains the range of the oscillations of the balance wheel.

Under the pallet and in the plane of the spur teeth is placed a small roller usually formed of ruby or other hard stone, having a notch on one side of it, into which the spur teeth successively fall. After any crown tooth, *a*, for example, passes the pallet, the corresponding spur tooth A falls into the notch of the roller, and this alternate action continues so long as the watch goes. It will be perceived therefore that the pallet and roller in the duplex escapement play the same part as the two edges of the semicylinder in the cylindrical escapement, and as the two pallets in the common recoil escapements.

The chief advantage claimed for this system is that there is but one pallet, and that the action does not require as perfect execution of the teeth of the escapement wheel as other arrangements.

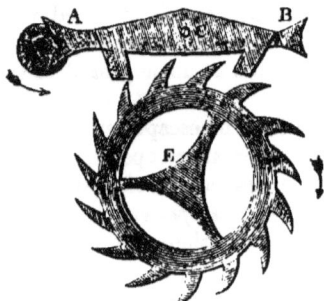

Lever escapement.—This is much used for English pocket watches. A lever A B (*fig* 340.), with a notch at one end, is attached to the anchor C. A pin at *a*, on a disc D, on the verge or arbor of the balance, enters this notch at each vibration, and first moves the dead part of the pallet off the tooth of

Fig. 340.

the scape wheel E, and then receives an impulse, which restores the force it has lost, leaving another tooth engaged in the opposite pallet. As the lever is detached from the balance, except for an instant at the middle of each vibration, the amount of friction is very small. Another advantage of this movement is, that it is but little liable to derangement, and when it is injured, is easily and cheaply repaired, while the duplex and cylindrical escapements are expensive to make, and can only be mended by such skilful workmen as are not often found, except in the metropolis or large towns.

694. **Detached escapement.** — In the class of portable timepieces used for the purposes of navigation, where the greatest attainable regularity of motion is required, an arrangement is adopted called the *detached escapement.* This system is represented in *fig.* 341.

Upon the arbor of the balance wheel is attached a disc, in which there is a notch *i* A smaller disc G, is also attached to it, from which a small pin *a* projects. By the oscillations of the balance wheel the notch *i* and the pin oscillate alternately right and left. A fine flexible spring A, attached to a fixed block B, carries upon it a projecting piece C. To the block D is attached another fine

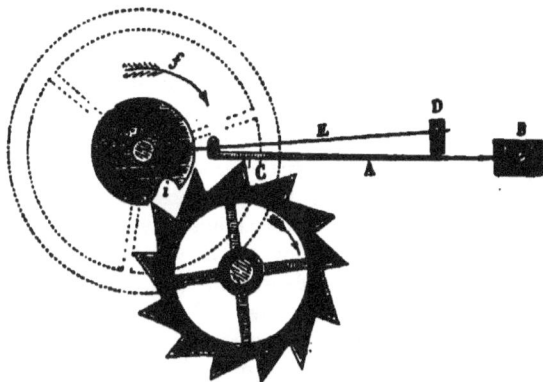

Fig. 341.

spring E, which extends to the edge of the small disc G. The projecting piece C, is so placed that when the spring A is not raised, it encounters a tooth of the scape wheel, but when slightly raised it allows the tooth to pass. The spring E rests in a small fork behind the extremity of A, presented downwards.

Now, let us suppose the balance wheel to swing from left to right. The pin *a*, projecting from the small disc G, coming against the end of the spring E, raises it; and this spring, acting in the fork behind it, raises the spring A, and therefore lifts the piece C, and liberates the tooth which that piece previously obstructed. The scape wheel therefore advances, but before the next tooth comes to the place occupied by the former one, the balance swings back from right to left. The spring E, no longer supported by the fork at the end of A, readily lets the pin pass, and the piece C, returning to its former position, comes in the way of the succeeding tooth and stops it.

At the moment that the balance is about to commence its swing from left to right, and when the piece C is about to liberate the tooth which rests against it, another tooth behind it rests against the side of the notch *i* in the disc G, and when the escapement wheel is liberated, and the swing of the balance from left to right is commencing, this tooth, pressing on the side of the notch *i*, gives it and the balance wheel an impulse which is sufficient to restore to it all the force which it lost in the previous oscillation. Except at this moment the balance wheel in this form of escapement is entirely free from all action of the mainspring.

695. **Maintaining power in clocks.**—While a clock or watch is being wound up, the weight or main-spring no longer presses upon the catch nor upon the ratchet wheel, through which the motion is imparted to the wheelwork. The motion of the hands is therefore sus-pended during the time occupied in wind-ing up; consequently, if the watch or clock keep regular time while it goes, it must lose just so much time as may be em-ployed in the process of winding it up. Al-though this, in common clocks and watches, does not produce any sensible effect, the errors incidental to their rates of going generally exceeding it, yet in clocks used in observatories and in chronometers used for the purposes of navigation, where the greatest degree of regularity is required, provisions are made by which the motion of the clock or watch is continued not-withstanding the process of winding up.

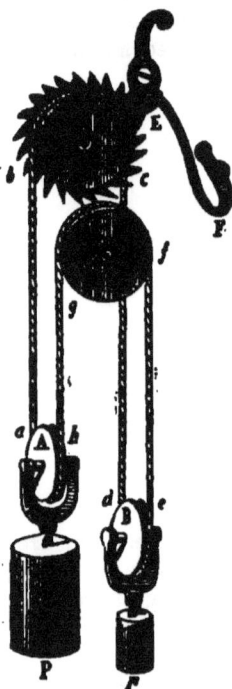

Such expedients are called the *main-taining power.*

One of the most simple arrangements by which this is accomplished in clocks moved by a weight is shown in *fig. 342.*

Fig. 342.

The weight F, which is the moving power, is connected with another much smaller weight *p*, by means of an endless cord, which passes over the grooves of a series of pulleys, A, B, C, and D, of which A and B are movable, and C and D fixed. The force with which F descends is the excess of its weight above that of *p*.

The pulley C, being prevented from revolving by the catch E during the descent of the weight F, and the cord being prevented from sliding upon its groove by the effects of its friction and adhesion, the parts *b a* and *c d*, which descend from the pulley C to the pulleys A and B, may be regarded as being virtually attached to fixed points at *b* and *c*, so that they cannot descend. This being the case, the weight F, descending by its prepon-derance over *p*, and consequently the weight *p* being drawn up, the part of the cord *c d* must pass over the pulley B, the part *e f* over the pulley D, and the part *g h* over the pulley A. In this way the parts *b a* and *g h* will be gradually lengthened as the weight F descends at the expense of the parts *c d* and *f e*, which will be shortened to an equal extent, so that the weight *p* will be raised through the same space as that through which the weight F is lowered. During this process the wheel D is kept in constant revolu-tion; and the first wheel of the train of clockwork being fixed on its axle, a motion is imparted by it through that train to the hands.

When it is desired to wind up the clock, the hand is applied to the cord *c d*, which is drawn downwards, so that the fixed pulley revolves, the catch E dropping from tooth to tooth until the weight P has been raised to the highest, and the weight p has fallen to the lowest point. The parts g h and f e of the cord not being at all affected by this, the pulley D continues to turn as before by the effect of the preponderance of P, which acts as powerfully while it ascends as it did when it descended.

It appears, therefore, that by this arrangement the motion of the works and of the hands is not suspended during the process of winding up.

696. **Maintaining power in watches.** — If the works of a watch be impelled by the force of a mainspring without a fusee, in the manner represented in *fig.* 331., it is evident that the movement must be suspended during the process of winding up; because the ratchet wheel B, by which the force of the spring is transmitted to the works, is then relieved from the action of the catch n o. This defect may, however, in such case be removed by a very simple expedient. Instead of connecting the external extremity of the mainspring with a fixed point, let it be attached to the inside of a barrel surrounding it, and let the wheel c be attached to this barrel. In that case, when the axle of the ratchet wheel is turned in winding up, and the ratchet wheel, therefore, relieved from the action of the catch, the wheel c will be acted upon by the barrel, which will itself be impelled by the reaction of the external extremity of the spring which is attached to it, the winding up being effected only by the internal extremity.

This is the arrangement generally adopted in chimney and table time-pieces, as constructed in France and Switzerland, and also in flat watches, in all of which the adoption of the cylindrical escapement (*fig.* 336.) enables the constructor to dispense with the fusee.

It will, of course, be understood that in such arrangements, while the wheel c is attached to the barrel, and by it to the external extremity of the mainspring, the ratchet wheel B is attached to the axle T B (*fig.* 331.), and by it to the internal extremity of the mainspring.

When a fusee is used, the ratchet wheel being fixed upon its axis, and not on that of the barrel containing the mainspring, this method of obtaining a maintaining power is not applicable. In such cases the object is attained by two ratchet wheels upon the axle of the fusee having their teeth and catches turned in opposite directions, one of them being impelled by a provisional spring, which only comes into play when the action of the mainspring is suspended during the process of winding up.

The fusee, with its appendages, as commonly constructed, without a maintaining power, is drawn in *fig.* 343., the grooved cone with the ratchet wheel attached to its base being raised from the cavity in the toothed wheel c D, in which it is deposited, to show the arrangement more clearly, and in the edge of which the catch *n* is placed, so that it shall fall into the teeth of the ratchet wheel. When the watch is being wound up, the chain passing from the barrel to the grooves of the fusee, the teeth of the ratchet wheel A B pass freely round the cavity, the catch *n* falling from tooth to tooth, and producing the clicking noise already noticed. But when the watch is going, the tension of the chain draws the fusee and the ratchet wheel attached to it round in the contrary direction, and, pressing the teeth of the ratchet wheel against the catch *n*, carries round the wheel C D, which gives motion to all the other wheels, and through them to the hands. Now, it will be evident that when the watch is being wound up, and the catch *n* relieved from the pressure of the teeth of the ratchet wheel, no motion will be imparted to C D, and consequently the movement of the entire works will be suspended.

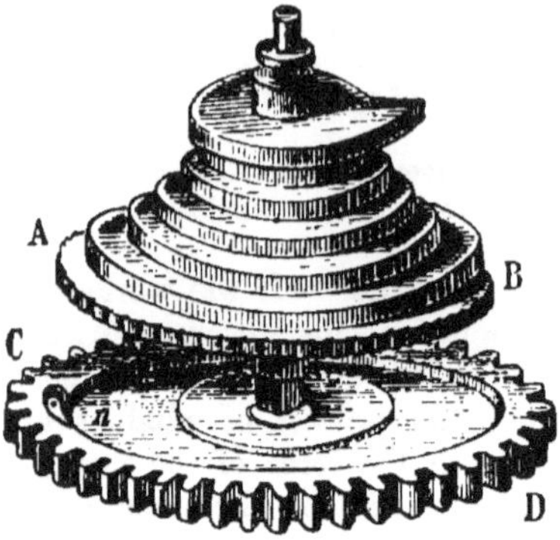

Fig. 343.

The modification by which a maintaining power is obtained by the combination of two contrary ratchet wheels is shown in *fig.* 344., where C D is the first wheel of the train from which the watch receives its motion. The ratchet wheel A is fixed upon the base of the fusee, so as to move with it. The catch *m*, which is pressed by a spring into the teeth of this ratchet wheel, is attached to the second ratchet wheel B, so that when A is carried round by the chain acting on the fusee, it must carry the wheel B round with it in the direction of the arrow *f*. The wheel B is connected with the wheel C D by a semicircular spring *a b c*, which is attached to the wheel C D at *c*, and to the wheel B at *a*. The catch *n*, which falls into the teeth of the ratchet wheel B, is attached to a fixed point on the bed of the watch.

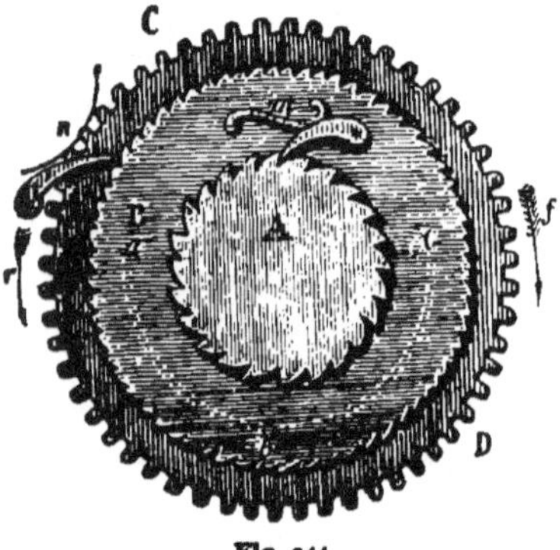

While the watch is going, the wheel B, driven by the fusee, draws after it the spring *a b c*, and this spring, acting at *c* on the wheel C D, draws it round in the direction of the arrow *f*. Now,

Fig. 344.

bending it round to a certain extent; and this spring, acting at *c* on the wheel C D, draws it round in the direction of the arrow *f*. Now,

B B 4

let us suppose that the watch is being wound up. The ratchet wheel A, being turned by the key in the direction of the arrow r, the catch m falls from tooth to tooth, and the force it before received from the ratchet wheel A is suspended. But the spring a b c has been drawn from its form of equilibrium, to a certain small extent, in drawing round after it the wheel C D, as already stated, and it has still a tendency to draw that wheel after it, so as to recover its form of equilibrium. In doing this it will continue to act upon the wheel C D, and to carry it round while the action of the fusee upon it is suspended during the process of being wound up. The spring a b c is so constructed as to act thus for an interval more than is necessary to wind up the watch.

697. **Cases in which weights and springs are used.** — From what has been explained, it will be observed that time-pieces, in general, are constructed with one or other of two moving powers — a descending weight or a mainspring ; and with one or other of two regulators — a pendulum or a balance wheel. These expedients are variously adopted and variously combined, according to the position and circumstances in which the time-piece is used, and the purpose to which it is appropriated.

A descending weight as a moving power, combined with a pendulum as a regulator, supply the best chronometrical conditions. But the weight can only be used where a sufficient vertical space can be commanded for its ascent and descent, and neither it nor the pendulum is applicable, except to time-pieces which rest in a fixed and stable position.

In the case of time-pieces whose position is fixed, but where the vertical space for the play of the weight cannot be conveniently obtained, the mainspring is applied as a moving power, combined with the pendulum as a regulator. Chimney and table clocks present examples of this arrangement. The height being limited, it is necessary also in these cases to apply short pendulums. The length of a pendulum which vibrates seconds being about 39 inches, such pendulums can only be used where considerable height can be commanded.

It has been shown that the lengths of pendulums are in the proportions of the squares of the times of their vibration. It follows, therefore, that the length of the pendulum which would vibrate in half a second, will be one fourth the length of one which vibrates in a second, and since the latter is 39 inches, the former must be 9¾ inches. Such a pendulum can, therefore, be conveniently enough applied to a chimney or a table-clock high enough to leave about ten inches for its play.

The pendulum is so good a regulator, and the anchor escapement renders it so independent of the variation of the moving power, that in time-pieces where it is combined with a mainspring, a fusee is found to be unnecessary. In such cases, therefore, the axis of the first wheel is placed in the centre of the mainspring, as represented in *fig.* 331.

698. **Watches and chronometers.** — The cylindrical escapement shown in *fig.* 336., is nearly as independent of the variation of the moving power as the pendulum, and therefore in common watches, where this escapement is used, the fusee is dispensed with.

In the class of time-pieces called chronometers, used for the purposes of navigation, and in general for all purposes where the greatest attainable perfection is required in a portable time-piece, all the expedients to insure regularity are united, and accordingly the detached escapement is combined with fusee and mainspring.

Besides the expedients above mentioned, for insuring uniformity of rate, provisions are made in the most perfect chronometers to prevent the variations of rate which would arise from the expansion and contraction of the metal composing the balance wheel by the variation of temperature.

Marine chronometers. — These are usually suspended in a box on gimbals, like those which support the ship's compass. The balance wheel usually vibrates in half seconds, being a much slower rate than that of common watches. They are of immense utility in navigation, and especially in long voyages.

699. **Observatory clocks.** — In observatories where stationary time-pieces can be used, the clock moved by weights and regulated by a pendulum is invariably adopted. The pendulum, in such cases, is always so constructed that its rate of vibration shall not be affected by variations of temperature.

700. **Striking apparatus.** — In clocks adapted for domestic and public use, it is found desirable that they should give notice of the time, not only to the eye, but to the ear; and for that purpose a bell is attached to them, which tolls at given intervals, the number of strokes being equal to that of the units in the number expressing the hour. This appendage is called the *striking train*.

The striking train, though connected with that which moves the hands, is quite independent of it, having its own moving and regulating power, and its own system of wheels by which the effect of the moving power is submitted to the regulator, and transmitted to the tongue of the bell.

Unlike the train which moves the hands, the striking train is not in continual motion. Its motion is always suspended, except at the particular moment at which the clock strikes. The mechanism partakes of the character of an alarum, being stopped by a certain catch until the hands point to some certain hour, and then being set free by the withdrawal of the catch. It remains free, however, only so long as is necessary for the tongue or hammer to make the necessary number of strokes upon the bell, after which the catch again engages itself in the striking mechanism, and stops it.

Some clocks only strike the hour. Others mark the half hours, and others the quarters, by a single stroke.

The general principle of the striking mechanism will be rendered intelligible by *fig.* 345., which represents it in the case of a common clock moved by a weight.

The weight suspended from the cord ε moves the train in the same manner as in the case of the train which moves the hands. The motion of the first wheel c is transmitted to the last wheel ι, which corresponds to the escapement wheel, by the intermediate wheels and pinions *f*, F, *g*, G, *h*, H, and *i*. The wheel ι drives the pinion *k*, upon the axle of which is fixed the regulator κ. This regulator is a fly, a side view of which, upon a larger scale, is shown in *fig.* 346.

The pinion, which is made to revolve by the wheel w, gives a motion of rotation to the fly A A′ B′ B, which consists of a thin rectangular plate of metal, along the centre of which the prolongation M L of the axle of the pinion is attached. The flat surfaces of the fly, revolving more or less rapidly, strike against the air, which resists them with a force which increases in the proportion of the square of the velocity of the rotation. Thus, if the velocity of rotation be increased in a two-fold ratio, the resistance to A A′ B′ B is increased in a four-fold ratio; if the rotation be increased in a three-fold ratio, the resistance is increased in a nine-fold ratio, and so on. It is evident, therefore, that by this very rapid increase the resistance to the motion of the train must soon become equal to the descending force of the weight, and then the motion will become uniform; for if it were to increase, the resistance would exceed the force of the weight, and would slacken the rate of motion; and if it were to decrease, the resistance being less than the force of the weight, the latter would accelerate the motion. In either case the motion would immediately be rendered uniform.

Projecting from the face of the wheel H (*fig.* 345.) there is a small pin which rests upon the end *m* of a lever *m* *n*, which turns upon the centre *n*. The lever *m* *n*, when in this position, stops the motion of the striking train. Behind the same lever *m* *n*, and projecting from it, there is another piece, which in the position represented in the figure rests in a notch of the wheel *o* *p*, lying behind the striking train, and indicated in the figure by dotted lines. Around the edge of this wheel there is a series of similar notches at unequal distances, determined in the manner which we shall presently explain.

Upon the face of the wheel G, at equal distances one from another surrounding it, a series of pins project, which, as the wheel turns, successively encounter a lever *b*, which plays upon a centre *d′*. Upon the same centre *a′* is fixed the handle *a a′* of the hammer A by which the bell v is struck. A spring fixed upon the same centre *a′* causes the lever *b* to rest in the position represented in the figure, and to return to that position if raised from it. The hammer handle *a a′* is made either elastic itself, or is provided with a like spring.

When the wheel G is made to revolve at a uniform rate by the weight ε,

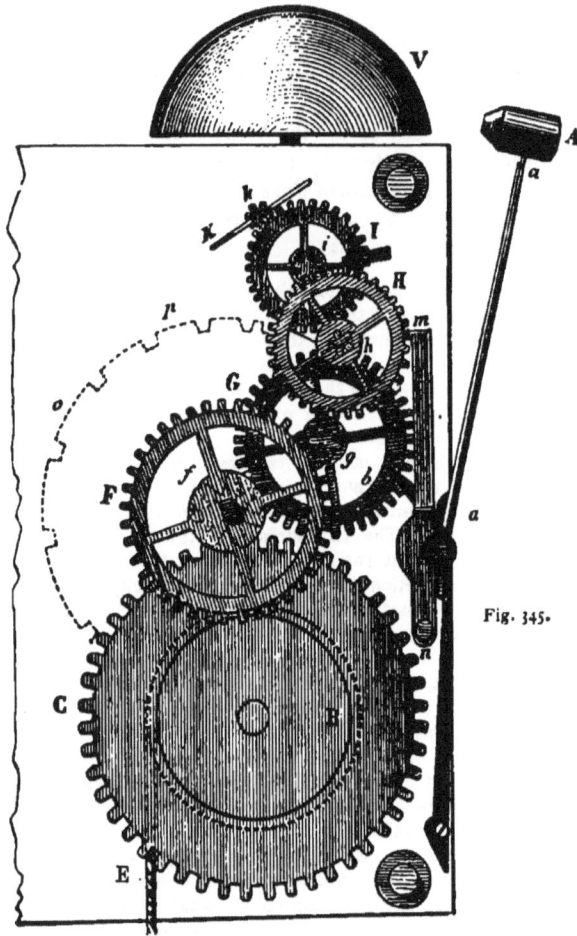

Fig. 345.

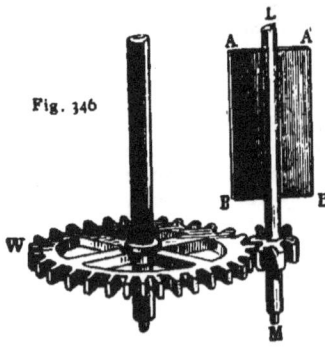

Fig. 346

regulated by the fly κ, the pins projecting from the face of the wheel G encounter successively the lever *b*, and raising it, throw back the handle *a a'* of the hammer which is in connection with the lever *b*. After the pin has passed the lever *b*, the latter is brought back with a jerk by the action of the spring; and the hammer A, receiving the same jerk, strikes upon the bell v, and instantly recoils from it; and if the wheel G continues to revolve, one pin after another upon it will encounter the lever *b*, and the ham mer A will make a stroke upon the bell each time that a pin passes the lever.

The wheel H is so constructed that it will make one revolution in the interval between two successive strokes of the bell; or, what is the same, in the interval between the moments at which two successive pins pass the lever b.

In the train which moves the hands, an expedient is provided by which each time that the minute-hand passes twelve upon the dial, the lever m n is thrown back from the position which it has in *fig.* 345.; and the top m being withdrawn from under the pin upon the wheel H, that wheel and the entire striking train is liberated and set in motion. At the same time the piece is taken out of the notch in which it rested on the wheel o p, and that wheel, in common with the other parts of the striking train, is put in motion.

For every complete revolution that the wheel H makes, the hammer A makes a stroke upon the bell, and the motion of H and of the entire striking train will continue until the end m of the lever m n again comes under the pin projecting from the face of H. During the motion the lever m n is kept back by the edge of the wheel o p acting against the piece projecting from the lever m n; but when the wheel o p has turned so as to bring the next notch to the projecting piece, it will be thrown into the notch; and the end m of the lever m n coming under the pin projecting from the wheel H, the motion of the train will be suspended.

Now it will be evident, from what has been stated, that when the lever m n is thrown back by the works, it is kept back by the edge of the wheel o p acting against the projecting piece; and so long as it is thus kept back, the striking train continues to move, and the hammer continues to strike the bell. But the duration of this motion will depend on the space between the notches on the wheel o p, since it is while that space upon the edge passes under the projecting piece on m n that the end m of the lever m n is kept back, so as to be out of the way of the pin projecting from the wheel H. These spaces between the notches are therefore so proportioned one to another that the lever m n at each hour is held back a sufficient time for the hammer to make upon the bell the number of strokes denoting the hour, and no more.

The arrangement for striking half hours and quarters is based upon similar principles.

CHAP. VI.

THE PRINTING PRESS.

WHETHER it be regarded as a mere piece of mechanism or viewed in relation to the vast influence it has exercised on the progress of the human race in knowledge and civilisation, the printing press must ever be an object of profound interest.

To render more easily intelligible the beautiful mechanical contrivances of which it consists, it will be necessary, in the first instance, to explain the succession of operations they are intended to execute

701. **Setting the types.**—The types are in the first instance put together or *composed*, as it is technically called, by persons therefore called *compositors*, who, standing with the manuscript before them, collect the letters one by one from an inclined desk, consisting of as many small compartments as there are letters, in which the types are assorted. The types thus put together are ranged in lines, pages, or columns, according to the sheet to be printed; and all the pages of type to be impressed upon the same side of a sheet are placed in juxtaposition, in a strong iron frame, corresponding in size and shape to the sheet, proper intervals being left to represent the margins between page and page, or column and column. The frame thus including the composed types to be impressed on one side of a sheet of paper is called a *form*.

Types are similarly composed and arranged, corresponding to the matter to be printed on the opposite side of the same sheet.

702. **Old method of printing.**—There are two classes of printing machines, *single* and *double*. By the former, the sheets are only printed on one side, and by the latter they are printed on both sides in the same operation.

The form to be printed being laid upon a horizontal table, with the faces of the types uppermost, the following operations are to be executed: — 1st, printing ink is to be applied to the faces of the type, so evenly that there shall be no blotting or inequalities in the printing; 2nd, the sheet of paper to be printed must be laid upon the form so as to receive the impression of the type in its proper position, and in the centre of it; 3rd, this paper must be urged upon the type by a sufficient pressure to enable it to receive the printed characters, such pressure, however, not being so great as to cause the type to penetrate or deface the paper; 4th, the paper is, in fine, when thus printed, to be withdrawn from the type and laid upon a table, where the printed sheets are collected.

In the old process these operations were performed by two men, one of whom was employed to ink the types, and the other to print. The former was armed with two bulky inking balls, consisting of a soft black substance of leathery appearance, spherical form, and about 12 inches in diameter. He flourished these with dexterity, dabbed them upon a plate smeared with ink, and then with both hands applied them to the faces of the types until the latter were completely charged with ink. This accomplished, the other functionary — the pressman — having prepared the sheet of paper while the type was being inked, turned it down upon the type, drew it under the press, and with a severe pull of the lever gave the necessary pressure by which the paper took the impression of the type. A contrary motion of the apparatus withdrew the type from under the press, and the pressman, removing the paper, now printed de-

posited it upon a table placed near him to receive it. The same
series of operations was then repeated for producing the impression
of another sheet, and so on. In this manner two men in ordinary
book work usually printed at the rate of 250 sheets per hour on
one side.

703. **Inking rollers.**—One of the first improvements which

Fig. 347.

took place upon this apparatus consisted in the substitution of a
cylindrical roller for the inking balls. This roller was mounted
with handles, so that the man employed to ink the type first rolled
it upon a flat surface smeared with ink (*fig.* 347.), and having
thus charged it, applied it to the type form, upon which he rolled
it in a similar manner, thus transferring the ink from the roller to
the faces of the type. The substitution of these inking rollers for
the inking balls constituted one of the most important steps in the
modern improvement of the art of printing. The rollers were
composed of a combination of treacle and glue, and closely resem-
bled caoutchouc in their appearance and qualities.

704. **Modern printing presses.** — The printing presses
which served the purposes of publication for some hundred years,
during which they received no other improvements than such as

might be regarded merely as modifications in the detail of their mechanism, have been entirely superseded by engines of vastly increased power and improved principles of construction. Although these admirable machines differ one from another in the details of their mechanism, according to the circumstances under which they are applied, and the power they are expected to exert, they are nevertheless characterised by certain common features. .

705. The form to be printed is laid in the usual manner upon a perfectly horizontal table, with the faces of the types uppermost; and upon the same table, in juxtaposition with the form, and level with the faces of the type, or nearly so, is placed a slab upon which a thin and perfectly regular stratum of printing ink is diffused ; the table thus carrying the form and inking slab is moved by proper machinery right and left horizontally, with a reciprocating rectilinear motion through a space a little greater than the length of the form.

Above the form and slab are mounted, also in juxtaposition, a large cylinder or drum which carries upon it the sheet of paper to be printed, and three or four inking rollers similar to that already described. There are also three or four other rollers in juxtaposition with the latter, one of which supplies ink to the others, which severally spread it in a uniform stratum upon the slab. The paper cylinder and the inking and diffusing rollers are so mounted that, when the horizontal table, carrying the form and inking slab, moves alternately backwards and forwards under them, they roll upon it.

In this way, when the table is moved towards the rollers, the form, passing under the inking rollers right and left, receives from them the ink upon the faces of the type ; and at the same time the slab, moving backwards and forwards under the diffusing rollers, receives from them, upon its surface, the proper stratum of ink to supply the place of that which was taken from it by the inking rollers.

706. **Single printing machines.** — When the table is moved alternately towards the other side, the form, with the types already inked, passes under the cylinder carrying the paper, the motion of which is so regulated as to correspond exactly with the rectilinear motion of the table carrying the form. The cylinder is urged upon the type with a regulated force, sufficient and not more than sufficient, to impress the type upon the paper.

The sheets of paper are supplied in succession to the cylinder, and held evenly upon it by bands of tape while they pass in contact with the type. After receiving the impression of the type, the tapes which bound them are separated, and the printed sheets discharged.

Such is the general principle of single printing machines

707. **Double printing machines.** — In these the table which
is moved alternately right and left carries two forms, one corresponding to the pages to be printed on one side of the sheet, and
the other to those to be printed on the other side. There are also
two inking slabs, one to the left of the left hand form, and the
other to the right of the right hand form. There are also two
paper cylinders, and two sets of inking and diffusing rollers. Each
sheet of paper to be printed, being held between tapes, as already
described, is carried successively round the two cylinders, being so
conducted, in passing through one to the other, that one side of it
passes in contact with, and is printed by, one form, and the opposite side by the other form. The proportion of the motions is so
nicely regulated, that the impression of each page or column made
on one side of the paper, corresponds exactly with that of the corresponding page or column on the other side.

This general description will be more clearly understood by
reference to the following illustrative diagrams : —

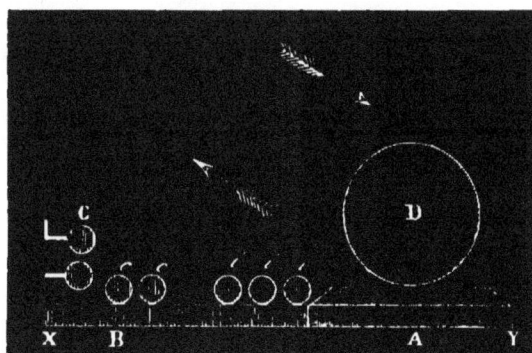

Fig. 348.

Fig. 348. illustrates the operation of a single printing machine. The form
A and the inking slab B, are placed on a horizontal table; above them is the.
paper cylinder D, the inking rollers *i i i*, the diffusing rollers *c c*, and the
rollers C, which supply the ink to the diffusing rollers. The first of these, C,
is called the ductor roller. When the table X Y is moved towards the left
from Y to X, the form A passes under the inking rollers *i i i*, and receives ink
from them on the faces of the type; at the same time the slab B passes under
the diffusing rollers *c c*, and receives from them a supply of ink to replace
what it has just given to the inking rollers.

When the table is moved in the contrary direction from X towards Y, the
form once more passed under the inking rollers, and afterwards under the
paper cylinder, which, being pressed upon it while it moves in exact accordance with it, the types discharge upon the paper the ink they have just
received from the rollers; and the printing of the paper being thus effected
on one side, the sheet is discharged from the tapes. The table is then again
moved to the left, and the types are again inked, and the same effects ensue as
have already been described.

In this manner sheet after sheet is printed.

The inking and diffusing rollers rest upon the slab and the types by their weight, the axes projecting from their ends being inserted in slits formed in upright supports attached to opposite sides of the frame which supports the moving table. The two upright pieces in which the axes of each roller are inserted, are not placed in exact opposition to each other; the consequence of which is, that the rollers are placed with their axes slightly inclined to the sides of the table. This arrangement is attended with a very important effect: for, in consequence of the friction or adhesion of the rollers with the slab, they are moved alternately in contrary directions with the longitudinal motion across the table. This motion, combined with their rolling motion upon the slab, aids materially in diffusing the ink in a perfectly uniform stratum.

An illustrative diagram of a double printing machine is shown in *fig.* 349. where D and D' are the two paper cylinders; A and A', the two forms; i i i

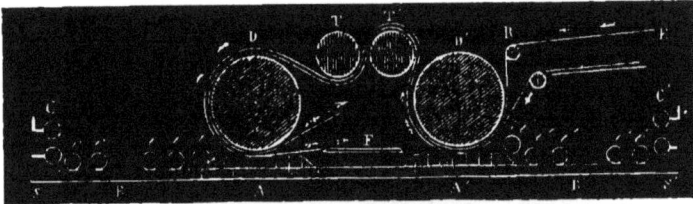

Fig. 349.

and i' i' i', the inking rollers; c c and c' c', the diffusing rollers; and c and c', the ductor rollers. The pile of paper placed on the table E, is supplied sheet by sheet to the tapes between which it is held; and being passed over the roller R, and under the cylinder D', it receives the impression of the types of the form A'; it then passes successively over the roller T', under T, and round the cylinder D, at the lower point of which its unprinted side comes into contact with the types of the form A, by which it is printed; after which, the tapes opening, it is thrown out by the centrifugal force upon the receiving table F.

It will be apparent by the figure, that while the sheet is printed on one side by the form A', the form A is passing between the inking rollers i i i and the slab B ; and on the contrary while the paper is printed on the other side by the form A, the form A' is passing between the inking rollers i' i' i', and the slab B'.

In this manner, by each alternate motion from right to left and from left to right, a sheet is printed on both sides.

708. A perspective view of a double-acting printing machine, as constructed by Messrs. Applegath and Cowper, is shown in *fig.* 350.

A boy, called the *layer on*, R, standing at an elevated desk, pushes the paper, sheet by sheet, towards the tapes, which, closing upon it, carry it over

C C

Fig. 160.   Applegath and Cowper's Double Printing Machine.

a roller R, passing under which it is carried to the right of the cylinder D under which it passes, and being carried up to the left of it, passes successively over the roller T, under the roller T', over the cylinder D, and drawn along its left side, after which it passes under it, and is flung into the hands of boy, F, called the *taker off*, seated before a table, placed between the two cylinders D and D', upon which he disposes the sheets as he receives them.

In this manner, the layer on feeds the machine in constant succession with blank sheets, which, being carried under the cylinder D, are printed on one side, and afterwards under the cylinder D', are printed on the other, when they are received by the taker off.

The doctor rollers, C and C', are kept in revolution by endless bands, carried over rollers at the lower parts of the frame, and then over grooved wheels fixed on the axes of the cylinders. The table carrying the forms and slabs is moved alternately right and left by means of a pair of bevelled wheels, w, under the frame, and a double rack and pinion above; one of the bevelled wheels, having a horizontal axis, receives motion from a steam-engine or other moving power; it imparts motion to the other bevelled wheel, having a vertical axis. This latter axis has a pinion fixed at its upper end, which works in a double rack attached to the movable type table, which is so constructed that the continuous rotation of the pinion imparts an alternate rectilinear motion right and left to the rack and to the table attached to it.

The manner in which the tapes lay hold of and conduct the paper successively round the cylinders, and finally discharge it upon the table of the *taker off*, will be easily understood by reference to *fig.* 351.

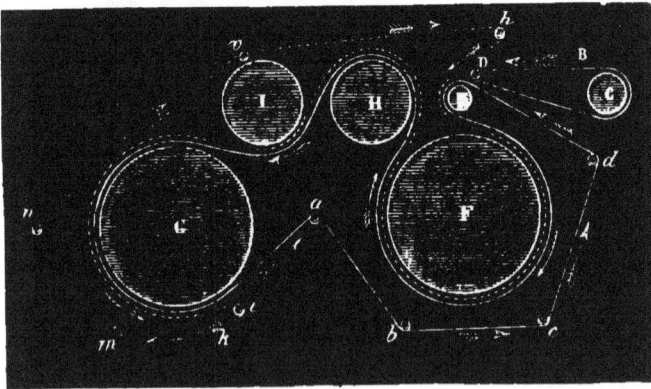

Fig. 351.

C and D are two grooved rollers, surrounded by an endless band which pushes the paper from the table of the *layer on* towards the tapes. The two endless tapes, between which the paper is held, are represented in the diagram by the continuous and dotted lines, and the direction of their motion round the rollers and cylinders is represented by the arrows. It will be perceived that opposite the table of the *layer on*, the tapes converge, from two small rollers d and h, and come into contact at the top of the roller E. The edges of the sheets of paper, being advanced from the table of the *layer on*, are caught between the tapes immediately above the roller E.

C C 2

It must be understood that there are two or three pairs of tapes parallel to each other, which correspond to the margins of the pages or columns; but only one pair is shown in the figure.

The paper, being thus seized between the tapes above the roller E, is carried successively, as shown in the figure, still held between the tapes, under and over F H I and G, until it arrives at i, where the tapes separate, that which is indicated by the continuous line being carried to the roller a, and that by the dotted line, over the roller i, to the roller k. The tapes being thus separated the printed sheet is discharged at i, upon the table of the *taker off*; meanwhile, the tape indicated by the continuous line is carried successively over the roller a, under b, under c, outside d, and is finally returned to the roller E.

In the same manner the tape indicated by the dotted line is carried successively under the rollers k and m, outside n, over v and h, from which it returns to E, where it again joins the other tape proceeding from d.

709. The first machines constructed upon this improved principle for printing newspapers, were erected at the printing office of *The Times* newspaper, and it was announced in that journal, on the 28th of November, 1814, that the sheet which was then placed in the hands of the reader, was the first printed by steam-impelled machinery.

By this, with some improvements which the apparatus received soon afterwards, the effective power of the printing press was augmented in a very high proportion. With the hand presses previously in use, not more, as we have seen, than 250 sheets could be printed on one side in an hour. Each of the two machines erected at *The Times* office produced 1800 impressions per hour.

710. **Further improvement.**— The power of the printing machine constructed upon this principle was soon after augmented, by increasing the number of printing cylinders to four, the principle of the machine, however, remaining the same.

The manner in which this was accomplished will be easily understood by the aid of the illustrative diagram (*fig.* 352.), where 1, 2, 3, and 4, are the printing cylinders; P P P' P', are the tables of the four *layers on*, and o o o' o', lead to those of the four *takers off*. The course followed by the sheets of paper, in passing to and from the cylinders, are indicated by the arrows. Inking rollers are in this case placed at R, between the printing cylinders; the two type forms are inked twice, while they move from right to left, and twice again, while they move from left to right. The printing cylinders are alternately let down upon the type and raised from them in pairs; while the type table moves from left to right, the cylinders 1 and 3 are in contact with the table, the cylinders 2 and 4 being raised from it, and, on the contrary, when it moves from right to left, the cylinders 2 and 4 are in contact with it, 1 and 3 being raised from it.

By this improvement, which was adopted in *The Times* office in 1827, the proprietors of that journal obtained the power, then unprecedented, of printing from 4000 to 5000 sheets per hour on one

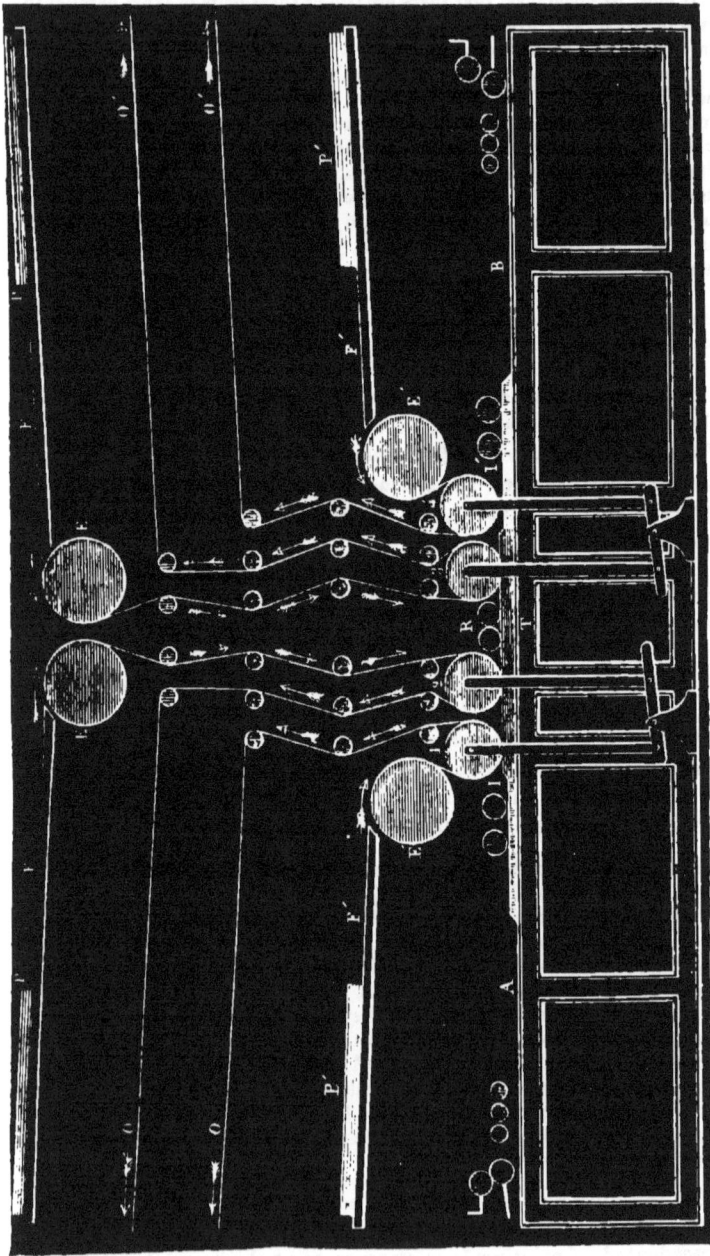

Fig. 252.

side of the paper. By this means they were enabled to satisfy the demands of a circulation amounting to 28000.

In reference to newspaper printing, it must be here observed that the great object to be attained, is to increase the celerity by which printing on one side only of the sheet can be augmented. It is found convenient so to arrange the letter-press that the matter appropriated to one side of the sheet shall be ready for press at an early hour, and may be printed before the contents of the other side, in which the most recent intelligence is included, can be prepared. Hence the advantage of using machines adapted to print one side only with the most extreme celerity for newspapers.

711. "**Times**" **printing machine.**—This machine continued to serve the purposes of *The Times* newspaper until a later epoch, when again the exigencies of the press exceeded even its immense powers, and another appeal was made to the inventive genius of Mr. Applegath. It was, in short, necessary to provide a machine by which *at least* 10000 *sheets an hour could be worked off from a single form!*

In considering the means of solving this problem, it is necessary to observe, that whatever expedient may be used, the sheets of paper to be printed must be delivered one by one to the machine by an attendant. After they once enter the machine they are carried through it and printed by self-acting machinery. But in the case of sheets so large as those of the newspapers, it is found that they cannot be delivered with the necessary precision by manipulation at a more rapid rate than two in five seconds, or twenty-five per minute, being at the rate of 1500 sheets per hour. Now, in this manner, to print at the rate of 10000 per hour, would require seven cylinders, to place which so as to be acted upon by a type form moving alternately in a horizontal frame, in the manner already described, would present insurmountable difficulties.

In the face of these difficulties, Mr. Applegath, to whom the world is indebted for the invention of *The Times* printing machine, decided on abandoning the reciprocating motion of the type form, arranging the apparatus so as to render the motion continuous. This necessarily involved circular motion, and accordingly he resolved upon attaching the columns of type to the sides of a large drum or cylinder, placed with its axis vertical, instead of the horizontal frame which had been hitherto used. A large central drum is erected, capable of being turned round its axis. Upon the sides of this drum are placed vertically the columns of type. These columns, strictly speaking, form the sides of a polygon, the centre of which coincides with the axis of the drum, but the breadth of the columns is so small compared with the diameter of the drum,

that their surfaces depart very little from the regular cylindrical
form. On another part of this drum is fixed the inking table.
The circumference of this drum in *The Times* printing machine
measures 200 inches, and it is consequently 64 inches in diameter.

The general form and arrangement of the machine are represented in *fig.* 353., where D is the great central drum which carries the type and inking tables.

This drum in *The Times* machine is surrounded by eight
cylinders, B, B, &c., also placed with their axes vertical, upon
which the paper is carried by tapes in the usual manner. Each
of these cylinders is connected with the drum by toothed wheels, in
such a manner that their surfaces respectively must necessarily
move at exactly the same velocity as the surface of the drum.
And if we imagine the drum, thus in contact with these eight
cylinders, to be put in motion, and to make a complete revolution,
the type form will be pressed successively against each of the eight
cylinders, and if the type were previously inked, and each of the
eight cylinders supplied with paper, eight sheets of paper would
be printed in one revolution of the drum.

It remains, therefore, to explain, first, how the type is eight times
inked in each revolution ; and, secondly, how each of the eight
cylinders is supplied with paper to receive their impression.

Beside the eight paper cylinders are placed eight sets of inking
rollers ; near these are placed two ductor rollers. These ductor
rollers receive a coating of ink from reservoirs placed above them.
As the inking table attached to the revolving drum passes each of
these ductor rollers, it receives from them a coating of ink. It
next encounters the inking rollers, to which it delivers over this
coating. The types next, by the continued revolution of the drum,
encounter these inking rollers, and receive from them a coating of
ink, after which they meet the paper cylinders, upon which they
are impressed, and the printing is completed.

Thus in a single revolution of the great central drum the inking
table receives a supply eight times successively from the ductor
rollers, and delivers over that supply eight times successively to
the inking rollers, which, in their turn, deliver it eight times successively to the faces of the type, from which it is conveyed finally
to the eight sheets of paper held upon the eight cylinders by the
tapes.

Let us now explain how the eight cylinders are supplied with
paper. Over each of them is erected a sloping desk, *h*, *h*, &c.,
upon which a stock of unprinted paper is deposited. Beside this
desk stands the " layer on," who pushes forward the paper, sheet
by sheet, towards the tapes.

These tapes, seizing upon it, first draw it down in a vertical

direction between tapes in the eight vertical frames, until its edges correspond with the position of the form of type on the printing cylinder. Arrived at this position, its downward motion is stopped by a self-acting apparatus provided in the machine, and it is then impelled by vertical rollers towards the printing cylinder, these rollers having upon them marginal tapes which carry the paper round the cylinder, from which it receives the impression of the types. After this the central and lower marginal tapes dismiss the sheet of paper, which the upper ones only become charged with, and carry it to its proper place, where the tapes are stopped with the paper suspended between them, until the "taker off" draws the sheets down, ranging them upon his table. These movements are continually repeated; the moment that one sheet passes from the hands of the "layer on," he supplies another, and in this manner he delivers to the machine at the average rate of two sheets every five seconds, and the same delivery taking place at each of the eight cylinders, there are 16 sheets delivered and printed every five seconds.

It is found that by this machine in ordinary work between 10000 and 11000 per hour can be printed; but with very expert men to deliver the sheets, a still greater speed can be attained. Indeed, the velocity is limited, not by any conditions affecting the machine, but by the power of the men to deliver the sheets to it.

In case of any misdelivery a sheet is spoilt, and, consequently, the effective performance of the machine is impaired. If, however, a still greater speed of printing were required, the same description of machine, without changing its principle, would be sufficient for the exigency; it would be necessary that the types should be surrounded with a greater number of printing cylinders.

It may be right to observe, that the cylinders and rollers are not uniformly distributed round the great central drum; they are so arranged as to leave on one side of that drum an open space equal to the width of the type form. This is necessary in order to give access to the type form so as to adjust it.

One of the practical difficulties which Mr. Applegath had to encounter in the solution of the problem, which he has so successfully effected, arose from the shock produced to the machinery by reversing the motion of the horizontal frame which, in the old machine, carried the type form and inking table, a moving mass which weighed 25 cwt. This frame had a motion of 88 inches in each direction, and it was found that such a weight could not be driven through such a space with safety at a greater rate than about 45 strokes per minute, which limited its *maximum* producing power to 5000 sheets per hour.

Another difficulty in the construction of this vast piece of machinery was so to regulate the self-acting mechanism that the impression of the type form should always be made in the centre of the page, and so that the space upon the paper occupied by the

printed matter on one side may coincide exactly with that occupied by the printed matter on the other side.

The type form fixed on the central drum moves at the rate of
about 80 inches per second, and the paper is moved in contact
with it of course at exactly the same rate. Now, if by any error
in the delivery or motion of a sheet of paper, it arrive at the printing cylinder 1-80th part of a second too soon or too late, the relative position of the columns will vary by 1-80th part of 80 inches
— that is to say, by one inch. In that case the edge of the printed
matter on one side would be an inch nearer to the edge of the
paper than on the other side.

This is an incident which rarely happens, but when it does, a
sheet, of course, is spoilt. In fact, the waste from that cause is
considerably less in the present vertical machine than in the former
less powerful horizontal one.

The vertical position of the inking rollers, in which the type is
only touched on its extreme surface, is more conducive to the
goodness of the work than the horizontal machine, where the
inking rollers act by gravity ; also any dust shaken out of the
paper, which formerly was deposited upon the inking rollers,
now falls upon the floor.

With this machine 50000 impressions have been taken without
stopping to brush the form or table.

712. **Marinoni's newspaper printing press.** — Messrs.
Marinoni and Co., of Paris, have within the last few years constructed improved printing presses for newspapers of large circulation, several of which have been erected and brought into operation in the printing office of the Paris journal *La Presse*. This
printing machine, which is capable of working off 6000 copies per
hour, printed on both sides of the paper, is represented in *fig.* 354.
It will be perceived that eight men are employed in the process,
four layers on and four takers off. The machine is double, the
parts at each side of a vertical line drawn through the axis of the
fly wheel being perfectly similar. The manner in which the sheets
pass to and from the printing rollers will be more readily understood by *fig.* 355., where A is the upper and A' the lower delivering board, and B the upper and B' the lower receiving board on
the right, the two delivering and receiving boards on the left being
similarly placed. The motion of the sheets, as they are conducted
to and from the rollers by the tapes, is indicated by arrows, and
the course followed by each sheet from the moment it leaves the
delivering board until it arrives at the receiving board, is indicated
by the numbers 1, 2, 3, 4, &c. Thus, the sheet delivered from
the board A is taken by the tapes which pass round the roller M,
and carried from 1 to 2. Arriving at the lower roller, it passes

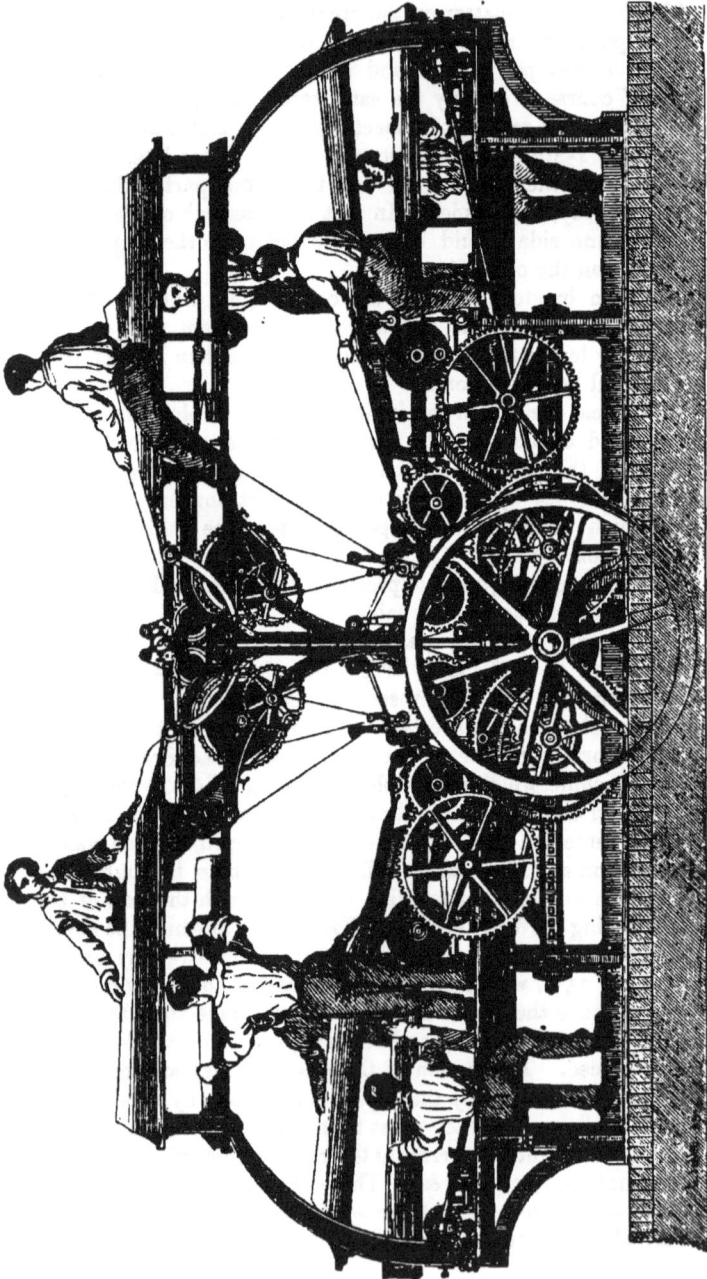

Fig. 354.

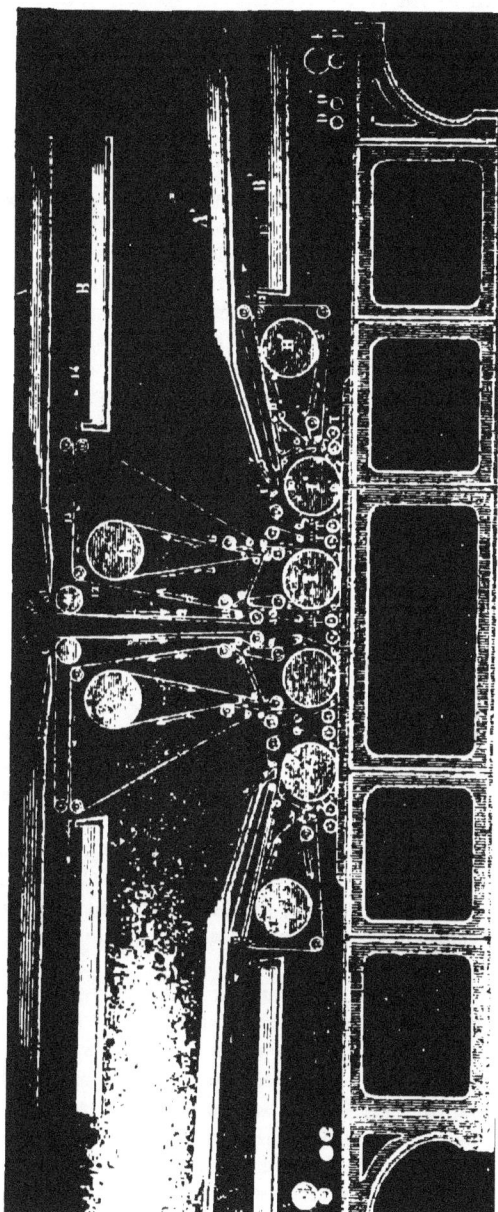

as shown by the arrow between the rollers, 3, and is carried from 4 under the printing roller 1, where it is printed on one side, after which it is carried up between the tapes to 5, from whence it is discharged between the tapes of 6, and carried up over the roller R at 7, from which it is carried down between the tapes 8, and thrown, as shown by the arrow, to the tapes 9, by which it is again carried under the roller and printed on the other side; after which, it is carried up successively between the tapes 10 and 11 to 12, and finally discharged from 13 at 14 upon the receiving board B.

The sheet delivered from the lower receiving board A', follows a course precisely similar, entering at 1' and passing round the printing roller at 2' 3', from which it passes between the tapes 4' round the roller R' at 5' 6', and thence from the tapes 7' 8' round the printing roller 1' at 9', by which it is printed on the other side; after which it is carried by 10', 11', 12' to the lower receiving board B' at 13'.

The inking rollers are shown at E P D and D', and are arranged in the usual manner, at T, T', T'', T''', and T'''', to spread the ink on the types.

It will be perceived that the power of this press is equal to that of *The Times*, the difference being that *The Times* prints 12000 sheets on one side only, while this prints 6000 on both sides. *The Times* machine requires 8 layers on and 8 takers off, being double the number required by Marinoni's press. It must, however, be observed that, in the practical management of newspaper printing, as conducted in *The Times* office, the power of Marinoni's press, though in a certain sense equal to that of *The Times*, would be altogether insufficient; for it is indispensably necessary for that journal to print 60000 copies on one side of the paper during the last 5 hours of the morning. The matter allotted to the other side of the paper is so selected that it can be composed and printed in the earlier part of the night, or even of the previous day; the pressure falling exclusively on the matter which occupies the other side of the paper, consisting chiefly of the latest intelligence and Parliamentary reports.

It may be asked, therefore, how the journal of *La Presse*, of which the circulation, though inferior to that of *The Times*, is still very large, being understood to be at this time (June, 1855) 40000 copies, can be printed with the necessary celerity? The answer is, that *La Presse* does not contain as much as the tenth part of the letter-press of a copy of *The Times*, and that, therefore, it is found practicable to compose the matter in type twice or oftener, so as to produce two or more distinct *forms*, as they are called, which are put to work at as many different presses. The expense of composition is further economised at the printing office

Fig. 356.

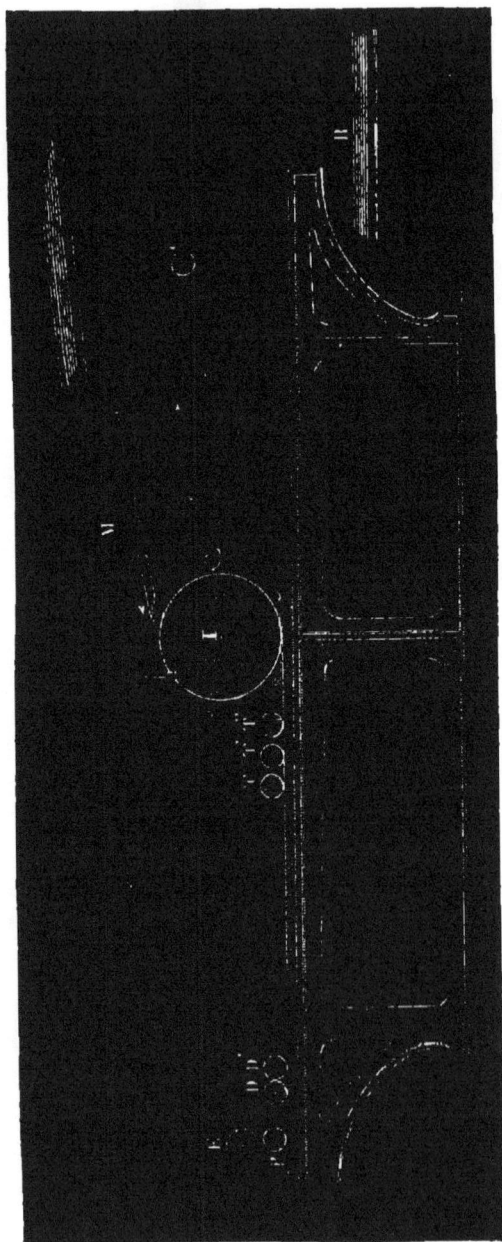

Fig. 157

of *La Presse* by stereotyping the matter, which is composed at a sufficiently early hour to admit of that process, the stereotype plates being melted down the next day. By this expedient double or triple composition is only necessary for the intelligence which comes too late to allow of being stereotyped.

713. **Marinoni's book printing machine.** — A convenient form of printing engine for books, constructed by the same engineers, is shown in *fig.* 356., by which, however, the sheets are printed on one side only. The layer on delivers the sheets upon the board M (*fig.* 357.), from which they pass round the printing roller I, and are discharged as indicated by the arrow upon the receiving board B. The rollers for delivering and spreading the ink on the types are arranged in the usual way.

INDEX.

NOTE. — This Index refers to the numbers of the paragraphs, and not to the pages.

A.

Action and reaction, 201.
Adhesion, 352 ; of wheels, 367.
Adjusting screw, 556.
Affinity, chemical, 354.
Air, resistance of, 589.
Alloys, strength of, 596.
Animalcules, minuteness of, 49.
Animal power, 621.
Annealing, 133.
Atoms, ultimate, 65.
Attraction, capillary, 353.
Atwood's machine, 243.
Axes, principal, 343.
Axle, wheel and, 432 ; of wheels, 554.

B.

Ball and socket, 550.
Balance, 414 ; sensibility of, 415 ; letter, 417 ; spring, 418 ; compensation, 688.
Balance wheel, 517. 686.
Barlow, table of strength of beams, 613.
Bevelled wheel, 450.
Bite of metals, 369.
Blood, its composition, 46 ; corpuscles in a drop of, 48.
Bodies, their parts, 12 ; falling, 237 ; strength of, 590.
Boring tools, 659.
Brakes, 582.
Bridges, suspension, 160.
Brittleness, 130.
Buoyancy, 81.

C.

Capillary attraction, 353. 648.
Capstan, 436. 624.
Centre of gravity, 268 ; of fluids, 308 ; of separate bodies, 309.
Centrifugal force, 311 ; rules for calculating, 314 ; examples, 318. 328 ; examples, 333.
Chemical affinity, 354.
Chemical agency, 645.
Chronometers, 698.
Clepsydra, 668.

Clamp, 556.
Cleavage, planes of, 63.
Clocks and watches, 671 ; maintaining power in, 695.
Cohesion, 262. 351 ; manifested, 356 ; examples, 357.
Coining press, 658.
Collision, 196 ; example, 199 ; effect of, 203 ; of equal masses, 205 ; example, 206 ; apparatus to illustrate, 215.
Combustion, 68.
Compensation balance, 688.
Compression, 87 ; of wood, 90 ; of stone, 91 ; of metal, 92 ; of liquids, 93 ; of gases, 94.
Composition, examples of, 170.
Congelation, 643.
Contractibility of liquids, 95.
Contraction, 109 ; effects of, 110.
Cordage, strength of, 599.
Couple, 155.
Coupling, 557.
Cradle joint, 551.
Crane, 649 ; movable, 650 ; building, 651 ; railway, 652 ; fixed, 653.
Crank, 529.
Crown wheel, 449.
Crystallisation, 61.

D.

Density, 76 ; example of, 80.
Diamond, polished surfaces of, 34.
Dilatation, 107 ; of liquids, 108 ; useful application of, 109 ; effects of, 110.
Distillation, destructive, 70.
Divisibility, unlimited, 28 ; minute, 51 shown by taste, 54.
Draught, line of, 579.
Drops for wharves, 654.
Drying machine, centrifugal, 326.
Ductility, 135.
Dynamical unit, 619.

E.

Earth once fluid, 360.
Earthwork, 628.

Elasticity of gases, 97; of liquids, 98; of solids, 100; examples, 101; of torsion, 106; effects of, 125; not proportionate to hardness, 127; of metals, 129; perfect and imperfect, 210.
Elastic force, limits of, 105.
Electro magnetic force, 644.
Endless screw, 495.
Engine, pile, 655; stamping, 656; screw cutting, 662; for cutting teeth of wheels, 663.
Equilibrium, 297; criterions of, 298; examples of, 301; not necessarily a state of rest, 385.
Erard's piano forte, 428.
Escapement, connection of pendulum with, 677; action of pendulum upon, 680; cylindrical, 692; recoil, ib.; duplex, 693; lever, ib.; detached, 694.
Evaporation, 69.
Expansibility of gases, 99.

F.

Falling body, 237.
Filtration, 83.
Fire escape, 459.
Flexibility, 130.
Flour mill, 664.
Fluids, resistance of, 585.
Fly, 518; position of, 540.
Fly wheel, 530.
Force, 141; resultant of, 145; in different directions, 147; composition of, 148; of moving mass, 185; examples of, 192; centrifugal, 311; rules to calculate centrifugal, 314; examples of centrifugal, 318. 328. 334; molecular, 350; resisting, 559; of animals, 634; electro magnetic, 644.
Friction, 560; sliding and rolling, 565.
Fuel, 639; effects of, 640.
Fusee, 527. 685.

G.

Gaseous state, 16.
Gases, elasticity of, 97; expansibility of, 99; may be reduced to liquids or solids, 362.
Gimbal, 549.
Glass, its filaments, 36.
Glue, effect of, 370.
Gold, visible on touchstone, 35; leaf, 43.
Governor, 526.
Gravity, 227.
Guinea and feather, 233.
Gun-cotton, 646.
Gunnery, 259.

H.

Hammer, effect of, upon a nail, 260. 534; sledge, 657.
Hardness, 122; may be modified, 122; of metals, 129.
Harrison's pendulum, 674.
Heat, effects of, 597; force produced by, 642.
Hinge, 552.
Hodgkinson, researches of, 603.
Horse power, 629.
Hunter's screw, 493.

I.

Inclined plane, 469; on railways, 472; double, 475.
Inclined roads, 471.
Inertia, 112; proof of, 116; examples of, 117.
Inking rollers, 703.
Insects, their wings, 42.
Iron, strength of, 596.
Isochronism, 687.

J.

Joint, knee, 289. 548; universal, 548; cradle, 551; telescope, 555.

K.

Knee joint, 289.
Knee lever, 427.

L.

Leaning towers, 282.
Level surface, 226.
Lever, 409; rectangular, 423; knee, 427.
Life preserver, 536.
Liquid state, 15.
Line of draught, 579.
Lubricants, 368; selection of, 572.

M.

Machine, Atwood's, 243; construction of a, 372; functions of, 390; simple, 404; weighing, 426; planing, 660; printing, 706; Times printing, 711; book printing, 713; Marinoni's printing, 712.
Magic clock, 307.
Magnitude, classified, 2; linear, 3; superficial, 4; solid, 5; unlimited, 10.
Mainspring, 684.
Maintaining power, 695.
Malleability, 131.
Marble, pulverised, 33.
Matter inert, 111.
Mechanical agents, 68.
Mercurial time measurer, 670; pendulum, 675.
Micrometer screw, 494.
Mill stones, 664.
Mirrors, 395.
Molecular forces, 350.
Molecules, 59; not in contact, 71.
Moments, 395.
Momentum of solids, 186; of liquids, 187; of air, 188; arithmetical expression for, 194; reservoir of, 222.
Morin's apparatus, 244.
Motion, 161; direction of, in a curve, 162; effect of force upon, 167; absolute and relative, 181; examples of, 182; laws of, 216; spontaneous, 220; down inclined plane, 253; of projectiles, 254; retarded, 352; modification of, 542; perpetual, 648.
Moving force, 190.
Musk, 52.

O.

Oblique impact, 212.
Observatory clock, 699.
Oscillation, 507.

P.

Pendulum, 278; simple, 497; indicates variation of gravity, 505; compensation, 510; a measure of force of gravity, 512; length of, 516. 672; compensation, 673; Harrison's, 674; mercurial, 675.
Penetrability, 25; examples of, 26.
Peron's table of human strength, 632.
Perpetual motion, 648.
Petrifaction, 84.
Piano forte, Erard's, 428.
Pile-engine, 655.
Pinion, mode of cutting, 682.
Plane, inclined, 469. 472. 475; self-acting, 476.
Planets, 360.
Planing machine, 660.
Plumb line, 224.
Pores, 73. 77. 78.
Porosity, examples of, 79. 80. 85, 86.
Power, moments of, 393; gained at expense of time, 399; animal, 621; horse, 629; water, 636; wind, 637; steam, 638.
Printing, old method of, 702.
Printing machine, single, 706; double, 707; Applegath and Cowper's double, 708; Times, 711; Marinoni's, 712.
Printing press, 701; modern, 704.
Projectiles, 254.
Pugilism, 207.
Pulley, 455; movable, 460; Smeaton's and White's, 465.

R.

Reaction, 201; example of, 202.
Rectangular lever, 423.
Regulators, 525.
Repose, 568.
Resistance of air, 589.
Resistance of fluids, 585.
Resisting force, 559.
Resolution of force, examples of, 170.
Retarded motion, 252.
Rollers, 575; inking, 703.
Rolling and punching mill, 539.
Ropes, 453.

S.

Salt of silver dissolved, 56.
Sand-glass, 669.
Saw mill, 661.
Screw, 484; application of, 460; mode of cutting, 491; Hunter's, 493; micrometer, 494; endless, 495; adjusting, 556.
Screw cutting engine, 662.
Screw press, 537.
Sheaves, 454.
Simple machine, 404.
Sledge, 574.
Sledge hammer, 657.
Soap bubble, its thinness, 41.
Solder, effect of, 370.

Solid state, 14.
Spade labour, 627.
Spider's web, 53.
Stability, 279; of vehicle, 284; of table, 286.
Stamping engine, 656.
Steam horse, 635.
Steam power, 638.
Steel-yard, 416.
Strength, 590; of timber, 598; of cordage, 599; of columns, 602; Peron's table of human, 632.
Strength of columns, 602.
Striking apparatus of clock, 700.
Strychnine dissolved, 55.
Sugar dissolved, 57.
Suspension, 508.
Sun dial, 666.
Swing, curious property of, 515.

T.

Teeth, 445; formation of, 446.
Teeth of wheels, 661.
Telescope joint, 555.
Tenacity, 137; of metals, 138; of fibrous texture, 139.
Timber, strength of, 598.
Time, 260.
Times printing machine, 709.
Torsion, elasticity of, 106.
Treadmills, 437.
Tredgold's table of strength of beams, 612; estimate of horse labour, 633.
Trunnions, 553.

U.

Universal joint, 546.

V.

Velocity, 163.
Vertical direction, 225.
Vis-inertiæ, 113.

W.

Watches, 698.
Water power, 636.
Wedge, 477; theory of, 479; use of, 481.
Weighing machine, 426.
Weight, moments of, 393.
Welding, 134.
Wheel, 432; crown, 449; bevelled, 450; balance, 517; fly, 530; position of fly, 540; use of, 575; mode of cutting, 682; balance, 686.
Whirling table, 316.
Windlass, 435.
Wind power, 637.
Wire in embroidery, 44.
Wollaston, his wire, 38; its minuteness, 39.

THE END.

www.ingramcontent.com/pod-product-compliance
Lightning Source LLC
Chambersburg PA
CBHW021348210326